STUDENT SURVIVAL MANUAL
A Conceptual Approach

Ken Seydel
San Bruno, California

CALCULUS

Gerald L. Bradley
Claremont McKenna College

Karl J. Smith
Santa Rosa Junior College

Prentice Hall

Englewood Cliffs, New Jersey 07632

Production Editor: Tina M. Trautz, PMI
Special Projects Manager: Barbara A. Murray
Editor-in-Chief: Jerome Grant
Associate Editor: Audra J. Walsh
Production Coordinator: Alan Fischer

Printed in the United States of America

10   9   8   7   6   5   4   3   2   1

ISBN: 0-13-305814-X

PRENTICE-HALL INTERNATIONAL (UK) LIMITED, LONDON
PRENTICE-HALL OF AUSTRALIA PTY. LIMITED, SYDNEY
PRENTICE-HALL CANADA INC. TORONTO
PRENTICE-HALL HISPANOAMERICANA, S.A., MEXICO
PRENTICE-HALL OF INDIA PRIVATE LIMITED, NEW DELHI
PRENTICE-HALL OF JAPAN, INC., TOKYO
SIMON & SCHUSTER ASIA PTE. LTD., SINGAPORE
EDITORA PRENTICE-HALL DO BRASIL, LTDA., RIO DE JANEIRO

Contents

PREFACE

This manual contains worked out solutions to every other odd-numbered exercise, with some notable exceptions. The "What Does This Say" questions are omitted since these questions are designed for you to put a concept into *your* words. The "Computational Window" problems are omitted as they will vary considerably depending upon the hardware and software used. The "Think Tank" problems are not included since I did not want to dictate the thought process. Your creativity and group interactions are important parts of these exercises. Occasionally, additional odd-numbered problems are solved to give better coverage of a particular type of problem. Finally, the section C problems are handled a bit differently. These problems are often not routinely assigned and are more of the nature of a challenge or extra credit problem. Since it would spoil that usefulness if complete solutions were given, only an analysis of the problem is presented. It is given, however, for all the odd-numbered problems.

Thirty years of teaching calculus has given me the ability to anticipate the types of errors commonly made. Some of the Survival Hints are directed toward those pitfalls. Others are an attempt to simplify and clarify concepts, and some are advice regarding what and how to study. Every student is a unique individual, so not all of these "hints" will be pertinent for everyone. I hope that most of them are helpful for you!

Regardless of how clear and lucid your professor's lecture on a particular topic may be, do not attempt to do the exercises without first reading the text and studying the examples. It will serve to reinforce and clarify the concepts and procedures.

Even though a solutions manual is basically a "how to" document, always ask *why* a particular approach was used, and understand the <u>concept</u> the problem is illustrating.

The concepts are more important than the answers!
The majority of the problems in this text can be solved
by a computer. Outside of the classroom the persons
with the concepts and problem-solving skills are the
ones needed to devise solutions and tell the machine
what to do.

Concentrate on the concepts and the process, not the
answer. Do not just solve the problem, learn *how* to
solve the problem. We all make arithmetic and
algebraic errors. Of course you should work carefully
to reduce these to a minimum. But do not spend too
much of your valuable study time looking for them.
Hopefully this Survival Manual will help!

Calculus is a collection of beautiful and powerful
ideas. I hope you enjoy them.

Ken Seydel
Skyline College

Chapter 1
Preview of Calculus: Functions and Limits

1.1 What Is Calculus?

(Due to the shortness of this section some additional problems are solved.)

SURVIVAL HINT

If your instructor does not assign the "WHAT DOES THIS SAY?" problems, take a few minutes to think about them anyway. These problems involve the *concepts* of the section. Could you explain this concept to another student? If not, it needs more work. Working with another student or a small study group is highly recommended!

5.   The limit of this sequence, which we will call L, is an infinite string of threes. You probably recognize this as $\frac{1}{3}$. So $L = \frac{1}{3}$.

7.   Unlike Problems 5 and 6 this sequence does not seem to have a repeating block. A nonrepeating decimal can not be written as a rational number. You probably recognize this sequence as the first seven digits of the irrational number π.

9.   For $n = 1, 10, 100, 100$; $L = \frac{2}{5}, \frac{20}{14}, \frac{200}{104}, \frac{2000}{1004}$.   It appears that $L = 2$. Notice that as n gets very large the 4 becomes negligible, and the numerator is always twice the denominator.

11.   For $n = 1, 10, 100, 1000$; $L = 1, \frac{30}{102}, \frac{300}{10002}, \frac{3000}{1000002}$.   As n gets very large the denominator grows faster than the numerator, so the fraction gets smaller. It appears that $L = 0$.

15.   By drawing radii to each of the vertices of a regular inscribed n-gon we get n congruent triangles. Each of these triangles has a central angle of $\frac{360}{n}$ degrees. A perpendicular to a side gives a right triangle with central angle $\frac{180}{n}$. So in this triangle $\frac{s}{2} = r \sin \frac{180}{n}$, and $h = r \cos \frac{180}{n}$. The area of each triangle will be $A = r^2 \sin \frac{180}{n} \cos \frac{180}{n}$ or more simply (using the double angle identity) $A = \frac{r^2}{2} \sin \frac{360}{n}$. The area of a regular inscribed n-gon will be:

$A_n = \frac{nr^2}{2} \sin \frac{360}{n}$. Now $A_3 \approx 5.196152$, $A_4 = 8.000000$, $A_5 \approx 9.510565$, $A_6 \approx 10.392305$, $A_{100} \approx 12.558104$, $A_{1000} \approx 12.566288$.

Since we are dealing with a circle, let's put these results in terms of π.

$A_6 \approx 3.307973\pi$, $A_{100} \approx 3.997369\pi$, $A_{1000} \approx 3.999974\pi$.

We will, therefore, assume that $L = 4\pi$.

17.   There are 16 rectangles, each with a width of $\frac{1}{16}$ and a height of the square of the right endpoint of the rectangle.

$A = \frac{1}{16}[(\frac{1}{16})^2 + (\frac{2}{16})^2 + (\frac{3}{16})^2 + ... + (\frac{16}{16})^2]$

$A = \frac{1}{16^3}(1^2 + 2^2 + 3^2 + ... + 16^2)$ Using the summation formula for squares we get:

$A = \frac{1}{16^3}\left(\frac{16(17)(33)}{6}\right) = \frac{187}{512}$

$A \approx 0.3652$

SURVIVAL HINT

This section is really worth a second reading after you have done the exercises. The entire course will be much more meaningful if you have an overall perspective of the basic concepts. Be certain you see how the tangent is the limit of the slopes of the secant, and be comfortable with the concept of area as an infinite sum of infinitely small pieces. These are very basic concepts.

1.2 Preliminaries

SURVIVAL HINT

Be careful in all your work to pay attention to the endpoints of intervals. Is the interval open, closed, or half-open? On most problems later in the text the interval is not specified, but may be implied by the function. For instance on $f(x) = \sqrt{4 - x^2}$ the endpoints are included, but on $f(x) = \dfrac{1}{\sqrt{4 - x^2}}$ the endpoints are not included.

3.   **a.**   This interval is closed on the left and open on the right.

     b.   This interval is closed on both the right and left.

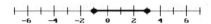

     c.   This interval is open on both the right and left.

     d.   This interval is closed on the left and open on the right.

7.   **a.** $d = \sqrt{(3+1)^2 + (-2-1)^2} = \sqrt{16+9} = 5$

         $M = (\frac{-1+3}{2}, \frac{1-2}{2}) = (1, -\frac{1}{2})$

     b. $d = \sqrt{(-1+2)^2 + (-2+1)^2} = \sqrt{2}$

         $M = (\frac{-2-1}{2}, \frac{-1-2}{2}) = (-\frac{3}{2}, -\frac{3}{2})$

11. $y^2 - 5y - 14 = 0$

       $(y - 7)(y + 2) = 0$

       $y = 7, \ -2$

SURVIVAL HINT

Is $-a$ positive or negative? Without further information we do not know. It might be neither. Be comfortable with the definition of absolute value. Especially the fact that $-a$ is positive when $a < 0$.

15. $2x + 4 = 16$ or $-(2x + 4) = 16$

       $2x = 12$ or $-2x = 20$

       $x = 6, \ -10$

19. ϕ; (The empty set.)

An absolute value can never be equal to a negative number.

SURVIVAL HINT

If your trigonometry is stale, it is essential that you do some review. If the basic identities and formulas are not fresh in your mind, make a list and put them in an easy-to-find place in your notebook for reference. Add other formulas as you find the need for them.

23.   By the zero product theorem:

$$2 \cos x + \sqrt{2} = 0 \quad \text{or} \quad 2 \cos x - 1 = 0$$
$$\cos x = -\frac{\sqrt{2}}{2} \quad \text{or} \quad \cos x = \frac{1}{2}$$
$$x = \frac{3\pi}{4}, \frac{5\pi}{4}, \frac{\pi}{3}, \frac{5\pi}{3}$$

27. $3x < -5$

$$x < -\frac{5}{3} \quad \text{or} \quad \left(-\infty, -\frac{5}{3}\right)$$

31. $2 < x \leq 7$ or $(2, 7]$

35.   Putting each of the three factors on a sign line:

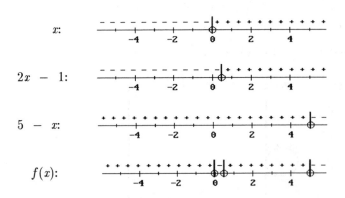

So we see the entire expression is positive on $(-\infty, 0)$ and $\left(\frac{1}{2}, 5\right)$

37.   Read this problem as a distance function: The distance between x and 8 is less than or equal to 0.001. The interval is $[7.999, 8.001]$

SURVIVAL HINT

Most work in calculus is done in radian measure. Set your calculator to radian mode and change it to degree mode only when you see the degree sign.

41.   **a.** $\frac{\pi}{3}$ or $60°$ **b.** $\frac{\pi}{6}$ or $30°$

43.   The length of arc is a fraction of the circumference of the circle.

In degree measure $L = \frac{\theta°}{360°} (2\pi r)$ or in radian measure $L = \frac{\theta}{2\pi} (2\pi r) = r\theta$.

In this case $L = 2(2) = 4$ cm

47.   These are radian measures. Be sure your calculator is in radian mode.

a. 1.5574   **b.** -0.0584 **c.** 0.9656

51. $x^2 + (y - 1.5)^2 = (0.25)^2$

SURVIVAL HINT

It is assumed that you have mastered the algebra skill of completing the square. If not, do some review. The detailed steps will usually not be given when it is used in an exercise.

55.   Completing the square:

$$(x^2 + 2x + 1) + (y^2 - 10y + 25) = -25 + 1 + 25$$
$$(x + 1)^2 + (y - 5)^2 = 1$$

Center: $(-1, 5)$ Radius: 1

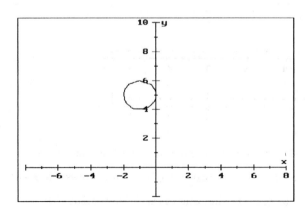

59. $\tan \frac{\pi}{12} = \tan \left(\frac{\pi}{4} - \frac{\pi}{6}\right) = \frac{\tan \frac{\pi}{4} - \tan \frac{\pi}{6}}{1 + \tan \frac{\pi}{4} \tan \frac{\pi}{6}} = \frac{1 - \frac{\sqrt{3}}{3}}{1 + \frac{\sqrt{3}}{3}} = 2 - \sqrt{3}$

69.   Analysis: Since absolute values only have an affect on signs, consider the three possible cases separately. What if x is a positive number, zero, or a negative number?

71.   Analysis: $|a|$ is the distance between a and the origin. Read this as "the distance between a and the origin is less than b".

73.   Analysis: This inequality says that the distance between x and y, $|x - y|$, is greater than or equal to the difference of their magnitudes. Is this true if they are both positive? If they are both negative? If one is positive and the other negative?

1.3 Lines in the Plane

SURVIVAL HINT

Many mistakes are made by good students, when doing material with which they are quite competent, because they try to do too much in their head. Get in the habit *NOW* of showing work on all problems which could be handed in to your instructor. It is *not* a waste of your time to be neat and well organized and write a sufficient number of steps so that someone else could follow your work. In a job situation this will be required. Your boss will not want to see a page of scratch-work with an answer circled at the bottom!

3.   Using the point-slope form :
$$y - 7 = \frac{2}{-1}(x + 1)$$
$$2x + y - 5 = 0$$

7.   A vertical line has form $x = a$. In this case, $x = -2$

11.   The given equation in slope-intercept form is: $y = -3x + 7$
A parallel line will have $m = -3$.
$$y - 8 = -3(x+1) \quad \text{or} \quad 3x + y - 5 = 0$$

15. $x - 4y + 5 = 0$ has $m = \frac{1}{4}$ so our line must have $m = -4$.
Solve the two given lines simultaneously to find their intersection at $(-1, 1)$.
Now use the point and slope to find our line:
$$y - 1 = -4(x + 1) \quad \text{or} \quad 4x + y + 3 = 0$$

17.   In slope-intercept form: $y = -\frac{3}{5}x - 3$

$m = -\frac{3}{5}$ y-intercept $= -3;$ $(0, -3).$

Let $y = 0$ to find x-intercept $= -5;$ $(-5, 0).$

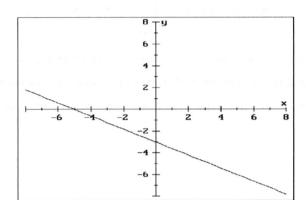

21.   This graph passes through the origin, $(0, 0)$, with $m = \frac{1}{5}$.

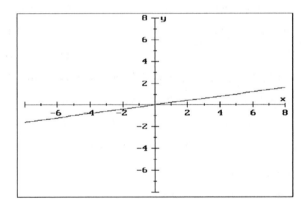

25.   The opposite sides of a parallelogram are parallel and equal.

$(3, -2)$ is 5 units down and 2 units right of $(1, 3)$, so the fourth vertex
is 5 units down and 2 units right of $(4, 11)$, which is $(6, 6)$, or 5 units
above and 2 units to the left of $(4, 11)$, which is $(2, 16)$.

Using the first choice:

Let A:$(1, 3)$, B:$(3, -2)$, C:$(4, 11)$, D:$(6, 6)$

$\overline{AB}$: $y - 3 = -\frac{5}{2}(x - 1)$ or $5x + 2y - 11 = 0$

$\overline{AC}$: $y - 3 = \frac{8}{3}(x - 1)$ or $8x - 3y + 1 = 0$

$\overline{CD}$: $y - 11 = -\frac{5}{2}(x - 4)$ or $5x + 2y - 42 = 0$

$\overline{BD}$: $y + 2 = \frac{8}{3}(x - 3)$ or $8x - 3y - 30 = 0$

25. (con't.)   Using the second choice:

Let A:(1, 3), B:(3, −2), C:(4, 11), D:(2, 16)

$\overline{AB}$: $y - 3 = -\frac{5}{2}(x - 1)$ or $5x + 2y - 11 = 0$

$\overline{AD}$: $y - 3 = 13(x - 1)$ or $13x - y - 10 = 0$

$\overline{CD}$: $y - 11 = -\frac{5}{2}(x - 4)$ or $5x + 2y - 42 = 0$

$\overline{BC}$: $y + 2 = 13(x - 3)$ or $13x - 3y - 41 = 0$

29. $C(0) = 5000$ and $m = 60$ so $C = 60x + 5000$

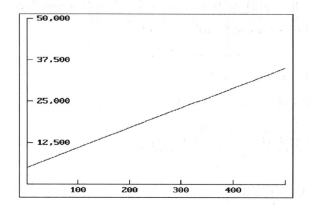

33.   **a.**   The three possible parallelograms are: $P_1P_2P_3A$, $P_1P_3P_2B$, $P_1P_3CP_2$.

Using the fact that the slopes and lengths of opposite sides are equal

(see Problem 25)   A:(3, 1),   B:(1, 11),   C:(−3, −5)

b.   For $P_1P_2P_3A$, P_1P_3 joins (2, 6) to (0, −2) and has midpoint (1 , 2)

$P_2P_4 = P_2A$ joins (−1, 3) to (3, 1) and has midpoint (1, 2).

Having the same midpoint they bisect each other.

37.   Analysis: Use the point-slope form of a line.

39.   Analysis: The distance from a point to a line is understood to be the shortest

distance, which would be on the line perpendicular to the given line.

We know how to find the distance between two points, so we need to know

where the line through $(x_0,\ y_0)$ and perpendicular to the given line intersects it.

We know the slope of the given line, so we know the slope of the perpendicular

line and can write its equation. Solve the equations of the two lines

simultaneously to find the needed second point. Then use the distance formula.

1.4 Functions and Their Graphs

1. D = all reals or $D = \mathbb{R}$ or $D = (-\infty, \infty)$

$f(-2) = 2(-2) + 3 = -1$

$f(1) = 2(1) + 3 = 5$

$f(0) = 2(0) + 3 = 3$

5. $f(x) = x - 2$; $x \neq -3$ So $D = \mathbb{R}$; $\neq -3$; or $(-\infty, -3) \cup (-3, \infty)$.

$f(2) = 2 - 2 = 0$, $f(0) = 0 - 2 = -2$, $f(-3)$ is undefined

9. $D = \mathbb{R}$ since the sine is defined for all values of x

$f(-1) = \sin[1 - 2(-1)] = \sin 3 \approx 0.1411$

$f(\tfrac{1}{2}) = \sin[1 - 2(\tfrac{1}{2})] = \sin 0 = 0$

$f(1) = \sin[1 - 2(1)] = \sin(-1) \approx -0.8415$

13. $\dfrac{f(x+h) - f(x)}{h} = \dfrac{[9(x+h) + 3] - (9x + 3)}{h} = \dfrac{9h}{h} = 9$

17. $\dfrac{f(x+h) - f(x)}{h} = \dfrac{|x+h| - |x|}{h}$

If $x < -1$ and $h < 1$ then $(x+h) < 0$ and $|x+h| = -(x+h)$

$= \dfrac{-(x+h) - (-x)}{h} = -1$

21.   a.   Not equal. $f(x) = g(x)$; $x \neq 0$

b.   Equal, since $g(x)$ restricts the domain.

SURVIVAL HINT

Values of x that make the denominator of a rational expression equal to 0 are not necessarily vertical asymptotes. They are only candidates and need to be tested. If the numerator is also 0 at that value, then you have a hole in the graph.

25.   a. $f_3(-x) = \dfrac{1}{3(-x)^3 - 4} = \dfrac{1}{-3x^3 - 3}$ Neither

b. $f_4(-x) = (-x)^3 + (-x) = -x^3 - x$ Odd

SURVIVAL HINT

Most useful "real" functions are compositions, and most in the text are compositions.
If you are not really comfortable with these problems, spend a little extra time on them
now and it will save you a lot of time later.

29. $f \circ g = \sin(1 - x^2);$ $g \circ f = 1 - \sin^2 x = \cos^2 x$

33. $f \circ g = \dfrac{1}{\tan x} = \cot x;$ $g \circ f = \tan \frac{1}{x}$

37.   **a.** $u(x) = \dfrac{x + 1}{2 - x};$ $g(u) = \sin x$

     b. $u(x) = \dfrac{2x}{1 - x};$ $g(u) = \tan x$

SURVIVAL HINT

"Graphing by composition" is one of the fastest and most useful methods. Many of the
problems in this text require a graph. If you have to graph by plotting and joining 23
points, it will take forever! Look for things in the equation that you recognize:
an absolute value, a cubic, a circle, a sine curve - and then translate, reflect, square,
or perform some operation on the known function. The next section discusses
"transformation of functions".

41.   This is the absolute value function reflected about the x-axis.

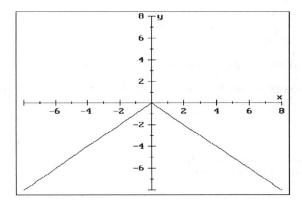

45.   This is the cube root function translated 2 units up.

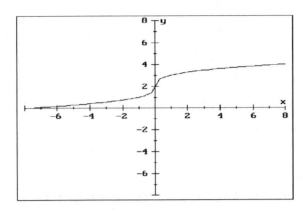

49.   This is the sine curve translated 2 units up.

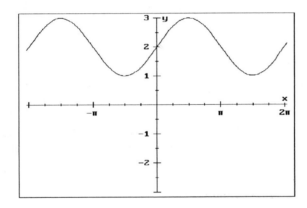

53. $R(a, \ g(a))$ S: $(x_0, \ g(x_0))$

57.   **a.** $I = f(t) = \dfrac{K}{s^2} = \dfrac{K}{(6t - t^2)^2} = \dfrac{30}{t^2(6 - t)^2}$

          b. $I(1) = \dfrac{30}{1^2(6 - 1)^2} = \dfrac{6}{5}$ candles

               $I(4) = \dfrac{30}{4^2(6 - 4)^2} = \dfrac{15}{32}$ candles

61.   **a.** $C(t) = (25t)^2 + (25t) + 900 = 625t^2 + 25t + 900$

          b. $C(3) = 75^2 + 75 + 900 = \$6600.$

61.   (con't.)

c. $11\,000 = 625t^2 + 25t + 900$

$$625t^2 + 25t - 10100 = 0$$

Dividing by 25: $25t^2 + t - 404 = 0$

Factoring: $(25t + 101)(t - 4) = 0$

$t = -\frac{101}{25}, 4$ Negative values of time are not in the domain

so $t = 4$ hours.

If the factoring is too difficult or the roots are likely irrational, solve with

the quadratic formula:

$$t = \frac{-1 \pm \sqrt{1^2 - 4(25)(-404)}}{2(25)} = \frac{-1 \pm \sqrt{40401}}{50} = \frac{-1 \pm 201}{50}$$

Rejecting the negative root $t = 4$ hours.

1.5 The Limit of a Function

SURVIVAL HINT

The limit is the basic concept underlying both the derivative and the definite integral.
Your intuitive concept of limit is correct, but you will find it essential to develop ability
to use the formal definition and learn some evaluation techniques. It is a spiral type
concept, in that you will come back to it time and time again; each time at a higher
level and with more understanding and greater perspective.

1. $\lim\limits_{x \to 3} f(x) = 0$ **5.** $\lim\limits_{x \to -1} g(x) = 7$

9. $\lim\limits_{x \to 2^+} t(x) = 2$

13. $\lim\limits_{x \to 5^-} f(x)$ for $f(x) = 4x - 5$ appears $= 15$

x	2	3	4	4.5	4.9	4.99
$f(x)$	3	7	11	13	14.6	14.96

17. $\lim\limits_{x \to 2} f(x) = 8$

21. $\lim\limits_{x \to \frac{3\pi}{2}} s(x) = -1$

25.

x	1	.5	.1	.01	.001
$f(x)$	.54030	.87758	.99500	.99995	1.00000

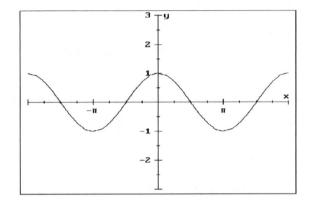

$f(x) = \cos x$

$$\lim_{x \to 0} f(x) = 1.00$$

29.

x	-2.5	-2.9	-2.99	-2.999	
$f(x)$	$-.28329$	$-.16949$	$-.16694$	$-.16669$	

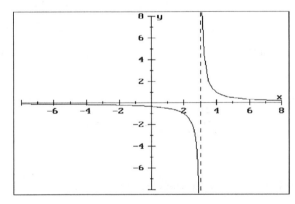

$f(x) = \dfrac{1}{x-3}$

$$\lim_{x \to -3^+} f(x) \approx -0.17$$

33.

x	3	3.1	3.14	3.141	3.1415
$f(x)$	$-.33$	$-.3223$	$-.31847$	$-.31837$	$-.31832$

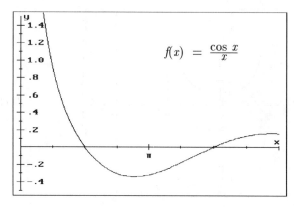

$$\lim_{x \to \pi} f(x) \approx -0.32$$

37.

x	3	3.1	3.14	3.141	3.1415
$f(x)$	.66333	.64488	.63694	.63674	.63664

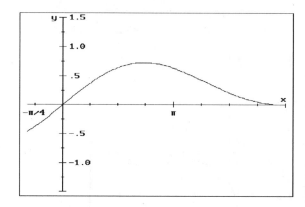

$$\lim_{x \to \pi} f(x) \approx 0.64$$

SURVIVAL HINT

If the limit of a rational expression has the form $\frac{0}{0}$ then it may have any value, L. (Check division by multiplying: $(L)(0) = 0$.) But if it has the form $\frac{a}{0}$, then the limit will be $\pm\infty$. (As the denominator of a fraction decreases, the quotient increases.)

41.

x	.5	.1	.01	.001
$f(x)$	1.04291	1.00167	1.00002	1.00000

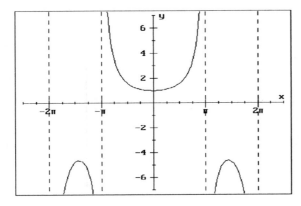

$$f(x) = \frac{x}{\sin x}$$

$$\lim_{x \to 0} f(x) = 1.00$$

45.

x	8	8.5	8.9	8.99	8.999
$f(x)$	$-.03431$	$-.01537$	$-.00283$	$-.00028$	$-.00003$

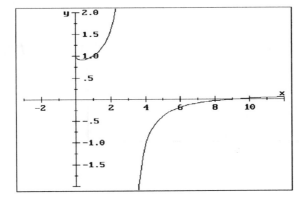

$$f(x) = \frac{\sqrt{x} - 3}{x - 3}$$

$$\lim_{x \to 9} f(x) = 0.00$$

49.

x	4.5	4.1	4.01	4.001	4.0001
$f(x)$	$-.05719$	$-.06135$	$-.06238$	$-.06249$	$-.06250$

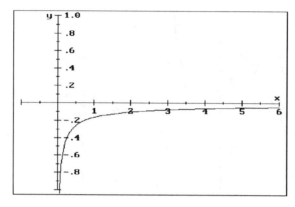

$$f(x) = \frac{\frac{1}{\sqrt{x}} - \frac{1}{2}}{x - 4}$$

$$\lim_{x \to 4} f(x) \approx -0.06$$

53.

x	.9	.99	.999	1.1	1.01	1.001
$f(x)$	1.1111	1.0101	1.0010	.9090	.9901	.9990

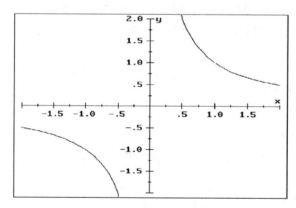

$$f(x) = \frac{1 - \frac{1}{x}}{x - 1}$$

$$\lim_{x \to 1} f(x) = 1.00$$

57.

x	.1	.01	.001	$-.1$	$-.01$	$-.001$
$f(x)$	.0335	.0033	.0003	$-.0335$	$-.0033$	$-.0003$

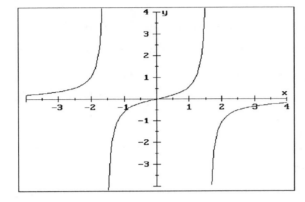

$$f(x) = \frac{\tan x - x}{x^2}$$

$$\lim_{x \to 0} f(x) = 0.00$$

SURVIVAL HINT

There is really no such thing as "instantaneous velocity" or "instantaneous rate of change". In an instant both time and motion are frozen. The concept of limit allows us to define what we mean by these terms.

65. Analysis: Follow the instructions in **a.** and **b.** and make your conclusion about the limits. How can they be different? A look at the graph might help:

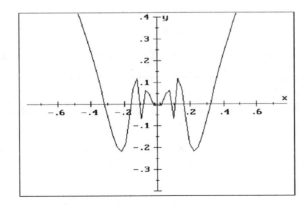

1.6 Properties of Limits

SURVIVAL HINT

When doing limit problems, use proper form and notation to avoid errors and to make each statement valid. Write the $\lim\limits_{x \to a}$ for each transformed expression until you actually take the limit. Do not be lazy in your notation.

3.   This is a polynomial, so $\lim\limits_{x \to 3} f(x) = f(3) = -8$

7.   This is a trigonometric function which is defined at the limiting value, so

$$\lim_{x \to \pi/3} \sec x = \sec \frac{\pi}{3} = 2$$

11.   Factor and reduce to eliminate the troublesome 0 in the denominator.

$$\lim_{x \to -2} \frac{(2+u)(2-u)}{(2+u)} = \lim_{x \to -2} (2-u) \quad \text{if } u \neq -2$$

Since x is *approaching* -2, $x \neq -2$ and $\lim\limits_{x \to -2} (2-u) = 4$

15.   Since $f(1)$ gives a denominator of 0, factor and reduce.

$$\lim_{x \to 1} \left(\frac{(x-2)(x-1)}{(x+2)(x-1)} \right)^2 = \lim_{x \to 1} \frac{(x-2)^2}{(x+2)^2} = \frac{1}{9}$$

19.   This is a rational trigonometric expression which is defined at the limiting value,

so $\lim\limits_{x \to 0} f(x) = f(0) = 1$

23.   Make use of the previously established $\lim\limits_{x \to 0} \frac{\sin x}{x} = 1$

$$\lim_{x \to 0} \frac{\frac{\sin x}{\cos x}}{x} = \lim_{x \to 0} \frac{\frac{\sin x}{x}}{\cos x} = \frac{\lim\limits_{x \to 0} \frac{\sin x}{x}}{\lim\limits_{x \to 0} \cos x} = \frac{1}{1} = 1$$

27.   To eliminate the troublesome x in the denominator there are several trigonometric identities that we could use. The numerator suggests the Pythagorean identity, $\tan^2 x = \sec^2 x - 1$.

$$\lim_{x \to 0} \frac{\sec x - 1}{x \sec x} = \lim_{x \to 0} \frac{(\sec x - 1)(\sec x + 1)}{x \sec x (\sec x + 1)} = \lim_{x \to 0} \frac{\sec^2 x - 1}{x \sec x (\sec x + 1)}$$

$$= \lim_{x \to 0} \frac{\tan^2 x}{x \sec x (\sec x + 1)} = \lim_{x \to 0} \left(\frac{\sin x}{x} \right) \left(\frac{\sin x}{\cos^2 x (\sec x)(\sec x + 1)} \right)$$

$$= \lim_{x \to 0} \left(\frac{\sin x}{x} \right) \lim_{x \to 0} \left(\frac{\sin x}{\cos x (\sec x + 1)} \right) = 1 \left(\frac{0}{1(2)} \right) = 0$$

35.   Note that 1 must be approached from the positive side for the radical to be

defined. $\lim\limits_{x \to 1^+} \dfrac{\sqrt{x-1}+x}{1-2x} = \dfrac{0+1}{1-2} = -1$

39. $x > 2$, so $f(x) = x^2 - 5$; $\lim\limits_{x \to 2^+} (x^2 - 5) = 4 - 5 = -1$

41. $s > 1$, so $g(s) = \sqrt{1-s}$; $\lim\limits_{s \to 1^+} \sqrt{1-s} = \sqrt{1-1} = 0$

53.   Factor and reduce: $\lim\limits_{x \to 1} \dfrac{\sqrt{x}-1}{(\sqrt{x}-1)(\sqrt{x}+1)} = \lim\limits_{x \to 1} \dfrac{1}{\sqrt{x}+1} = \dfrac{1}{2}$

55.   Doing the subtraction: $\lim\limits_{x \to 0} \left(\dfrac{x-1}{x^2}\right) = \dfrac{-1}{0}$

This fraction approaches $-\infty$ as $x \to 0$, so the limit does not exist.

59. $\lim\limits_{x \to 3^-} f(x) = \lim\limits_{x \to 3^-} 2(x+1) = 8$

$\lim\limits_{x \to 3^+} f(x) = \lim\limits_{x \to 3^+} (x^2 - 1) = 8$

Since the limit from the left is the same as the limit from the right,

$\lim\limits_{x \to 3} f(x) = 8$. The fact that $f(x) = 4$ does not matter.

65.   Set up the suggested identity by first introducing the constants into the

numerator and denominator: $\lim\limits_{x \to 0} \left(\dfrac{\frac{a\sin ax}{a}}{\frac{b\sin bx}{b}}\right)$

Then divide numerator and denominator by x:

$\lim\limits_{x \to 0} \left(\dfrac{\frac{a\sin ax}{ax}}{\frac{b\sin bx}{bx}}\right) = \dfrac{a\lim\limits_{x \to 0}\frac{\sin ax}{ax}}{b\lim\limits_{x \to 0}\frac{\sin bx}{bx}} = \dfrac{a}{b}$

67.   Analysis: We need to solve the inequality $\dfrac{1}{x^2} > 100L$ for x. In doing this,

remember that $\sqrt{x^2} = |x|$.

71.   Analysis: Use the formal definition of limit and apply the suggested identity.

1.7 Continuity

SURVIVAL HINT

Once again, your intuitive notion of "continuity" is correct. But think about all the
possible situations that could make a function discontinuous. The formal definition
takes care of all of these. Do not just memorize the definition, understand the concept.

1.   Temperature is continuous, so TEMPERATURE $= f(\text{time})$ would be a continuous

function. The domain would be midnight to midnight.

5.   The charges (range of the function) consist of rational numbers only (dollars
and cents to the nearest cent), so the function CHARGE $= f($MILEAGE$)$ would
be a step function. The domain would consist of the mileage from the beginning
of the trip to its end.

9.   The denominator factors to $x(x - 1)$, so suspicious points would be $x = 0, 1$.
There will be a hole discontinuity at $x = 0$ and a pole discontinuity at $x = 1$.

13.   We have suspicious points where the denominators are 0 at $t = 0, -1$.
There are pole discontinuties at each of these points.

17.   The sine and cosine are continuous on the reals, but the tangent is discontinuous
at $x = \frac{\pi}{2} + n\pi$. Each of these values will have a pole type discontinuity.

21.   For continuity $f(2)$ must equal $\lim_{x \to 2} f(x) = \lim_{x \to 2} \dfrac{(x - 2)(x + 1)}{x - 2}$
$= \lim_{x \to 2} (x + 1) = 3$

25.   The function is not defined at 2, and since $\lim_{x \to 2^-} f(x) = 11$ and $\lim_{x \to 2^+} f(x) = 9$
no value can be assigned to $f(2)$ to "tie together" the two pieces. This is
sometimes called an "essential" discontinuity. Only hole type discontinuities
are "removable".

29.   The function has a pole type discontinuity at $t = 0$, but 0 is not in the interval.
The function is continuous on $[-3, 0)$.

33.   Both $y = x$ and $y = \sin x$ are continuous on the reals, so $f(x) = x \sin x$ will be
continuous on $(0, \pi)$.

SURVIVAL HINT

A common error is to apply a theorem when it really is not applicable. Theorems are
if − then statements, and the conclusion is not justified unless the hypothesis is met.
When learning a theorem pay careful attention to the "if" part. When using a theorem
always be certain the hypothesis has been met.

39. $f(x) = \sqrt[3]{x} - x^2 - 2x + 1$ is continuous on $(-\infty, \infty)$ and
$f(0) = 1, \quad f(1) = -1$ so the hypotheses of the intermediate value theorem
are met, and we are guaranteed that there is at least one number c on $[0, 1]$
such that $f(c) = 0$.

43. $f(x) = \cos x - \sin x - x$ is continuous on the reals and $f(0) = 1$,

$f(\frac{\pi}{2}) = -1 - \frac{\pi}{2}$.

Since the hypotheses of the intermediate value theorem have been met, we are

guaranteed that there exists at least one number c on $[0, \frac{\pi}{2}]$ such that $f(c) = 0$.

47.   At 12:00 the hands coincide. For any other hour h the minute hand at $f(h)$ is

at a position $5h$ minutes *less than* the position of the hour hand, and at $f(h+1)$

the minute hand is at a position $5(12 - h)$ minutes *greater than* the position of

the hour hand. Since the position of the minute hand is a continuous function,

we have met the hypotheses of the intermediate value theorem and are guaranteed

of the existence of some time t on the interval $[h, h+1]$ where the distance

between the hands is 0.

51.   This function is continuous on $[0, \infty)$ except possibly at $x = 1$.

For continuity it is required that $\lim_{x \to 1} f(x) = f(1)$. So we need:

$\lim_{x \to 1} \dfrac{\sqrt{x} - a}{x - 1} = b$ Since this expression is undefined, we rationalize the

numerator:

$\lim_{x \to 1} \dfrac{(\sqrt{x} - a)(\sqrt{x} + a)}{(x - 1)(\sqrt{x} + a)} = \lim_{x \to 1} \dfrac{x - a^2}{(x - 1)(\sqrt{x} + a)}$.   Now if we let $a = \pm 1$,

this will reduce to $\lim_{x \to 1} \dfrac{1}{\sqrt{x} \pm 1} = \dfrac{1}{1 \pm 1} = \dfrac{1}{2} = b$.

Therefore $a = 1$, $b = \frac{1}{2}$ will make this function continuous at $x = 1$.

55.   For the bisection method we will bisect each interval for which $f(x)$ has a change

of sign:

x	1	2	3/2	7/4	13/8	25/16
$f(x)$	-1	1	$-.25$	.3125	.015625	$-.12109375$

So there is a change in sign, and therefore a root, between $\frac{25}{16}$ and $\frac{26}{16}$.

The *nearest* sixteenth would be $\frac{26}{16}$, or $1\frac{5}{8}$.

Using a calculator to continue to the nearest $\frac{1}{1000}$:

x	1.625	1.563	1.594	1.610	1.618	1.619
$f(x)$	.0156	$-.1200$	$-.0532$	$-.0378$	$-.00008$	.002

In the last step we abandoned the bisection of the interval because one of

the values was so near to 0. The root to the *nearest* $\frac{1}{1000}$ will be 1.618.

59. For the bisection method we will bisect each interval for which $f(x)$ has a change

of sign:

x	1	2	3/2	5/4	9/8	17/16
$f(x)$	-1	13	4.375	1.328	.080	$-.480$

So there is a change in sign, and therefore a root, between $\frac{18}{16}$ and $\frac{17}{16}$.

The *nearest* sixteenth would be $\frac{18}{16}$ or $1\frac{1}{8}$.

Using a calculator to continue to the nearest $\frac{1}{1000}$:

x	1.094	1.110	1.117	1.114	1.119	1.116	1.117
$f(x)$	$-.203$	$-.058$	.006	$-.022$	.024	$-.003$	.006

The root to the *nearest* $\frac{1}{1000}$ will be 1.116 .

63. Analysis: Since $f(a)$ and $f(b)$ have opposite signs, 0 is an intermediate value.

Verify that the hypotheses of the intermediate value theorem are met and then

apply it.

65. Analysis: Note that $f(0) = 1$, while $\lim\limits_{x \to 0} f(x) = 0$.

1.8 Introduction to the Theory of Limits

SURVIVAL HINT

Although it is possible to teach calculus without doing the exercises in this section,

you *should* understand the concept of the "epsilon-delta game". The key phrase in the

definition is "for each". Which means for *any* epsilon, no matter how small, we can

find a delta. This is the key to our confidence that L really is the limit.

1. For any chosen value of ϵ we want the distance between $f(x)$ and L to be less

than ϵ. In absolute value notation this is $|f(x) - L| < \epsilon$. For our function

and limit we need $|(2x - 5) - (-3)| < \epsilon$ or $|2x - 2| < \epsilon$ or $2|x - 1| < \epsilon$.

So the distance between x and 1 needs to be $< \frac{\epsilon}{2}$. Therefore, let $\delta = \frac{\epsilon}{2}$.

5. For any chosen value of ϵ we want the distance between $f(x)$ and L to be less

than ϵ. In absolute value notation this is $|f(x) - L| < \epsilon$. For our function

and limit we need $|(x^2 + 2) - (6)| < \epsilon$ or $|x^2 - 4| < \epsilon$

or $|x - 2||x + 2| < \epsilon$. As we approach 2, $|x + 2|$ is less than 5.

If the distance between x and 2, written as $|x - 2|$, is to be less than δ,

we need $|x - 2|(5) < \epsilon = \delta$. Therefore, let $\delta = $ minimum of $\{1, \frac{\epsilon}{5}\}$.

9.   For any chosen value of ϵ we want the distance between $f(x)$ and L to be less
than ϵ. In absolute value notation this is $|f(x) - L| < \epsilon$.
For our function and limit we need $|(3x + 7) - (1)| < \epsilon$ or $|3x + 6| < \epsilon$
or $3|x+2| < \epsilon$. This can be written as $|x - (-2)| < \frac{\epsilon}{3}$. Therefore,
let $\delta = \frac{\epsilon}{3}$ and for any given ϵ this value of δ will guarantee that $f(x)$
on the interval $[-2-\delta, -2+\delta]$ will be within ϵ units of L.

11.   For any chosen value of ϵ we want the distance between $f(x)$ and L to be less
than ϵ. In absolute value notation this is $|f(x) - L| < \epsilon$. For our function
and limit we need $|(x^2 + 2) - (6)| < \epsilon$ or $|x^2 - 4| < \epsilon$ or
$|x - 2| \, |x + 2| < \epsilon$. As we approach 2, we will arbitrarily require that δ be
less than 1. (Other values could be chosen, but 1 is nice to work with.)
This now means that x will be within 1 unit of 2 : $1 < x < 3$, so $3 < x + 2 < 5$.
We now replace $|x + 2|$ in our epsilon inequality with 5:
$|x - 2| 5 \ < \epsilon$, giving $|x - 2| < \frac{\epsilon}{5}$. Therefore, let $\delta = \min\{1, \frac{\epsilon}{5}\}$.
We chose the 5 to replace $|x + 2|$ instead of the 3 because our proof goes
from the selected δ value to the ϵ inequality:
$|x - 2| < \delta \Rightarrow |x - 2| < \frac{\epsilon}{5} \Rightarrow 5|x - 2| < \epsilon \Rightarrow |x + 2| \, |x - 2| < \epsilon$
(Replacing the 5 with a smaller value guarantees the inequality will hold.)
$\Rightarrow |x^2 - 4| < \epsilon \Rightarrow |(x^2 + 2) - (6)| < \epsilon \Rightarrow |f(x) - L| < \epsilon.$

SURVIVAL HINT

Many errors are made in calculus by students mis-applying a theorem. Review a bit of
the logic of if $-$ then statements. Be able to distinguish between the statement,
the converse, the inverse, and the contrapositive. Iff means that both the statement and
its converse are true. If a statement (theorem) is true then so is its contrapositive.
It is an <u>error</u> to assume the converse or to assume the inverse.

13.   In order for $f(x)$ to be continuous at $x = 0$, $\lim\limits_{x \to 0} f(x)$ must equal $f(0)$.
To show that $\lim\limits_{x \to 0} \sin \frac{1}{x} = 0$ we need to show that for any $\epsilon > 0$
there exists a $\delta > 0$ such that $|f(x) - L| < \epsilon$ when $|x - 0| < \delta$.
Arbitrarily letting $\epsilon = .5$ we need to find a δ-interval about 0 such that
$\left|\sin \frac{1}{x}\right| < .5$. However any interval about 0 contains a point $x = \frac{2}{\pi n}$
(with n odd) where $\sin \frac{1}{x} = \sin \frac{\pi n}{2} = \pm 1$.
Therefore, there does not exist a δ-interval about 0 such that $\left|\sin \frac{1}{x}\right| < .5$,
and $f(x)$ must be discontinuous at $x = 0$.

15.　Analysis: If the limits exist for f and g, then for any given ϵ there exists a δ. Instead of choosing ϵ for each function, choose $\frac{\epsilon}{2}$. For f there will be a δ_1, and for g a δ_2. The limit of the sum will be less than $\frac{\epsilon}{2} + \frac{\epsilon}{2} = \epsilon$ if you choose δ as the minimum of δ_1 and δ_2.

17.　Analysis: If the limits for f and g exist and are equal to zero, then there is an ϵ_1 and ϵ_2 that satisfy the formal definition. The lesser of these two values will cause both f and g to be sufficiently small for any specified δ.

19.　Analysis: Show that the function $h(x)$ meets the hypotheses of the limit limitation theorem. Then $\lim_{x \to c} h(x) \geq 0$, which makes $\lim_{x \to c} [f(x) - g(x)]$ $= \lim_{x \to c} f(x) - \lim_{x \to c} g(x) \geq 0$. So $\lim_{x \to c} f(x) \geq \lim_{x \to c} g(x)$.

21.　Analysis: If from problem 20 there exists a δ such that $|x - c| < \delta$ implies that $\big||f(x) - |L|\big| \leq |f(x) - L|$, let $\epsilon = \frac{1}{2}|L|$. Then the distance between $f(x)$ and L will be less than $\frac{1}{2}|L|$, which gives the required inequality.

23.　Analysis: To prove that LM is the limit of fg we need to find a δ such that $|fg - LM| < \epsilon$. Use the suggested identity and the results of problem 22 to simplify the expression. Your required δ will be the minimum of other deltas.

Chapter 1 Review

PRACTICE PROBLEMS

22.　**a.**　Using the point-slope formula: $y - y_1 = m(x - x_1)$

$y - 5 = -\frac{3}{4}\left[x - \left(-\frac{1}{2}\right)\right]$

$y - 5 = -\frac{3}{4}\left(x + \frac{1}{2}\right)$　Multiply both sides by 4,

$4y - 20 = -3x - \frac{3}{2}$

$3x + 4y = \frac{37}{2}$　or　$6x + 8y - 37 = 0$

b.　Use the two points to find the change in y and the change in x to determine the slope, then use either of the points in the point-slope formula.

$m = \frac{-3}{10}$ and using the point $(-3, 5)$.

$y - 5 = -\frac{3}{10}[x - (-3)]$

$10y - 50 = -3x - 9$

$3x + 10y - 41 = 0$

c.　Use the two given points to find the slope, and then use the slope-intercept form of an equation: $y = mx + b$

$m = \frac{3}{28}$, and $b = $ the y-intercept $= -\frac{3}{7}$.

$y = \frac{3}{28}x - \frac{3}{7}$ or $3x - 28y - 12 = 0$

22. (con't.)

    d.   If we write the given equation in slope-intercept form, $y = -\frac{2}{5}x + \frac{11}{5}$,

        we see that the slope is $-\frac{2}{5}$. A parallel line must have the same slope.

        Now use the point-slope form.

        $y - 5 = -\frac{2}{5}\left(x + \frac{1}{2}\right)$

        $5y - 25 = -2x - 1$

        $2x + 5y - 24 = 0$

    e.   Find the slope of $\overline{PQ}$. A perpendicular line will have a slope which is the

        negative reciprocal. Find the midpoint of $\overline{PQ}$. Then use the point-slope

        form for the equation of the line.

        The slope of $\overline{PQ}$ is $-\frac{3}{4}$. The midpoint of $\overline{PQ}$ is $(1, 4)$.

        $y - 4 = \frac{4}{3}(x - 1)$

        $3y - 12 = 4x - 4$

        $4x - 3y + 8 = 0$

23.   **a.**   Probably the easiest way to graph this equation is by finding the intercepts.

        If $x = 0$, then $y = 6$, and if $y = 0$, then $x = 4$. Using $(0, 6)$ and $(4, 0)$:

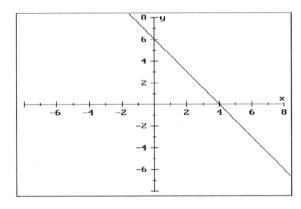

    b.   Complete the square in order to find the vertex of the parabola:

        $y = (x - 2)^2 - 14$, so the vertex is at $(2, -14)$.

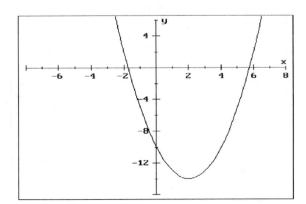

23. (con't.)

    c.   This is an absolute value function that has been translated one unit to the left and 3 units up. It has a vertex at $(-1, 3)$.

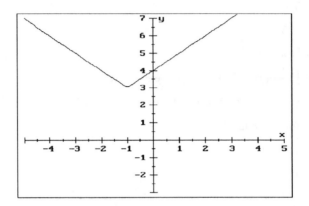

    d.   This is a cosine curve that has an amplitude of 2 and has been translated 1 unit to the right.

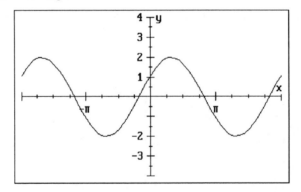

    e.   Remove the x coefficient from the parentheses in order to determine the phase shift and the change in amplitude: $y + 1 = \tan 2(x + \frac{3}{2})$. This is a tangent curve that has been translated $\frac{3}{2}$ units to the left, 1 unit down, and has a change in frequency from π to $\frac{\pi}{2}$.

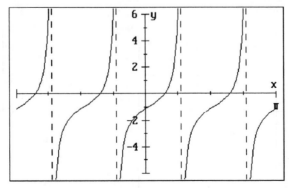

24. If $f(x) = \dfrac{1}{x+1}$, then $f\left(\dfrac{1}{x+1}\right) = \dfrac{1}{\dfrac{1}{x+1}+1}$, and $f\left(\dfrac{2x+1}{2x+4}\right) = \dfrac{1}{\dfrac{2x+1}{2x+4}+1}$

So we need to find the values of x for which:

$$\dfrac{1}{\dfrac{1}{x+1}+1} = \dfrac{1}{\dfrac{2x+1}{2x+4}+1}.$$ Multiply numerator and denominator on

the left by $(x+1)$, and on the right by $(2x+4)$:

$$\dfrac{x+1}{1+x+1} = \dfrac{2x+4}{2x+1+2x+4}$$

$$\dfrac{x+1}{x+2} = \dfrac{2(x+2)}{4x+5}$$
$$4x^2 + 9x + 5 = 2x^2 + 8x + 8$$
$$2x^2 + x - 3 = 0,$$
$$(2x+3)(x-1) = 0, \quad x = -\tfrac{3}{2}, 1$$

25. The root of a quotient equals the quotient of the roots when they have the same domain. The domain of $f(x)$ is $(-\infty, 0) \cup (1, \infty)$. The domain of $g(x)$ is $(1, \infty)$. The two functions are not the same.

26. $f \circ g = \sin\left(\sqrt{1-x^2}\right)$, and $g \circ f = \sqrt{1 - \sin^2 x} = |\cos x|$

27. Since the cost is the number of worker-hours times the rate per hour,
$$C = 25\ T = \dfrac{25(3N+4)}{2N-5}.$$

28.   **a.**   Since this function is defined at the limiting value, $\lim\limits_{x \to 3} f(x) = f(3) = \tfrac{3}{2}$

       b.   Factor the denominator and reduce: $\lim\limits_{x \to 4} \dfrac{\sqrt{x}-2}{(\sqrt{x}-2)(\sqrt{x}+2)}$

$$= \lim\limits_{x \to 4} \dfrac{1}{\sqrt{x}+2} = \tfrac{1}{4}$$

       c.   Factor and reduce: $\lim\limits_{x \to 2} \dfrac{(x-2)(x-3)}{(x+2)(x-2)} = \lim\limits_{x \to 2} \dfrac{x-3}{x+2} = -\tfrac{1}{4}$

       d. $\lim\limits_{x \to 0} \dfrac{1 - \cos x}{2\tan x} = \lim\limits_{x \to 0} \dfrac{(1-\cos x)(1+\cos x)}{2\dfrac{\sin x}{\cos x}(1+\cos x)}$

$$= \lim\limits_{x \to 0} \dfrac{\sin^2 x}{2\dfrac{\sin x}{\cos x}(1+\cos x)}$$

$$= \lim\limits_{x \to 0} \dfrac{\sin x \cos x}{1 + \cos x} = \tfrac{0}{2} = 0$$

29. Polynomials are everywhere continuous, so the only problem is at $x = 1$.

We need $\lim\limits_{x \to 1^-} (Ax+3) = 2$, and $\lim\limits_{x \to 1^+} (x^2 + B) = 2$.
$A + 3 = 2, \ A = -1$ and $1 + B = 2, \ B = 1$

SURVIVAL HINT

Develop the habit of verifying the hypothesis before using a theorem.

30.   Let $f(x) = \dfrac{1}{\sqrt{x} + 3} - x - \sin x$. This function is continuous on $[0, \infty)$, and

$f(0) = \frac{1}{3}$, $f(\pi) = \dfrac{1}{\sqrt{\pi} + 3} - \pi \approx -2.93$

So by the root location theorem there must be some value c on $[0, \pi]$ where

$f(c) = 0$

Chapter 2
Techniques of Differentiation with Selected Applications

2.1 An Introduction to the Derivative: Tangents

SURVIVAL HINT

Memorize the definition of the derivative. Recognize the concept of slope as the change in y divided by the change in x: $\dfrac{\Delta y}{\Delta x}$.

3. $f'(x) = \lim\limits_{\Delta x \to 0} \dfrac{f(x + \Delta x) - f(x)}{\Delta x} = \lim\limits_{\Delta x \to 0} \dfrac{(3) - (3)}{\Delta x} = 0 \quad f'(-5) = 0$

7. $f'(x) = \lim\limits_{\Delta x \to 0} \dfrac{f(x + \Delta x) - f(x)}{\Delta x} = \lim\limits_{\Delta x \to 0} \dfrac{[2 - (x + \Delta x)^2] - (2 - x^2)}{\Delta x}$

$ = \lim\limits_{\Delta x \to 0} \dfrac{-2x\Delta x - (\Delta x)^2}{\Delta x} = \lim\limits_{\Delta x \to 0} (-2x - \Delta x) = -2x$

$ f'(0) = 0$

11. $f'(t) = \lim\limits_{\Delta t \to 0} \dfrac{f(t + \Delta t) - f(t)}{\Delta t} = \lim\limits_{\Delta t \to 0} \dfrac{[3(t + \Delta t) - 7] - (3t - 7)}{\Delta t}$

$ = \lim\limits_{\Delta t \to 0} \dfrac{3\Delta t}{\Delta t} = 3; \quad f(t) \text{ is differentiable for all real numbers.}$

15. $f'(x) = \lim\limits_{\Delta x \to 0} \dfrac{f(x + \Delta x) - f(x)}{\Delta x}$

$ = \lim\limits_{\Delta x \to 0} \dfrac{[(x + \Delta x)^2 - (x + \Delta x)] - (x^2 - x)}{\Delta x}$

$ = \lim\limits_{\Delta x \to 0} \dfrac{2x\Delta x + (\Delta x)^2 - \Delta x}{\Delta x} = \lim\limits_{\Delta x \to 0} (2x + \Delta x - 1) = 2x - 1$

$f'(x)$ is defined for all real values, so $f(x)$ is differentiable for all reals.

19. $f'(x) = \lim\limits_{\Delta x \to 0} \dfrac{f(x + \Delta x) - f(x)}{\Delta x}$

$\qquad = \lim\limits_{\Delta x \to 0} \dfrac{\sqrt{5(x + \Delta x)} - \sqrt{5x}}{\Delta x}$ *Rationalizing the numerator:*

$\qquad = \lim\limits_{\Delta x \to 0} \dfrac{\left(\sqrt{5(x + \Delta x)} - \sqrt{5x}\right)\left(\sqrt{5(x + \Delta x)} + \sqrt{5x}\right)}{\Delta x \left(\sqrt{5(x + \Delta x)} + \sqrt{5x}\right)}$

$\qquad = \lim\limits_{\Delta x \to 0} \dfrac{5(x + \Delta x) - 5x}{\Delta x \left(\sqrt{5(x + \Delta x)} + \sqrt{5x}\right)}$

$\qquad = \lim\limits_{\Delta x \to 0} \dfrac{5}{\sqrt{5(x + \Delta x)} + \sqrt{5x}} = \dfrac{5}{2\sqrt{5x}} = \dfrac{\sqrt{5x}}{2x}$ for $x > 0$

23. $f'(s) = 3s^2$ (see Example 7) $f'(-\frac{1}{2}) = 3(-\frac{1}{2})^2 = \frac{3}{4}$

$f(-\frac{1}{2}) = (-\frac{1}{2})^3 = -\frac{1}{8}$

Using the point-slope formula: $y - (-\frac{1}{8}) = \frac{3}{4}[x - (-\frac{1}{2})]$ or $3x - 4y + 1 = 0$

27. Since $f(x)$ has a slope of 5, the required equation will have a slope of $-\frac{1}{5}$

and must pass through $(3, 13)$. Using the point-slope equation:

$y - 13 = -\frac{1}{5}(x - 3)$ or $x + 5y - 68 = 0$

31. $\dfrac{dy}{dx} = \lim\limits_{\Delta x \to 0} \dfrac{f(x + \Delta x) - f(x)}{\Delta x} = \lim\limits_{\Delta x \to 0} \dfrac{2(x + \Delta x) - 2x}{\Delta x} = \lim\limits_{\Delta x \to 0} 2 = 2$

$\dfrac{dy}{dx}\bigg|_{x = -1} = 2$

35.   **a.** $m_{\text{sec}} = \dfrac{f(x + \Delta x) - f(x)}{\Delta x}$

$\qquad\qquad = \dfrac{f(-2 + .1) - f(-2)}{.1}$

$\qquad\qquad = \dfrac{(-1.9)^2 - (-2)^2}{.1} = -3.9$

  b. $f'(x) = 2x$ (see Example 2) so $f'(-2) = -4$

Since Δx is "small" with respect to x the slope of the secant (average change)

is close to the value of the slope of the tangent (instantaneous change).

SURVIVAL HINT

Do not confuse zero slope with no slope. Zero slope is a horizontal line, no slope may

mean the function is discontinuous at the given value, is continuous but has a cusp or

39.   **a.** $f'(x) = \lim\limits_{\Delta x \to 0} \dfrac{f(x + \Delta x) - f(x)}{\Delta x}$

$$= \lim\limits_{\Delta x \to 0} \dfrac{[4 - 2(x + \Delta x)^2] - (4 - 2x^2)}{\Delta x}$$

$$= \lim\limits_{\Delta x \to 0} \dfrac{-4x\Delta x - 2(\Delta x)^2}{\Delta x} = -4x$$

b.   A horizontal line has $m = 0$, so $f'(x) = 0$, $-4x = 0$, $x = 0$.

At $x = 0$, $f(0) = 4$, so $\qquad y - 4 = 0(x - 0)$,

$y = 4$ is the equation.

c. $8x + 3y = 4$ has a slope of $-\frac{8}{3}$ so our line must have the same slope in

order to be parallel. $f'(x) = -4x = -\frac{8}{3}$, $x = \frac{2}{3}$. $f\left(\frac{2}{3}\right) = \frac{28}{9}$.

So the point at which the tangent is parallel to the given line is $\left(\frac{2}{3}, \frac{28}{9}\right)$.

If we wanted the equation of this tangent line we would use the

point-slope equation: $\qquad y - \frac{28}{9} = -\frac{8}{3}\left(x - \frac{2}{3}\right)$ or $24x + 9y - 44 = 0$

43.   Let $\Delta x = .1$, $c = 1$, $f(x) = (2x - 1)^2$

$f'(x) \approx \dfrac{f(x + \Delta x) - f(x)}{\Delta x} = \dfrac{f(1.1) - f(1)}{.1} = \dfrac{1.44 - 1}{.1} = 4.4$

Let $\Delta x = .01$

$f'(x) \approx \dfrac{f(x + \Delta x) - f(x)}{\Delta x} = \dfrac{f(1.01) - f(1)}{.01} = \dfrac{1.0404 - 1}{.01} = 4.04$

It appears that $f'(x) = 4$.

47. $\dfrac{dy}{dx} = 2Ax$, $\dfrac{dy}{dx}\Big|_{x = c} = 2Ac$. The point is $(c, f(c)) = (c, Ac^2)$

So the equation of the tangent line will be:

$y - Ac^2 = 2Ac(x - c)$. The x-intercept occurs when $y = 0$:

$-Ac^2 = 2Ac(x - c)$, $-c = 2(x - c)$, $x = \frac{c}{2}$.

The y-intercept occurs when $x = 0$: $y - Ac^2 = 2Ac(0 - c)$, $y = -Ac^2$.

51.

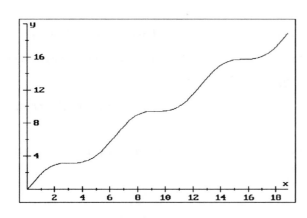

$f(x) = x + \sin x$

55.   **a., b.**

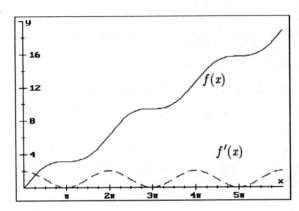

The derivative is 0 at π, 3π, 5π . This agrees with the horizontal

tangents on the graph of $f(x)$.

b.   The graph is increasing where the derivative is positive and the graph is

decreasing where the derivative is negative.

59.   Analysis: From problem 47 we have more than sufficient information.

Only two points are necessary to determinine a line and we have three:

both intercepts and the given point P.

2.2 Techniques of Differentiation

SURVIVAL HINT

Addition is commutative. So the product rule can be used as $(fg)' = fg' + gf'$ or

$gf' + fg'$. Subtraction is *not* commutative so in the quotient rule you must always

begin with the denominator times the derivative of the numerator.

3.   2.1, #15 $f'(x) = 2x - 1$ *Linearity and Power rules*

2.1, #16 $g'(t) = -2t$ *Linearity and Power rules*

2.1, #17 $f(s) = s^2 - 2s + 1$

$f'(s) = 2s - 2$ *Linearity and Power rules*

2.1, #18 $f(x) = \frac{1}{2} x^{-1}$

$f'(x) = -\frac{1}{2}x^{-2}$ *Power rule*

2.1, #19 $f(x) = \sqrt{5}\, x^{1/2}$

$f'(x) = \frac{\sqrt{5}}{2\, x^{1/2}}$ *Power Rule*

3. (con't.)   2.1, #20 $f(x) = (x - 1)^{1/2}$

$$f'(x) = \frac{1}{2(x - 1)^{1/2}} \qquad \textit{Power Rule}$$

7.   **a.**   By the Power Rule: $f'(x) = 3x^2$

b. C^2 is a constant with derivative of 0: $g'(x) = 1$

11.   Write $f(x)$ as $7x^{-2} + x^{2/3} + C$ then $f'(x) = -14x^{-3} + \frac{2}{3}x^{-1/3}$

or $-\frac{14}{x^3} + \frac{2}{3\sqrt[3]{x}}$

15.   By the Product Rule: $(2x + 1)(-12x^2) + (1 - 4x^3)(2) = -32x^3 - 12x^2 + 2$

19.   Use the Product Rule letting $f(x) = x^2$

and $h(x) = (x + 2)^2 = (x + 2)(x + 2)$

$f'(x) = 2x$ and $h'(x) = (x + 2)(1) + (1)(x + 2) = 2x + 4$

Now, $g'(x) = f(x)h'(x) + f'(x)h(x)$

$$= x^2(2x + 4) + 2x(x + 2)^2 = 4x^3 + 12x^2 + 8x$$

Or in factored form: $4x(x + 2)(x + 1)$

You also could have expanded $g(x)$ to a polynomial first.

$g(x) = x^4 + 4x^3 + 4x^2$ then by linearity rule $g'(x) = 4x^3 + 12x^2 + 8x$.

23. $f(x) = -2x^{-2}$, $f'(x) = 4x^{-3}$, $f''(x) = -12x^{-4}$,

$f'''(x) = 48x^{-5}$, $f^{(4)}(x) = -240x^{-6}$.

27. $f(x) = x^2 - 3x - 5$, $f(-2) = 5$, so the point of tangency is $(-2, 5)$.

$f'(x) = 2x - 3$ and $f'(-2) = -7 = m_{\tan}$. Using the point-slope formula:

$y - 5 = -7[x - (-2)]$, $7x + y + 9 = 0$.

31. $f(x) = \frac{x^2 + 5}{x + 5}$; $f(1) = 1$, so the point of tangency is $(1, 1)$.

Using the Quotient Rule: $f'(x) = \dfrac{(x + 5)2x - (x^2 + 5)(1)}{(x + 5)^2}$

Simplifying: $f'(x) = \dfrac{x^2 + 10x - 5}{(x + 5)^2}$,   and $f'(1) = \frac{1}{6} = m_{\tan}$

Using the point-slope formula: $y - 1 = \frac{1}{6}(x - 1)$ or $x - 6y + 5 = 0$

35.   For a horizontal tangent we must have $f'(t) = 0$.

Write $f(t) = t^{-2} - t^{-3}$ and use the Power Rule: $f'(t) = -2t^{-3} + 3t^{-4}$

so $f'(t) = \dfrac{-2}{t^3} + \dfrac{3}{t^4} = \dfrac{-2t + 3}{t^4}$; $f'(t) = 0$ when $t = \frac{3}{2}$.

Now, $f(\frac{3}{2}) = \dfrac{1}{\frac{9}{4}} - \dfrac{1}{\frac{27}{8}} = \frac{4}{9} - \frac{8}{27} = \frac{4}{27}$.

So the point at which there is a horizontal tangent is: $(\frac{3}{2}, \frac{4}{27})$

39. For a horizontal tangent we must have $h'(x) = 0$. Using the Quotient Rule:

$$h'(x) = \frac{(2x + 3)(8x + 12) - (4x^2 + 12x + 9)(2)}{(2x + 3)^2}$$

$$= \frac{8x^2 + 24x + 18}{(2x + 3)^2} = \frac{2(2x + 3)^2}{(2x + 3)^2} = 2$$

Therefore, this is a line for which the slope is always 2; no horizontal tangents.

If you were alert and noticed in the beginning that the numerator of $h(x)$ was a trinomial square

$$h(x) = \frac{(2x + 3)^2}{2x + 3} = 2x + 3 \text{ then } h'(x) = 2.$$

Again we see that this is a line (with a hole at $-\frac{3}{2}$) with a slope of 2.

43. The given line $2x - y - 3 = 0$ has a slope of 2, so we want the points on $y = x^4 - 2x + 1$ that have a slope of 2. That is, $y' = 2$ so $4x^3 - 2 = 2$. $x^3 = 1 \Rightarrow$ only one real root: $x = 1$. $f(1) = 0$, so we want the equation of a line through the point $(1, 0)$ with a slope of 2:

$$y - 0 = 2(x - 1) \text{ or } 2x - y - 2 = 0.$$

47. We are looking for particular points (x_0, y_0) on the graph of $y = 4x^2$ which have a tangent at that point which will pass through the point $(2, 0)$.

$y' = 8x$, $f(x_0) = 4x_0^2$, $f'(x_0) = 8x_0$. So we have a point $(x_0, 4x_0^2)$ and the slope of the line at that point: $8x_0$. We can now write the equation of the line:

$y - 4x_0^2 = 8x_0(x - x_0)$. This line must pass through the point $(2, 0)$, so

$$0 - 4x_0^2 = 8x_0(2 - x_0)$$

$$4x_0^2 - 16x_0 = 0$$

$$4x_0(x_0 - 4) = 0$$

$$x_0 = 0, 4.$$

Therefore there are two points on the curve at which the tangent line will pass through $(2, 0)$: $(0, 0)$ and $(4, 64)$.

51. $y = \frac{1}{2}x^2 + 3$ $y' = x$ $y'' = 1$ $y''' = 0$.

$y''' + y'' + y' = x + 1$

$0 + 1 + x = x + 1$. This function satisfies the given equation.

53. Analysis: Use a little inductive reasoning:

For a 1st degree polynomial, P, the 1st derivative is a constant and the 2nd derivative will be 0. For P second degree, the 1st derivative is linear, the 2nd derivative is a constant, and the 3rd derivative is 0. To generalize, the $(k + 1)$th derivative is 0.

55.   Analysis: Start with the definition for $(f + g)'$ then commute and associate
the numerator to have the definition for f' and g'.

57.   Analysis: f^2 is just the product of two functions: $[f(x)][f(x)]$.
The proof will follow the same pattern as the proof for the product rule,
$[f(x)][g(x)]$.

59.   Analysis: In looking at the definition of $g'(x)$, in addition to $g(x) \neq 0$, we need
to verify that $g(x + \Delta x) \neq 0$. Show that this is true.

61.   Analysis: Treat this as the quotient of two functions, where the function in the
numerator is a constant, and use the quotient rule.

63.   Analysis: Use brute force. Let $g(x) = [f(x)f(x)]f(x)$. Then do it one more time
for part b.

2.3 Derivatives of the Trigonometric Functions

SURVIVAL HINT

If your trigonometry is rusty, it will save you time in the long run to do some serious
review right now.   About the second or third time you have to look up a graph, exact
value, or identity, add it to your list of formulas and try to memorize it.

3.   Multiply and divide by $\frac{3x}{2x}$ to set up the identity:

$$\lim_{x \to 0} \frac{3}{2} \frac{\frac{\sin 3x}{3x}}{\frac{\sin 2x}{2x}} = \frac{3}{2} \frac{\lim_{x \to 0} \frac{\sin 3x}{3x}}{\lim_{x \to 0} \frac{\sin 2x}{2x}} = \frac{3}{2}$$

7.   Divide numerator and denominator by x to set up the identities:

$$\lim_{x \to 0} \frac{\frac{1 - \cos x}{x}}{\frac{\sin x}{x}} = \frac{\lim_{x \to 0} \frac{1 - \cos x}{x}}{\lim_{x \to 0} \frac{\sin x}{x}} = \frac{0}{1} = 0$$

11. $\lim_{x \to 0} \frac{\sin^2 x}{x^2} = \lim_{x \to 0} \frac{\sin x}{x} \frac{\sin x}{x} = \lim_{x \to 0} \frac{\sin x}{x} \lim_{x \to 0} \frac{\sin x}{x} = (1)(1) = 1$

15.   Notice that $\cos \frac{\pi}{4}$ is a constant. $g'(t) = 2t - \sin t$

19.   Let $f(t) = (\sin t)(\sin t)$. Now use the product rule:

$$f'(t) = (\sin t)(\cos t) + (\sin t)(\cos t) = 2 \sin t \cos t = \sin 2t$$

21.   Each of the terms needs the product rule:

$$f'(x) = \sqrt{x}\,(-\sin x) + (\cos x)\tfrac{1}{2} x^{-1/2} + x(-\csc^2 x) + (\cot x)(1)$$
$$f'(x) = -\sqrt{x} \sin x + \tfrac{1}{2} x^{-1/2} \cos x - x \csc^2 x + \cot x$$

23. This is *not* a limit. Do not use the identity. Use the quotient rule:

$$f'(x) \; = \; \frac{x\,(\cos x) \; - \; \sin x(1)}{x^2} \; = \; \frac{x\cos x \; - \; \sin x}{x^2}$$

27. Use the quotient rule: $f'(x) \; = \; \dfrac{(1 \; - \; 2x)\,(\sec^2 x) \; - \; \tan x(-2)}{(1 \; - \; 2x)^2}$

$$f'(x) \; = \; \frac{\sec^2 x \; - \; 2x\sec^2 x \; + \; 2\tan x}{(1 \; - \; 2x)^2}$$

31. Using the quotient rule: $f'(x) \; = \; \dfrac{(1 \; - \; \cos x)(\cos x) \; - \; \sin x\,(\sin x)}{(1 \; - \; \cos x)^2}$

$$f'(x) \; = \; \frac{\cos x \; - \; \cos^2 x \; - \; \sin^2 x}{(1 \; - \; \cos x)^2} \; = \; \frac{\cos x \; - \; (\cos^2 x \; + \; \sin^2 x)}{(1 \; - \; \cos x)^2} \; = \; \frac{\cos x \; - \; 1}{(1 \; - \; \cos x)^2}$$

$$= \frac{-1}{1 - \cos x} \quad \text{or} \quad \frac{1}{\cos x - 1}$$

37. Write as $g(x) = (\sec x)(\sec x) \; - \; (\tan x)(\tan x) \; + \; \cos x$ and use product rule:

$$g'(x) \; = \; [(\sec x)(\sec x \tan x) + (\sec x)(\sec x \tan x)]$$
$$- \; [(\tan x)(\sec^2 x) \; + \; (\tan x)(\sec^2 x)] \; - \; \sin x$$

$$g'(x) \; = \; 2\sec^2 x \tan x \; - \; 2\sec^2 x \tan x \; - \; \sin x \; = \; -\sin x$$

Or, if you remember your trigonometric identities, $\sec^2 x \; - \; \tan^2 x \; = \; 1$, so

$$g(x) \; = \; 1 \; + \; \cos x, \quad g'(x) \; = \; -\sin x$$

39. $f(\theta) \; = \; \sin \theta; \quad f'(\theta) \; = \; \cos \theta; \quad f''(\theta) \; = \; -\sin \theta$

43. $f(\theta) \; = \; \sec \theta; \quad f'(\theta) \; = \; \sec \theta \tan \theta;$

$$f''(\theta) \; = \; \sec \theta(\sec^2 \theta) \; + \; \tan \theta(\sec \theta \tan \theta)$$

$$f''(\theta) \; = \; \sec^3 \theta \; + \; \sec \theta \tan^2 \theta \; = \; \sec \theta(\sec^2 \theta \; + \; \tan^2 \theta)$$

47. $g(y) \; = \; \csc y \; - \; \cot y; \quad g'(y) \; = \; -\csc y \cot y \; - \; (-\csc^2 y)$

$$= \; \csc y(\csc y \; - \; \cot y)$$

$$g''(y) \; = \; \csc y[-\csc y \cot y \; - \; (-\csc^2 y)] \; + \; (\csc y \; - \; \cot y)(-\csc y \cot y)$$

$$= \; -\csc^2 y \cot y \; + \; \csc^3 y \; - \; \csc^2 y \cot y \; + \; \csc y \cot^2 y$$

$$= \; \csc y(\csc^2 y \; - \; 2\csc y \cot y \; + \; \cot^2 y)$$

$$= \; \csc y(\csc y \; - \; \cot y)^2$$

49.

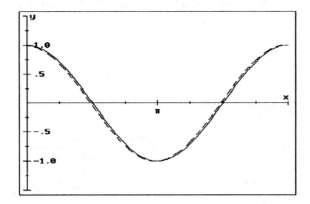

The two graphs almost coincide; that is, for "small" values of Δx, the difference quotient approximates the derivative. The *limit* as Δx approaches 0 of the difference quotient *is* $\cos x$.

SURVIVAL HINT

Exact values are preferred to decimal approximations. Your calculator and decimal approximations are acceptable in applications in which you know how many significant digits you are entitled to. Almost everything done with a calculator is approximate and not exact. If you use a calculator be sure to specify $\approx$, and do not say $=$.

53. $f(x) = \sin x, \quad f(\frac{\pi}{6}) = \frac{1}{2}, \quad f'(x) = \cos x, \quad f'(\frac{\pi}{6}) = \frac{\sqrt{3}}{2}$

Given a point, $(\frac{\pi}{6}, \frac{1}{2})$, and the slope at that point, $\frac{\sqrt{3}}{2}$, we can

find the equation of the line: $y - \frac{1}{2} = \frac{\sqrt{3}}{2}(x - \frac{\pi}{6})$

or $\sqrt{3}x - 2y + \left(1 - \frac{\sqrt{3}\pi}{6}\right) = 0$.

57. $y = x\cos x, \quad y(\frac{\pi}{3}) = \frac{\pi}{3}(\frac{1}{2}) = \frac{\pi}{6}$. Using the product rule:

$y' = x(-\sin x) + \cos x(1) = \cos x - x\sin x,$

$y'(\frac{\pi}{3}) = \frac{1}{2} - (\frac{\pi}{3})(\frac{\sqrt{3}}{2}) = \frac{3 - \pi\sqrt{3}}{6}$

Given a point, $(\frac{\pi}{3}, \frac{\pi}{6})$, and the slope at that point, $\dfrac{3 - \pi\sqrt{3}}{6}$, we can find the

equation of the line: $y - \frac{\pi}{6} = \dfrac{3 - \pi\sqrt{3}}{6}(x - \frac{\pi}{3})$

$$6y - \pi = (3 - \pi\sqrt{3})(x - \frac{\pi}{3})$$

or in standard form: $(\pi\sqrt{3} - 3)x + 6y - \pi^2\sqrt{3} = 0$.

59. $y = 4 \cos x - 2 \sin x;$ $\dfrac{dy}{dx} = -4 \sin x - 2 \cos x;$

$\dfrac{d^2 y}{dx^2} = -4 \cos x + 2 \sin x$

So $\dfrac{d^2 y}{dx^2} + y$ becomes: $-4 \cos x + 2 \sin x + (4 \cos x - 2 \sin x) = 0$

63. Analysis: $\cos h - 1$ suggests multiplication by $\cos h + 1$ in the numerator and denominator. Use the $\displaystyle \lim_{h \to 0} \dfrac{\sin h}{h} = 1$ identity to take care of the h in the denominator.

65. Analysis: Write as $\dfrac{\cos t}{\sin t}$ and use the quotient rule.

67. Analysis: Treat this as the quotient of two functions: $\dfrac{1}{\sin x}$. After applying the quotient rule your answer will be in terms of $\sin x$ and $\cos x$. Use the basic reciprocal identities to change these to $\csc x$ and $\cot x$.

69. Analysis: **a.** Area of sector $AOB = \dfrac{\pi}{2}$. Area of triangle $AOC = \dfrac{1}{2}(OC)\sin x$.

  $(OC) < 1$ and the area of the triangle is less than that of the sector.

  b. In the first quadrant x and $\sin x$ are ≥ 0.

  c. Because of **b.** we can approach 0 from both sides.

  d. If $\sin x$ and $\cos x$ are continuous on $[0, \dfrac{\pi}{2}]$, then any real number, x, can be factored into nx_1 where $x_1 < \dfrac{\pi}{2}$. The sums and products of continuous functions are continuous, so repeated use of the addition identity will also be continuous.

71. Analysis: Use the same area inequality that was used in the proof of the squeeze theorem, but express the areas in terms of $\tan h$.

  Area of $\triangle AOD = \dfrac{1}{2} \cos^2 h \tan h$, area of $\triangle BOC = \dfrac{1}{2} \tan h$.

  Invert the inequalities, multiply by $\dfrac{1}{2} \tan h$, and take the limit as $h \to 0$.

2.4 Rates of Change: Rectilinear Motion

1. The instantaneous rate of change, or rate of change at a point, is given by $f'(x)$.

  $f'(x) = 2x - 3,$ $f'(2) = 1$

5. By the quotient rule: $f'(x) = \dfrac{(3x + 5)(2) - (2x - 1)(3)}{(3x + 5)^2} = \dfrac{13}{(3x + 5)^2},$

  $f'(-1) = \dfrac{13}{4}$

9. $\quad f'(x) = 1 + \dfrac{(2-4x)(0)-3(-4)}{(2-4x)^2} = 1 + \dfrac{12}{(2-4x)^2} = \dfrac{16x^2-16x+16}{(2-4x)^2}$

$\qquad = \dfrac{4(x^2-x+1)}{(1-2x)^2}; \qquad f'(0) = 4$

Note that all of the algebra could have been avoided by evaluating the first
form of the derivative at $x = 0$. However it is desirable to have the
general form.

13. $\quad$ Write $f(x) = \left(x - \dfrac{2}{x}\right)\left(x - \dfrac{2}{x}\right)$; then use the product rule:

$\qquad f'(x) = \left(x - \dfrac{2}{x}\right)\left(1 + \dfrac{2}{x^2}\right) + \left(x - \dfrac{2}{x}\right)\left(1 + \dfrac{2}{x^2}\right)$

$\qquad = 2\left(\dfrac{x^2-2}{x}\right)\left(\dfrac{x^2+2}{x^2}\right) = \dfrac{2(x^4-4)}{x^3},$

$\qquad f'(1) = -6$

17. $\quad$ **a.** $\quad s(t) = t^3 - 9t^2 + 15t + 25$

$\qquad\qquad v(t) = s'(t) = 3t^2 - 18t + 15$

$\quad$ **b.** $\quad a(t) = s''(t) = v'(t) = 6t - 18$

$\quad$ **c.** $\quad v(t) = 3(t-1)(t-5) = 0 \quad$ at $\ t = 1, 5.$

$\qquad\qquad$ Checking the sign of $v(t)$ on $[0, 6]$, we find $v(t)$ is

$\qquad\qquad\qquad$ positive on $[0, 1)$, so object is advancing

$\qquad\qquad\qquad$ negative on $(1, 5)$, so object is retreating

$\qquad\qquad\qquad$ positive on $(5, 6]$, so object is advancing

$\qquad\qquad$ Note that the total distance covered is *not* $s(6) - s(0)$, but

$\qquad\qquad D = \big|s(1) - s(0)\big| + \big|s(5) - s(1)\big| + \big|s(6) - s(5)\big|$

$\qquad\qquad D = |32 - 25| + |0 - 32| + |7 - 0| = 46$

$\quad$ **d.** $\quad a(t) = 6(t - 3)$ is negative on $[0, 3)$, so object is decelerating

$\qquad\qquad$ and positive on $(3, 6]$, so object is accelerating

21. $\quad$ **a.** $\quad s(t) = 3\cos t, \quad v(t) = s'(t) = -3\sin t$

$\quad$ **b.** $\quad a(t) = v'(t) = -3\cos t$

$\quad$ **c.** $\quad v(t) = -3\sin t$ is negative on $[0,\pi)$, so object is retreating,

$\qquad\qquad$ and positive on $(\pi, 2\pi]$, so object is advancing.

$\qquad\qquad D = \big|s(\pi) - s(0)\big| + \big|s(2\pi) - s(\pi)\big| = |-3-3| + \big|3 - (-3)\big| = 12$

21. (con't.)

    d. $a(t) = -3 \cos t$ is negative on $[0, \frac{\pi}{2})$, so object is decelerating,

        positive on $(\frac{\pi}{2}, \frac{3\pi}{2})$, so object is accelerating,

        and negative on $(\frac{3\pi}{2}, 2\pi]$, so object is decelerating

 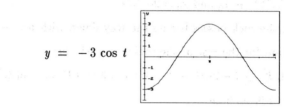

$$y = -3 \cos t$$

25. $x(t) = t^3 - 9t^2 + 24t + 20.$ $x'(t) = 3t^2 - 18t + 24 = 0$ at $t = 2, 4.$

$x(t)$ is advancing on $[0, 2)$ and $(4, 8]$ and retreating on $(2, 4)$. So the total distance

traveled $= |x(2) - x(0)| + |x(8) - x(4)| + |x(4) - x(2)|$

$D = |40 - 20| + |148 - 36| + |36 - 40|$

   $= 20 + 112 + 4 = 136$ units

27. **a.** $s(t) = 10t + \dfrac{5}{t+1},$ $v(t) = 10 - \dfrac{5}{(t+1)^2},$

        $v(4) = 10 - \frac{5}{25} = 9.8$ m/s

    b. $v(t) > 0$ so the distance traveled during the 5th minute is $s(5) - s(4),$

        $D = \left(50 + \frac{5}{6}\right) - \left(40 + \frac{5}{5}\right) = 9.8$ m

SURVIVAL HINT

The formula for the position of a falling object is used frequently enough that it is
worth the effort to remember it.

31.   The general equation for position, s, of a falling object (measured in feet) is

$s(t) = -16t^2 + v_0 t + s_0.$ In this case the object was dropped so $v_0 = 0,$

and $s_0 = 90.$ Solving this equation for the time it takes for the rock to hit the

ground: $s(t) = 0 : 0 = -16t^2 + 90$

                $t \approx 2.37$ sec.

Now the second rock has to travel a distance H in one second less. $v_0 = 0,$

    $s_0 = H,$ $0 = -16(1.37)^2 + H$

        $H \approx 30$ ft

35.   On Mars $g = 12$ ft/s^2, so $s(t) = -6\,t^2 + v_0 t + s_0$,

$v(t) = -12t + v_0$.

Since the rock goes up and back to ground level in 4 s,

it reaches its maximum height in 2 s. At that time $v = 0$.

So $0 = -12(2) + v_0$ and $v_0 = 24$ ft/s

Therefore, the rock passes her on the way down with $v = -24$ ft/s.

The equation for the rest of the rock's trip will be:

$s(t) = -6\,t^2 + (-24)t + 0$,   for $t = 3$, $s(3) = -6(3)^2 - 24(3) = -126$ ft

The cliff is 126 ft. high.

39. $q(t) = 0.05t^2 + 0.1t + 3.4$

    a. $q'(t) = 0.1t + 0.1$; $q'(1) = 0.2$ ppm/yr

    b.   Change in first year $= q(1) - q(0)$

$$= (0.05 + 0.1 + 3.4) - (3.4)$$

$$= 0.15 \text{ ppm}$$

    c.   Change in second year $= q(2) - q(1)$

$$= (0.2 + 0.2 + 3.4) - (3.55)$$

$$= 0.25 \text{ ppm}$$

43. $P(x) = 2x + 4x^{3/2} + 5000$

    a. $P'(x) = 2 + 6x^{1/2}$

$P'(9) = 2 + 18 = 20$ persons per month

    b. $\dfrac{P'(9)}{P(9)}(100) = \dfrac{20}{5126}(100) \approx 0.39\%$ per month

47. $N(t) = 5 - t^2(t - 6) = -t^3 + 6t^2 + 5$

    a. $N'(t) = -3t^2 + 12t$.

Percentage change, $C(t) = (100\,)\dfrac{-3t^2 + 12t}{-t^3 + 6t^2 + 5}$

47. (con't.)

    b.   We want to find values of $C(t)$ such that $(100)\dfrac{-3t^2+12t}{-t^3+6t^2+5} > 30$, so $\dfrac{3t^2-12t}{t^3-6t^2-5} > 0.3$. Evaluating $C(t)$ for various weeks:

$C(0) = 0$, $C(1) = 0.90$, $C(2) \approx .57$, $C(3) \approx .28$, $C(4) = 0$,

$C(5) = -.5$, $C(6) = -7.2$.

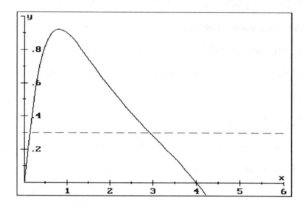

    From early in the first week until almost the third week the disease has epidemic level. Epidemic will be declared for the second and third weeks.

51.   Analysis: The volume of a cube is given by $V = s^3$, and the surface by $S = 6s^2$. The change in volume with respect to the change in the length of an edge is given by $\dfrac{dV}{ds}$. Find this value and express it in terms of S.

53.   Analysis: What is asked for is $\dfrac{dP}{dV}$. We could expand the equation and solve explicitly for P, and then differentiate with respect to V, or differentiate implicitly (Sect. 2.6) with respect to V, remembering that P is a function of V and A, B, and k are constants.

2.5 The Chain Rule

SURVIVAL HINT

The chain rule is probably the most used of all the differentiation rules. Most interesting and/or useful functions are compositions, and their derivatives require the chain rule. When you are finding $f'[u(x)]$, identify the u function, and remember to include the $\dfrac{du}{dx}$. Write out intermediate steps; many mistakes are made by trying to do too much "in your head".

3. $\dfrac{dy}{dx} = \dfrac{dy}{du}\dfrac{du}{dx} = \dfrac{d}{du}(u^2+1)\dfrac{d}{dx}(3x-2) = 2u(3) = 6(3x-2)$

7. $\dfrac{dy}{dx} = \dfrac{dy}{du}\dfrac{du}{dx} = \dfrac{d}{du}(\cos u)\,\dfrac{d}{dx}(x^2 + 7) = (-\sin u)(2x) = -2x\sin(x^2 + 7)$

11. **a.** $g'(u) = 15u^{14}$

      b. $u'(x) = 6x + 5$

      c. $f'(x) = 15(3x^2 + 5x - 7)^{14}(6x + 5)$

15. $s'(\theta) = [\cos(4\theta + 2)](4) = 4\cos(4\theta + 2)$

19. Use the product rule and the chain rule:

$$p'(x) = (\sin x^2)\dfrac{d}{dx}(\cos x^2) + (\cos x^2)\dfrac{d}{dx}(\sin x^2)$$

$$= (\sin x^2)(-\sin x^2)(2x) + (\cos x^2)(\cos x^2)(2x)$$

$$= 2x(\cos^2 x^2 - \sin^2 x^2) \qquad \textit{Recognize this as the double angle identity}$$

$$= 2x(\cos 2x^2)$$

If we applied the double angle identity to the original function our work would have been easier: $p(x) = \tfrac{1}{2}(2\sin x^2\cos x^2) = \tfrac{1}{2}\sin 2x^2$

$$p'(x) = \tfrac{1}{2}(\cos 2x^2)(4x) = 2x\cos 2x^2$$

23. Use the product rule and the chain rule:

$$f'(t) = (1 - t^2)\dfrac{d}{dx}(\sin t^2) + (\sin t^2)\dfrac{d}{dx}(1 - t^2)$$

$$= (1 - t^2)(\cos t^2)(2t) + (\sin t^2)(-2t)$$

$$= 2t(\cos t^2 - t^2\cos t^2 - \sin t^2)$$

27. We need the power rule, the quotient rule and the chain rule:

$$f'(x) = \dfrac{1}{2}\left(\dfrac{2x^2 - 1}{3x^2 + 2}\right)^{-\frac{1}{2}}\dfrac{d}{dx}\left(\dfrac{2x^2 - 1}{3x^2 + 2}\right)$$

$$= \dfrac{1}{2}\left(\dfrac{2x^2 - 1}{3x^2 + 2}\right)^{-\frac{1}{2}}\left\{\dfrac{(3x^2 + 2)(4x) - (2x^2 - 1)(6x)}{(3x^2 + 2)^2}\right\}$$

$$= \dfrac{1}{2}\left(\dfrac{2x^2 - 1}{3x^2 + 2}\right)^{-\frac{1}{2}}\left\{\dfrac{14x}{(3x^2 + 2)^2}\right\}$$

$$= \dfrac{14x}{2(2x^2 - 1)^{1/2}(3x^2 + 2)^{3/2}}$$

$$= \dfrac{7x}{(2x^2 - 1)^{1/2}(3x^2 + 2)^{3/2}}$$

31. We need the power rule and the chain rule.

$$f'(x) = \tfrac{1}{3}\left(x^2 + 2\sqrt{x}\right)^{-2/3}\dfrac{d}{dx}(x^2 + 2\sqrt{x})$$

$$= \tfrac{1}{3}\left(x^2 + 2\sqrt{x}\right)^{-2/3}\left(2x + \dfrac{1}{\sqrt{x}}\right)$$

$$= \dfrac{2x\sqrt{x} + 1}{3\sqrt{x}(x^2 + 2\sqrt{x})^{2/3}}$$

33. $f'(x) = \frac{1}{2}(x^2 + 5)^{-1/2}(2x)$

$f'(2) = \frac{2}{\sqrt{9}} = \frac{2}{3}$ *Using the point-slope formula:*

$y - 3 = \frac{2}{3}(x - 2)$ or $2x - 3y + 5 = 0$

37. $f'(x) = x^2 \frac{d}{dx}[\tan(4 - 3x)] + [\tan(4 - 3x)]\frac{d}{dx}(x^2)$

$= x^2[-3\sec^2(4 - 3x)] + 2x\tan(4 - 3x)$

$f'(0) = 0$ and $f(0) = 0$.   The line through $(0, 0)$ with a slope of 0 is the x-axis: $y = 0$

41. $f'(x) = \dfrac{(x + 2)^3 2(x - 1) - (x - 1)^2 3(x + 2)^2}{(x + 2)^6}$

$= \dfrac{(x - 1)(x + 2)^2[2(x + 2) - 3(x - 1)]}{(x + 2)^6}$

$= \dfrac{(x - 1)(x + 2)^2(-x + 7)}{(x + 2)^6}$

$= \dfrac{(x - 1)(-x + 7)}{(x + 2)^4}$

$f'(x) = 0$ and therefore has a horizontal tangent when $x = 1, 7$

43.   **a.** $f'(x) = (x + 3)^2 2(x - 2) + (x - 2)^2 2(x + 3)$

$= 2(x + 3)(x - 2)(2x + 1)$

There are horizontal tangents where $f'(x) = 0$: $x = -3, -\frac{1}{2}, 2$

b.   To find $f''(x)$ write $f'(x)$ as a polynomial:

$f'(x) = 4x^3 + 6x^2 - 22x - 12$

$f''(x) = 12x^2 + 12x - 22 = 2(6x^2 + 6x - 11)$

Use the quadratic formula to find when this is 0.

$x = \dfrac{-6 \pm \sqrt{36 - 4(6)(-11)}}{2(6)} = \dfrac{-6 \pm 10\sqrt{3}}{12} = \dfrac{-3 \pm 5\sqrt{3}}{6}$

47.   Here we have a function of a function, so the chain rule is required.

$C'(t) = \dfrac{dC}{dq}\dfrac{dq}{dt} = (0.4q + 1)(2t + 100)$.   Since $q = t^2 + 100t$,

$C'(t) = [0.4(t^2 + 100t) + 1](2t + 100)$

$C'(1) = [0.4(101) + 1](102) = \$4,222.80$ per hour

51. **a.** $f(x) = L(x^2), \quad f'(x) = L'(x^2)(2x) = \frac{1}{x^2}(2x) = \frac{2}{x}$

b. $f(x) = L\left(\frac{1}{x}\right), \quad f'(x) = L'\left(\frac{1}{x}\right)\left(-\frac{1}{x^2}\right) = x\left(-\frac{1}{x^2}\right) = -\frac{1}{x}$

c. $f(x) = L\left(\frac{2}{3\sqrt{x}}\right), \quad f'(x) = L'\left(\frac{2}{3\sqrt{x}}\right)\left(-\frac{1}{3x^{3/2}}\right) = \left(\frac{3\sqrt{x}}{2}\right)\left(-\frac{1}{3x^{3/2}}\right) = -\frac{1}{2x}$

d. $f(x) = L\left(\frac{2x+1}{1-x}\right),$

$$f'(x) = L'\left(\frac{2x+1}{1-x}\right)\left(\frac{(1-x)2 - (2x+1)(-1)}{(1-x)^2}\right)$$

$$= \left(\frac{1-x}{2x+1}\right)\left(\frac{3}{(1-x)^2}\right) = \frac{3}{(2x+1)(1-x)}$$

55. **a.** From the figure: $\tan\theta = \frac{s(t)}{2}$, but $\theta = 6\pi t$, so $s(t) = 2\tan 6\pi t$.

b. $s'(t) = 2(\sec^2 6\pi t)(6\pi) = 12\pi\sec^2 6\pi t$. Although we are not given t, we are given the distance from the lighthouse is 4. So $s = 2\sqrt{3}$ and $2\sqrt{3} = 2\tan 6\pi t$, $\tan 6\pi t = \sqrt{3}$, $6\pi t = \arctan\sqrt{3}$, $t = \frac{\arctan\sqrt{3}}{6\pi}$,

$t = \frac{1}{18}$ min.

We can now find $s'(t) = 12\pi\sec^2 6\pi t = 12\pi\sec^2\frac{\pi}{3}$

$$= 12\pi(4) = 48\pi \approx 150.8 \text{ km/min}$$

59. Analysis: To find $g'(2)$ we will need to know $g'(x)$. We can find this since $g(x)$ is given as the product of two functions. The function $g'(x)$ will involve both $f\left(\frac{x}{x-1}\right)$ and $f'\left(\frac{x}{x-1}\right)$. We do not know $f\left(\frac{x}{x-1}\right)$, but we do know $f(2)$. $f'\left(\frac{x}{x-1}\right)$ can be found since we were given $f'(x)$. Remember to use the chain rule when finding the derivative of $f\left(\frac{x}{x-1}\right)$.

61. Analysis: No difficulty here. Just carefully apply the chain rule carefully here.

2.6 Implicit Differentiation

SURVIVAL HINT

Since y is assumed to be a function of x, implicit differentiation is simply the chain rule with $y = u(x)$.

1. $2x + 2y\dfrac{dy}{dx} = 0, \quad \dfrac{dy}{dx} = -\dfrac{x}{y}$

5. $2x + 3(x\frac{dy}{dx} + y) + 2y\frac{dy}{dx} = 0$

$(3x + 2y)\frac{dy}{dx} = -2x - 3y$

$\frac{dy}{dx} = -\frac{2x + 3y}{3x + 2y}$

9. $-\frac{1}{y^2}\frac{dy}{dx} - \frac{1}{x^2} = 0, \quad \frac{dy}{dx} = -\frac{y^2}{x^2}$

13. We need to use the chain rule and the product rule on the argument of cosine:

$(-\sin xy)\frac{d}{dx}(xy) = -2x, \quad (-\sin xy)(x\frac{dy}{dx} + y) = -2x$

$(-x\sin xy)\frac{dy}{dx} = y\sin xy - 2x, \quad \frac{dy}{dx} = \frac{y\sin xy - 2x}{-x\sin xy} = \frac{2x - y\sin xy}{x\sin xy}$

17. **a.** $1 - \frac{1}{y^2}\frac{dy}{dx} = 0, \quad \frac{dy}{dx} = y^2$

   b. Multiplying by y: $xy + 1 = 5y, \quad (x - 5)y = -1, \quad y = \frac{-1}{x - 5}$

   $\frac{dy}{dx} = -\frac{-1}{(x - 5)^2} = \frac{1}{(x - 5)^2}.$ You can see this is also y^2.

21. For the slope we need $\frac{dy}{dx}$, so: $\frac{d}{dx}[\sin(x - y)] = \frac{d}{dx}(xy)$

$[\cos(x - y)]\frac{d}{dx}(x - y) = x\frac{dy}{dx} + y$

$[\cos(x - y)](1 - \frac{dy}{dx}) = x\frac{dy}{dx} + y$

$\cos(x - y) - [\cos(x - y)]\frac{dy}{dx} = x\frac{dy}{dx} + y$

$[-x - \cos(x - y)]\frac{dy}{dx} = y - \cos(x - y)$

$\frac{dy}{dx} = \frac{y - \cos(x - y)}{-x - \cos(x - y)} = \frac{\cos(x - y) - y}{\cos(x - y) + x}$

$\frac{dy}{dx}(0, \pi) = \frac{\cos(-\pi) - \pi}{\cos(-\pi)} = \frac{-1 - \pi}{-1} = \pi + 1.$ So the line is:

$y - \pi = (\pi + 1)(x - 0), \quad (\pi + 1)x - y + \pi = 0$

25. For the slope we need $\frac{dy}{dx}$, so: $3x^2 + 3y^2\frac{dy}{dx} - \frac{9}{2}(x\frac{dy}{dx} + y) = 0$

$(3y^2 - \frac{9}{2}x)\frac{dy}{dx} = \frac{9}{2}y - 3x^2$

$\frac{dy}{dx} = \frac{9y - 6x^2}{6y^2 - 9x} = \frac{3y - 2x^2}{2y^2 - 3x}$

$\frac{dy}{dx}(2, 1) = \frac{3 - 8}{2 - 6} = \frac{5}{4}.$ This looks reasonable in the illustration.

29. $7x + 5y^2 = 1;$ $7 + 10yy' = 0;$

$y' = -\dfrac{7}{10y}.$ For y'' we again differentiate:

$$\frac{d}{dx}(y') = \frac{d}{dx}\left(-\frac{7}{10y}\right), \quad y'' = \frac{7}{10y^2}\,y', \quad y'' = \left(\frac{7}{10y^2}\right)\left(-\frac{7}{10y}\right) = -\frac{49}{100y^3}$$

33.   **a.**   Considering that u is some function of v we will differentiate both sides of the equation with respect to v: $\dfrac{d}{dv}\left(\dfrac{u^2}{a^2}\right) + \dfrac{d}{dv}\left(\dfrac{v^2}{b^2}\right) = \dfrac{d}{dv}(1)$

$$\frac{2u}{a^2}\frac{du}{dv} + \frac{2v}{b^2} = 0, \quad \frac{du}{dv} = -\frac{\dfrac{2v}{b^2}}{\dfrac{2u}{a^2}} = -\frac{a^2 v}{b^2 u}$$

b.   This time differentiate with respect to u : $\dfrac{d}{du}\left(\dfrac{u^2}{a^2}\right) + \dfrac{d}{du}\left(\dfrac{v^2}{b^2}\right) = \dfrac{d}{du}(1)$

$$\frac{2u}{a^2} + \left(\frac{2v}{b^2}\right)\frac{dv}{du} = 0, \quad \frac{dv}{du} = -\frac{\dfrac{2u}{a^2}}{\dfrac{2v}{b^2}} = -\frac{b^2 u}{a^2 v}$$

37.   **a.**   Differentiating with respect to x: $2x + 2g(x)[g'(x)] = 0$

$g'(x) = -\dfrac{x}{g(x)}$

b. $g(x) = -\sqrt{10 - x^2}$ is a differentiable function of x for which $g(x) < 0$.

Using the chain rule: $g'(x) = -\dfrac{1}{2\sqrt{10 - x^2}}\dfrac{d}{dx}(10 - x^2) = \dfrac{x}{\sqrt{10 - x^2}}$

Verifying the result of (a): $\dfrac{x}{\sqrt{10 - x^2}} = \dfrac{-x}{-\sqrt{10 - x^2}} = \dfrac{-x}{g(x)}$

41. For a vertical tangent we need $y' = \infty$. This occurs when the denominator of an expression is 0, and the numerator is not 0. Finding y' by implicit differentiation:

$$\frac{3}{2}(x^2 + y^2)^{1/2}\frac{d}{dx}(x^2 + y^2) = \frac{1}{2}(x^2 + y^2)^{-1/2}\frac{d}{dx}(x^2 + y^2) + 1$$

$$\frac{3}{2}(x^2 + y^2)^{1/2}(2x + 2yy') = \frac{1}{2}(x^2 + y^2)^{-1/2}(2x + 2yy')$$

$$\left(3y(x^2 + y^2)^{1/2} - y(x^2 + y^2)^{-1/2}\right)y' = x(x^2 + y^2)^{-1/2} - 3x(x^2 + y^2)^{1/2} + 1$$

$$y' = \frac{x(x^2 + y^2)^{-1/2} - 3x(x^2 + y^2)^{1/2} + 1}{3y(x^2 + y^2)^{1/2} - y(x^2 + y^2)^{-1/2}}.$$

Multiplying numerator and denominator by $(x^2 + y^2)^{1/2}$ we get:

$$y' = \frac{x - 3x(x^2 + y^2) + (x^2 + y^2)^{1/2}}{3y(x^2 + y^2) - y}$$

$$= \frac{x(1 - 3x^2 - 3y^2) + (x^2 + y^2)^{1/2}}{y(3x^2 + 3y^2 - 1)}$$

Now the denominator, $y(3x^2 + 3y^2 - 1) = 0$, when $y = 0$, or $3x^2 + 3y^2 - 1 = 0$.

To find the x-coordinate of the point let $y = 0$ in the equation of the cardioid:

At $y = 0$, $\qquad x^3 = |x| + x.$

$\qquad$ For $x < 0$, $\quad x^3 = 0$, $\quad x = 0.$

$\qquad$ For $x > 0$. $\quad x^3 = 2x$, $\quad x = 0, \pm\sqrt{2}$

At $(0, 0)$ the derivative is undefined. At $(-\sqrt{2}, 0)$ the function is undefined.

So this factor of the denominator gives one vertical tangent at $(\sqrt{2}, 0)$.

Now consider the other factor: $3x^2 + 3y^2 - 1 = 0$.

$x^2 + y^2 = \frac{1}{3}$ is a circle with radius of $\frac{\sqrt{3}}{3}$.

We need to find the points where it intersects the cardioid.

Substituting $x^2 + y^2 = \frac{1}{3}$ into the cardioid equation:

$\left(\frac{1}{3}\right)^{3/2} = \left(\frac{1}{3}\right)^{1/2} + x$, $\quad x = -\frac{2\sqrt{3}}{9}$. Using the circle to find y at this point:

$$\left(-\frac{2\sqrt{3}}{9}\right)^2 + y^2 = \frac{1}{3}, \quad y^2 = \frac{5}{27}, \quad y = \pm\frac{\sqrt{15}}{9}$$

There are two additional vertical tangents, at $\left(-\frac{2\sqrt{3}}{9}, \frac{\sqrt{15}}{9}\right)$ and $\left(-\frac{2\sqrt{3}}{9}, -\frac{\sqrt{15}}{9}\right)$. This seems to agree with the shape of a cardioid.

45.   Analysis: To find the equation of a line we can use the point-slope formula. We have been given the point (x_0, y_0), but we will need the slope. This can be found by differentiating implicitly the equation for the hyperbola, and finding $y'(x_0, y_0)$.

47.   Analysis: Find the equation of the tangent to the curve at any point (x_0, y_0) in the same manner as problem 45. Find the x intercept by letting $y = 0$, and the y intercept by letting $x = 0$. The sum of these intercepts can then be shown to be the square of the original function, which equals C.

49.   Analysis: To find the angle between the curves we need the slopes of the tangent lines for each curve at their intersection. Differentiate each curve implicitly to find y'. Solve the equations simultaneously to find the points of intersection.

2.7 Related Rates

SURVIVAL HINT

The most common error on related rate problems is to use constants in place of variables. For instance: if a 12 ft ladder is sliding down a wall and we wish to know how fast the top is decending when the bottom is 4 ft from the wall, *do not* put the 4 ft on your figure. The 12 ft is a constant and belongs on the figure. The vertical and horizontal distances are in motion and should be given variable names. The 4 ft is a point at which you wish to evaluate your solved equation. Draw a careful figure, label constants and variables, then look for a *general* relationship. Then find a specific equation by substituting the given values. Do not try to write the specific equation in the first step.

3.   Differentiating $5x^2 - y = 100$ with respect to t; $10x \dfrac{dx}{dt} - \dfrac{dy}{dt} = 0$.

Now, evaluate when $\dfrac{dx}{dt} = 10$, $x = 10$ and $y = 400$.

$10(10)(10) - \dfrac{dy}{dt} = 0$, $\dfrac{dy}{dt} = 1{,}000$

7.   Differentiating $xy = 10$ with respect to t; $x \dfrac{dy}{dt} + y\dfrac{dx}{dt} = 0$

If $x = 5$ then $y = 2$ and $\dfrac{dx}{dt} = -2$. So $5(\dfrac{dy}{dt}) + 2(-2) = 0$, $\dfrac{dy}{dt} = \dfrac{4}{5}$

11.   Using Hooke's law, with $k = 12$, $F(x) = -12x$.

We are asked to find $\dfrac{dF}{dt}$ when $x = 3$ and $\dfrac{dx}{dt} = \dfrac{1}{4}$.

Differentiating with respect to t: $\dfrac{dF}{dt} = -12 \dfrac{dx}{dt}$; $\dfrac{dF}{dt} = -3$.

Notice that since F is a linear function of x, the change in F is a constant, and does not depend upon the value of x.

15.   We are asked to find $\frac{dr}{dt}$, given specific values for r and $\frac{dA}{dt}$, so we need a general equation relating A and r. Of course this is $A = \pi r^2$. So

$$\frac{dA}{dt} = 2\pi r\, \frac{dr}{dt}; \quad 4 = 2\pi(1)\frac{dr}{dt}; \quad \frac{dr}{dt} = \frac{2}{\pi}\ \text{ft/s} \ \approx \ 0.637\ \text{ft/s}$$

19.   Given $PV = C$, we are asked to find $\frac{dP}{dt}$ for the specific values of $V = 30$, $P = 90$, and $\frac{dV}{dt} = 10$. Differentiating Boyle's law with respect to t: $P\frac{dV}{dt} + V\frac{dP}{dt} = 0$. Using our specified values: $90(10) + 30\frac{dP}{dt} = 0$; $\frac{dP}{dt} = -30\ \text{lb/in}^2/\text{s}$. The negative indicates the pressure is decreasing.

23.   We are asked to find $\frac{dr}{dt}$ at a specific instant when $\frac{dV}{dt} = -5$, and $r = 4$. We need a general relationship between r and V for a sphere : $V = \frac{4}{3}\pi r^3$. So $\frac{dV}{dt} = 4\pi r^2\, \frac{dr}{dt}$. Substituting our specific values: $-5 = 4\pi(4)^2\frac{dr}{dt}$; $\frac{dr}{dt} = -\frac{5}{64\pi} \approx -0.025\ \text{in/min}$

Next we are asked to find $\frac{dS}{dt}$ at that same instant, so we need a generality relating S and r : $S = 4\pi r^2$. So $\frac{dS}{dt} = 8\pi r\, \frac{dr}{dt}$; $\frac{dS}{dt} = 8\pi(4)\Big(-\frac{5}{64\pi}\Big)$; $\frac{dS}{dt} = -2.5\ \text{in}^2/\text{min}$.

27.   Draw a figure and put in the constant 13, and the variables x and y:

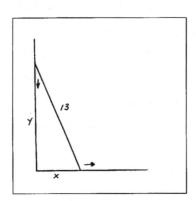

We are asked to find $\frac{dx}{dt}$ at the instant when $\frac{dy}{dt} = -3$ and $x = 5$.

The Pythagorean theorem is the general relationship here: $x^2 + y^2 = 13^2$.

So $2x\frac{dx}{dt} + 2y\frac{dy}{dt} = 0$. Now we need to have a value for y. Find y at this instant by $5^2 + y^2 = 13^2$; $y = 12$. Now find $\frac{dx}{dt}$:

$$2(5)\frac{dx}{dt} + 2(12)(-3) = 0; \quad \frac{dx}{dt} = 7.2\ \text{ft/s}$$

31.   Draw a figure and put in the constant 12 and the variables x and D:

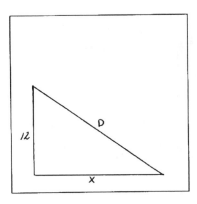

We are asked to find $\frac{dx}{dt}$ at the instant when $\frac{dD}{dt} = -6$ and $x = 16$.

The Pythagorean theorem relates these variables: $12^2 + x^2 = D^2$.

So, $0 + 2x\frac{dx}{dt} = (2D)\frac{dD}{dt}$; $2(16)\frac{dx}{dt} = 2D(-6)$.   Use the Pythagorean

relationship at the instant when $x = 16$ to find $D = 20$.

So, $\frac{dx}{dt} = -\frac{240}{32} = -7.5$ ft/min. Notice that the boat is moving toward

the pier faster than the rate at which the rope is being pulled in. This is not

intuitively obvious — but it is correct!

35.   Do not fall into the trap of putting 16 ft. and 14 ft. on your figure. These are

not constants. We do have similar triangles giving the proportion:

$\frac{s}{10+L} = \frac{6}{L}$.   We could differentiate the general expression in this form,

but it will be nicer if we solve for L: $L = \frac{60}{s-6}$. Remembering that s and L

are both functions of time, we differentiate implicitly: $\frac{dL}{dt} = -\frac{60}{(s-6)^2}\frac{ds}{dt}$.

We now need values for s and $\frac{ds}{dt}$. Since $s = 30 - 16t^2$ and $t = 1$; $s = 14$,

and $\frac{ds}{dt} = -32t$. Therefore $\frac{dL}{dt} = -\frac{60}{64}(-32) = 30$ ft/s.

39. Draw a figure using the variables of θ, x and D, and the constant of 3:

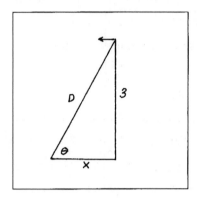

We are asked to find $\frac{d\theta}{dt}$ at the instant when $x = 4$ and $\frac{dx}{dt} = -500$.

We need a general relationship between θ and x: $\tan \theta = \frac{3}{x}$.

Since all of these variables are functions of time we can differentiate implicitly:

$(\sec^2\theta) \ \frac{d\theta}{dt} \ = \ -\frac{3}{x^2} \ \frac{dx}{dt}$. Now we need some instantaneous values:

$x = 4$, $\frac{dx}{dt} = -500$, and from our figure we see that $D = 5$ when $x = 4$ so,

$\sec \theta = \frac{5}{4}$. Substituting: $(\frac{5}{4})^2 \ \frac{d\theta}{dt} \ = \ -\frac{3}{4^2} \ (-500);$ $\frac{d\theta}{dt} = 60$ rad/hr $= 1$ rad/min

If for some reason you need degrees, 60 rad $= \dfrac{10,800}{\pi}$ degrees and

1 hour $= 3600$ sec, so 60 rad/hr $= \frac{3}{\pi}$ deg/s ≈ 1 deg/s.

43. Draw a figure using the variables of A for the distance traveled by the first

ship, B for the distance traveled by the second ship, D for the distance between

them, and the constant angle of $60°$.

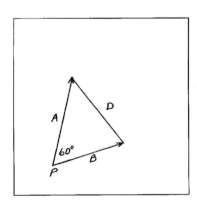

43. (con't.)

We are asked to find $\frac{dD}{dt}$ at $t = 2$ and $t = 5$. As the problem suggests, these variables are all generally related by the law of cosines:

$D^2 = A^2 + B^2 - 2AB \cos \theta$.   Differentiating with respect to t:

$2D \frac{dD}{dt} = 2A \frac{dA}{dt} + 2B \frac{dB}{dt} - \left(A \frac{dB}{dt} + B \frac{dA}{dt} \right)$

(Note that $\cos \theta = \frac{1}{2}$ is a constant.)

Now we need some instantaneous values: A at 2 hours $= 8(2) = 16$,

B at 2 hours $= 1(12) = 12$ (It left 1 hour later.) $\frac{dA}{dt} = 8$, $\frac{dB}{dt} = 12$,

and we need to find D: $D^2 = 16^2 + 12^2 - 2(16)(12)(\frac{1}{2})$; $D = \sqrt{208}$.

Substituting: $2 \sqrt{208} \frac{dD}{dt} = 2(16)(8) + 2(12)(12) - [16(12) + 12(8)]$

$\frac{dD}{dt} = \frac{128}{\sqrt{208}} = \frac{32 \sqrt{13}}{13} \approx 8.875$ knots.

At 5 pm $A = 5(8) = 40$, $B = 4(12) = 48$, $\frac{dA}{dt} = 8$, $\frac{dB}{dt} = 12$, and

$D^2 = 40^2 + 48^2 - 2(40)(48)(\frac{1}{2})$; $D = \sqrt{1984}$.

Substituting: $2 \sqrt{1984} \frac{dD}{dt} = 2(40)(8) + 2(48)(12) - [40(12) + 48(8)]$

$\frac{dD}{dt} = \frac{464}{\sqrt{1984}} = \frac{58 \sqrt{31}}{31} \approx 10.417$ knots.

47.   Analysis: We are asked to find $\frac{d\theta}{dt}$. To express θ as a function of the given information and the variable x, use the tangent. Implicit differentiation with respect to t will give an expression that can be solved for $\frac{d\theta}{dt}$. Take $\lim\limits_{x \to 0}$ of this expression to answer part **b**.

2.8 Differentials and Tangent Line Approximations

SURVIVAL HINT

The differential can be thought of as the change in f along the tangent. If Δx is "small" and the function is reasonably "well behaved", then the difference between the actual value of f and the corresponding value on the tangent will be relatively small.

It is most useful when extrapolating data, that is, when we really do not know how the function will behave beyond a given point, and our best bet is the tangent at that point.

3. **a.** 1.967 989 671

    b. Let $x_0 = 16$ and $\Delta x = -1$ and $f(x) = x^{1/4}$.

$$(15)^{1/4} = f(x_0 + \Delta x) \approx f(x_0) + f'(x)\Delta x$$
$$= (16)^{1/4} + \tfrac{1}{4}(x)^{-3/4}(-1)$$
$$= 2 - \tfrac{1}{4}(16)^{-3/4} = 2 - \tfrac{1}{32} \approx 1.97$$

7. **a.** 1.4641

    b. Let $x_0 = 1$ and $\Delta x = 0.1$ and $f(x) = x^4$.

$$(1.1)^4 = f(x_0 + \Delta x) \approx f(x_0) + f'(x)\Delta x$$
$$= (1)^4 + 4x^3(.1) = 1 + (.4) = 1.4$$

11. **a.** $-0.009\ 999\ 833\ 33$

    b. Let $x_0 = \frac{\pi}{2}$ and $\Delta x = 0.01$ and $f(x) = \cos x$.

$$\cos\left(\tfrac{\pi}{2} + 0.01\right) = f(x_0 + \Delta x) \approx f(x_0) + f'(x)\Delta x$$
$$= \cos\left(\tfrac{\pi}{2}\right) + (-\sin \tfrac{\pi}{2})(0.01) = 0 - 0.01 = -0.01$$

15. $f(x) = x^2 + x - 2;\quad f'(x) = 2x + 1;\quad f'(1) = 3;\quad \Delta x = b - a = 0.001$

$$\Delta f = f(x_0 + \Delta x) - f(x_0) \approx f'(x_0)\Delta x = f'(1)(0.001) = 0.003$$

19. $f(x) = \sqrt{x} + x^{3/2};\quad f'(x) = \frac{1}{2\sqrt{x}} + \frac{3}{2}\sqrt{x};\quad \Delta x = b - a = 0.001$

$$\Delta f = f(x_0 + \Delta x) - f(x_0) \approx f'(x_0)\Delta x = f'(4)(0.001)$$
$$= \left(\tfrac{1}{4} + 3\right)0.001 = 0.00325$$

21. $d(2x^3) = 6x^2\,dx$ (Do not forget the dx — a differential must always equal another differential.)

25. $d(x \cos x) = (\cos x - x \sin x)\,dx$

29. $d\left(x\sqrt{x^2 - 1}\right) = \left(x(d\sqrt{x^2 - 1}) + \sqrt{x^2 - 1}\,d(x)\right)$

$$= \left(\frac{x^2}{(x^2 - 1)^{1/2}} + (x^2 - 1)^{1/2}\right)dx$$

$$= \frac{2x^2 - 1}{(x^2 - 1)^{1/2}}\,dx$$

33. Let $f(x) = x^5 - 2x^3 + 3x^2 - 2;\quad f'(x) = 5x^4 - 6x^2 + 6x;$

$x_0 = 3$ and $\Delta x = dx = 0.01$. Now, $f(x_0 + \Delta x) \approx f(x_0) + f'(x_0)dx$,

so $f(3.01) \approx f(3) + f'(3)dx$

$$= [(3)^5 - 2(3)^3 + 3(3)^2 - 2] + [5(3)^4 - 6(3)^2 + 6(3)](0.01)$$
$$= 214 + 3.69 = 217.69$$

Comparing this to a calculator value of $f(3.01) = 217.715\ 588\ 2$ we see an error of $\approx 0.025\ 588\ 2$

37. $V = \frac{4}{3}\pi r^3;$ $V' = 4\pi r^2;$ $r_0 = 6;$ $dr = 0.01(6)$

$\Delta V \approx V'(r_0)\,dr = 4\pi(6)^2(0.06) = 8.64\,\pi \approx \pm 27.143 \text{ in}^3.$

So the volume is $904.779 \pm 27.143 \text{ in}^3.$

$\dfrac{27.143}{904.779} \approx 0.03 \text{ or } 3\%$

41. $Q(L) = 60{,}000L^{1/3};$ $Q'(L) = 20{,}000L^{-2/3};$ $Lo = 1000;$ $\Delta L = -60$

$\Delta Q \approx Q'(L_0)\Delta L = 20{,}000(1000)^{-2/3}(-60) = -12{,}000 \text{ units}$

45. $P(x) = \dfrac{596}{\sqrt{x}};$ $P'(x) = -\dfrac{298}{x^{3/2}};$ $x_0 = 59;$ $\Delta x = 1$

$\Delta P \approx P'(x_0)\,\Delta x = -\dfrac{298}{59^{3/2}}(1) \approx -0.658 \text{ beats/min; or about 2 beats}$

every 3 minutes.

49. Be careful with the differentiation as the chain rule is needed twice:

$N(\theta) = \dfrac{1}{\sin^4\left(\frac{\theta}{2}\right)};$ $N'(\theta) = -4[\sin^{-5}\left(\frac{\theta}{2}\right)\cos\left(\frac{\theta}{2}\right)]\frac{1}{2} = -2\sin^{-5}\left(\frac{\theta}{2}\right)\cos\left(\frac{\theta}{2}\right);$

$\theta_0 = 1 \text{ rad},$ $\Delta\theta = 0.1 \text{ rad}.$

$\Delta N \approx N'(\theta)\Delta\theta = -2[\sin^{-5}(0.5)\cos(0.5)](0.1) \approx -6.93 \text{ particles/unit area.}$

SURVIVAL HINT

Any time you see the adjective "marginal" in an economics problem you can translate it as derivative. It designates the rate of change.

53. $C(x) = \frac{2}{5}x^2 + 3x + 10;$ $p(x) = \frac{1}{5}(45 - x).$

a. Marginal cost $= C'(x) = \frac{4}{5}x + 3.$

b. $\frac{4}{5}x + 3 = 23;$ $x = 25.$ So $p(25) = \frac{1}{5}(20) = \$4.00$

c. The cost of producing the 11th unit $\approx C'(10) = \frac{4}{5}(10) + 3 = \$11.00$

d. The actual cost of producing the 11th unit $= C(11) - C(10)$

$= [\frac{2}{5}(11)^2 + 3(11) + 10] - [\frac{2}{5}(10)^2 + 3(10) + 10]$

$= 91.4 - 80 = \$11.40$

57. Analysis: "Approximately equal to" suggests the use of the approximation formula.

a. Let $f(x) = \sqrt{x},$ $f'(x) = \dfrac{1}{2\sqrt{x}},$ $f(1) = 1,$ $f'(1) = \frac{1}{2},$ and the formula says for small values of $h,$ $f(1 + h) \doteq f(1) + f'(1)h.$

b. Let $f(x) = \frac{1}{x},$ $f'(x) = -\dfrac{1}{2x^2},$ $f(1) = 1,$ $f'(1) = -\frac{1}{2},$ and the formula says for small values of $h,$ $f(1 + h) \approx f(1) + f'(1)h.$

59. Analysis: By using 81 instead of 100 you will see that $\triangle x$ is 16 instead of 3, which will make the approximation less accurate by about a factor of 40.

2.9 The Newton-Raphson Method for Approximating Roots

SURVIVAL HINT

There are other difficulties, in addition to the ones listed in the text, that will cause the Newton-Raphson method to fail. If the function is not too complex, or a graphing calculator is available, it is worth the time to sketch the region of the root and see if your results are converging and reasonable.

1. $f(x) = x^3 - 29 \quad f'(x) = 3x^2 \quad x_{n+1} = x_n - \dfrac{f(x_n)}{f'(x_n)} = x_n - \dfrac{x_n^3 - 29}{3x_n^2}$

So $x_{n+1} = \dfrac{2x_n^3 + 29}{3x_n^2}$. Letting $x_0 = 3$, $x_1 \approx 3.074\,074\,074$,

$x_2 \approx 3.072\,317\,830$, $x_3 \approx 3.072\,316\,825$. At this point we can be confident of the value to the nearest 0.00001 as 3.072 31. Compare this to the exact root of $\sqrt[3]{29}$.

5. $f(x) = x^4 - 16.2 \quad f'(x) = 4x^3 \quad x_{n+1} = x_n - \dfrac{f(x_n)}{f'(x_n)}$

$= x_n - \dfrac{x_n^4 - 16.2}{4x_n^3} = \dfrac{3x_n^4 + 16.2}{4x_n^3}$. Letting $x_0 = 2$, $x_1 \approx 2.006\,25$

$x_2 \approx 2.006\,220\,915$, $x_3 \approx 2.006\,220\,915$. Since x_2 and x_3 agree we have accuracy to 9 decimal places. Compare this with the exact answer of $\sqrt[4]{16.2}$.

9. $f(x) = x^3 + 2x - 1 \quad f'(x) = 3x^2 + 2 \quad x_{n+1} = x_n - \dfrac{f(x_n)}{f'(x_n)}$

$= x_n - \dfrac{x_n^3 + 2x_n - 1}{3x_n^2 + 2} = \dfrac{2x_n^3 + 1}{3x_n^2 + 2}$. Letting $x_0 = 0$, $x_1 \approx 0.5$

$x_2 \approx 0.454\,545\,455$, $x_3 \approx 0.453\,398\,337$

13. $f(x) = x^3 + 3x - 1 \quad f'(x) = 3x^2 + 3 \quad x_{n+1} = x_n - \dfrac{f(x_n)}{f'(x_n)}$

$= x_n - \dfrac{x_n^3 + 3x_n - 1}{3x_n^2 + 3} = \dfrac{2x_n^3 + 1}{3x_n^2 + 3}$. Now since $f(0) = -1$, and $f(1) = 3$,

and polynomials are continuous, the intermediate value theorem guarantees a root on this interval. Using linear interpolation, a first guess would be $x_0 = 0.25$.

$x_1 \approx 0.323\,529\,512$, $x_2 \approx 0.322\,185\,883$, $x_3 \approx 0.322\,185\,355$ which gives us a root accurate to 0.000001 of $x \approx 0.322\,185$

17. $f(x) = \sqrt[3]{x-3} - x - 1$ $f'(x) = \frac{1}{3}(x-3)^{-2/3} - 1$

$$x_{n+1} = x_n - \frac{f(x_n)}{f'(x_n)} = x_n - \frac{\sqrt[3]{x_n-3} - x_n - 1}{\frac{1}{3}(x_n-3)^{-2/3} - 1}.$$ Since $f(x)$ is continuous

on the interval, and $f(-3) \approx 1.3066$, and $f(-2) \approx -0.7099$, the intermediate

value theorem guarantees a root on this interval. Using linear interpolation

a first guess would be $x_0 = -2.4$, $x_1 \approx -2.719\ 779\ 498$,

$x_2 \approx -2.781\ 904\ 986$, $x_3 \approx -2.793\ 619\ 243$, $x_4 \approx -2.795\ 815\ 698$,

$x_5 \approx -2.796\ 227\ 107$, $x_6 \approx -2.796\ 304\ 151$. So to the nearest 0.001

our root is $x \approx -2.796$

21. $f(x) = 1 - \frac{1}{x}$ $f'(x) = \frac{1}{x^2}$ $x_{n+1} = x_n - \frac{f(x_n)}{f'(x_n)}$

$$= x_n - \frac{1 - \frac{1}{x_n}}{\frac{1}{x_n^2}} = -x_n^2 + 2x_n$$

If we let $x_0 = 2$ then $x_1 = 0$ and all other iterations $= 0$. This would seem

to indicate that $x = 0$ is a root, but $f(x)$ is not defined at $x = 0$

25. $f(x) = x^3 - 3x - \cos x + \sin x$ $f'(x) = 3x^2 - 3 + \sin x + \cos x$

$$x_{n+1} = x_n - \frac{f(x_n)}{f'(x_n)} = x_n - \frac{x_n^3 - 3x_n - \cos x_n + \sin x_n}{3x_n^2 - 3 + \sin x_n + \cos x_n}$$

$f(0) = -1$, $f(\frac{\pi}{2}) \approx 0.163\ 395\ 6$

Since this is a continuous function the intermediate value theorem guarantees

a root on the interval $[0, \frac{\pi}{2}]$. A good estimate would be $x_0 = 1.5$.

$x_1 \approx 1.541\ 144\ 180$, $x_2 \approx 1.539\ 807\ 364$, $x_3 \approx 1.539\ 805\ 927$,

$x_4 \approx 1.539\ 805\ 927$

We have a root at $x \approx 1.539\ 805\ 927$

Of course, a different x_0 will give a different sequence of estimates,

but they will converge to the same value.

29. $f(x) = 4x^3 + 3x^2 - 12x + 7 \qquad f'(x) = 12x^2 + 6x - 12$

$$x_{n+1} = x_n - \frac{f(x_n)}{f'(x_n)} = x_n - \frac{4x_n^3 + 3x_n^2 - 12x_n + 7}{12x_n^2 + 6x_n - 12}$$

$$= \frac{8x_n^3 + 3x_n^2 - 7}{12x_n^2 + 6x_n - 12}$$

After some trial values you will find $f(-1) = 2$ and $f(-2) = -2.458333$ so there is a root on $[-2, -1]$. Let $x_0 = -2.5$

$x_1 \approx -2.359\,375, \quad x_2 \approx -2.346\,512, \quad x_3 \approx -2.346\,407,$

$x_4 \approx -2.346\,407.$ So there is a root at $x \approx -2.346\,407$

33. $f(x) = -2x^4 + 3x^2 + \frac{11}{8} \qquad f'(x) = -8x^3 + 6x$

a. $f(0) = \frac{11}{8}$ and $f(2) = -\frac{149}{8}$ so there is a root on $[0, 2]$.

Also note that $f(x)$ is an even function and, therefore, symmetric about the y-axis. So it must have at least two roots.

b. $x_{n+1} = x_n - \frac{f(x_n)}{f'(x_n)} = x_n - \frac{-2x_n^4 + 3x_n^2 + \frac{11}{8}}{-8x_n^3 + 6x_n}$

$$= \frac{-6x_n^4 + 3x_n^2 - \frac{11}{8}}{-8x_n^3 + 6x_n}$$

If $x_0 = \frac{1}{2}$ then $x_1 = -0.5$, $x_2 = 0.5$ and the values will continue to oscillate. This is due to the symmetry about the y-axis.

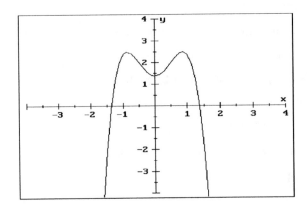

39.   Analysis: If $f(x_n) = 0$, then x_n is a root. If we put this value in the formula what do we get for x_{n+1}?

41.   Analysis: If we want a value $a = \frac{1}{x}$, then we need to find a root to the function $f(x) = \frac{1}{x} - a$. The Newton-Raphson formula will give $x_{n+1} = 2x_n - ax_n^2$.

Chapter 2 Review

PRACTICE PROBLEMS

15. Write the middle term with fractional exponents, and remember to use the chain

rule on the third term: $y = x^3 + x^{3/2} + \cos(2x)$

$$\frac{dy}{dx} = 3x^2 + \tfrac{3}{2}x^{1/2} - 2\sin(2x)$$

16. This is a function of a function, so the chain rule is necessary.

$$\frac{dy}{dx} = -4\left(\sqrt{3}x + \frac{3}{x^2}\right)^{-5}\left(\sqrt{3} + (-2)\frac{3}{x^3}\right)$$

$$= -4\left(\sqrt{3}x + \frac{3}{x^2}\right)^{-5}\left(\sqrt{3} - \frac{6}{x^3}\right)$$

$$= \frac{-4x^{10}}{(\sqrt{3}x^3 + 3)^5} \cdot \frac{\sqrt{3}x^3 - 6}{x^3}$$

$$= \frac{-4\,x^7(\sqrt{3}x^3 - 6)}{(\sqrt{3}x^3 + 3)^5}$$

17. The chain rule must be used several times here.

$$\frac{dy}{dx} = \tfrac{1}{2}[\sin(3 - x^2)]^{-\frac{1}{2}}[\cos(3 - x^2)](-2x)$$

$$= -x\frac{[\cos(3 - x^2)]}{\sqrt{\sin(3 - x^2)}}$$

18. Here we can not solve explicitly for y, so implicit differentiation is used.

$$x\frac{dy}{dx} + y + 3y^2\frac{dy}{dx} = 0$$

$$(x + 3y^2)\frac{dy}{dx} = -y$$

$$\frac{dy}{dx} = \frac{-y}{x + 3y^2}$$

19. Think before you compute! $\sin^2 a + \cos^2 a = 1$. So $y = 1$, $\frac{dy}{dx} = 0$

20.   The equation is quadratic in y, and could be solved explicitly for y. But it

will probably be easier to use implicit differentiation.

$$2x + xy' + y - 4yy' = 0$$

$$(x - 4y)y' = -(2x + y)$$

$$y' = \frac{2x + y}{4y - x} \qquad \text{Now find the second derivative with the quotient rule.}$$

$$y'' = \frac{(4y - x)(2 + y') - (2x + y)(4y' - 1)}{(4y - x)^2}$$

$$y'' = \frac{(4y - x)\left(2 + \dfrac{2x + y}{4y - x}\right) - (2x + y)\left(4\dfrac{2x + y}{4y - x} - 1\right)}{(4y - x)^2}$$

$$= \frac{(4y - x)\left(\dfrac{9y}{4y - x}\right) - (2x + y)\left(\dfrac{9x}{4y - x}\right)}{(4y - x)^2}$$

$$= \frac{9[(4y^2 - xy) - (2x^2 + xy)]}{(4y - x)^3}$$

$$= \frac{9(4y^2 - 2xy - 2x^2)}{(4y - x)^3}$$

$$= \frac{18(2y^2 - xy - x^2)}{(4y - x)^3}$$

$$= \frac{18(x + 2y)(x - y)}{(x - 4y)^3}$$

21. $f(x) = x - 3x^2$, $f(x + h) = (x + h) - 3(x + h)^2$

$$= x + h - 3x^2 - 6xh - 3h^2$$

$$\frac{dy}{dx} = \lim_{h \to 0} \frac{f(x + h) - f(x)}{h}$$

$$= \lim_{h \to 0} \frac{(x + h - 3x^2 - 6xh - 3h^2) - (x - 3x^2)}{h}$$

$$= \lim_{h \to 0} \frac{(h - 6xh - 3h^2)}{h} = \lim_{h \to 0} (1 - 6x - 3h) = 1 - 6x$$

22. We will use the point-slope form to get the equation. We have been given a point, (1, 8), and we can find $\frac{dy}{dx}$ to determine the slope.

$$\frac{dy}{dx} = (x^2 + 3x - 2)(-3) + (7 - 3x)(2x + 3)$$

$$\frac{dy}{dx}(1) = (2)(-3) + (4)(5) = 14$$

$y - 8 = 14(x - 1)$ which is standard form is $14x - y = 6$

23. We will use the point-slope form to get the equation. $y = f(1) = \frac{1}{2}$, so the point is $(1, \frac{1}{2})$.

$$\frac{dy}{dx} = 2\sin\left(\frac{\pi x}{4}\right)\cos\left(\frac{\pi x}{4}\right)\left(\frac{\pi}{4}\right). \quad \frac{dy}{dx}(1) = 2\left(\frac{\sqrt{2}}{2}\right)\left(\frac{\sqrt{2}}{2}\right)\left(\frac{\pi}{4}\right) = \frac{\pi}{4}$$

$y - \frac{1}{2} = \frac{\pi}{4}(x - 1)$ or in slope-intercept form: $y = \frac{\pi}{4}x - \left(\frac{\pi}{4} - \frac{1}{2}\right)$.

The equation of the normal line at that same point must have a slope that is the negative reciprocal of the slope of the tangent, $m = -\frac{4}{\pi}$.

The equation of the normal: $y - \frac{1}{2} = -\frac{4}{\pi}(x - 1)$ or $y = -\frac{4}{\pi}x + \frac{4}{\pi} + \frac{1}{2}$.

24. Express the area as a function of the radius, then differentiate with respect to time to find the change in area with respect to time.

$$A = \pi r^2 \text{ so } \frac{dA}{dt} = 2\pi r \frac{dr}{dt}$$

$$\frac{dr}{dt} = 0.5 \text{ when } r = 2 \text{ so}$$

$$\frac{dA}{dt} = 2\pi(2)(0.5) = 2\pi \text{ ft}^2/\text{s}$$

25. $f(x) = x^3 - x^2 - x - 13 \quad f'(x) = 3x^2 - 2x - 1$

$f(0) = -13$, $f(3) = 2$, and since polynomials are continuous, the hypotheses of the intermediate value theorem are met, and we are guaranteed at least one point, c, in the interval such that $f(c) = 0$.

By the Newton-Raphson method $\quad x_{n+1} = x_n - \dfrac{f(x_n)}{f'(x_n)}$

$$x_{n+1} = x_n - \frac{x_n^3 - x_n^2 - x_n - 13}{3x_n^2 - 2x_n - 1}$$

$$= \frac{2x_n^3 - x_n^2 + 13}{3x_n^2 - 2x_n - 1}$$

Using a little linear interpolation for a first guess: $\frac{x}{3} = \frac{13}{15}$, $x = \frac{39}{15} \approx 2.6$

$x_1 \approx 2.93977$

$x_2 \approx 2.89650$

$x_3 \approx 2.89571$

$x_4 \approx 2.89571$

Chapter 3
Additional Applications of the Derivative

3.1 Extreme Values of a Continuous Function

SURVIVAL HINT

It is likely that you have had a test covering Chapter 2. The concepts of limit and
differentiation, and skill in finding derivatives accurately and quickly, are essential
to success in the remainder of the course. It is essential that your exam not be filed
away until you have mastered the material of Chapter 2! Analyze your mistakes. They
are usually one of three types: you did not understand the concept, you made
a computational or algebraic error, you did not have time to finish the exam. For the
first type ask the instructor, a tutor, or another student to explain the concept to you.
It is a good idea to form a small group of 3 to 5 students to do a "post mortem" on the
exam. For the second type error attempt to organize your work better. Put more steps
on paper and do less in your head. Stay composed on the exam and do not rush. It is
usually better to do fewer problems and get them right, than to turn in more completed
problems full of errors. The solution for the third type of difficulty, assuming most
other students completed the exam, is practice. The more problems you solve, the faster
you become. Do all of the chapter review problems, and some of the supplementary
problems before the next exam. This takes **time**, but the standard rule-of-thumb is two
hours out of class for every hour in class.

SURVIVAL HINT

Remember that the *candidates* for extrema are the endpoints, and points where the derivative is either zero or does not exist.

3.   The extreme value theorem guarantees that a continuous function on a closed interval will have an absolute maximum and minimum. The candidates for these extrema are the endpoints and the critical points. The critical points are where $f'(x)$ is undefined or equal to 0. For this function, $f'(x) = 3x^2 - 6x = 0$ at $x = 0, 2$. Testing these points and the endpoints:
$$f(0) = 0, \ f(2) = -4, \ f(-1) = -4, \ f(3) = 0.$$
So we have a maximum of 0 at $(0, 0)$ and $(3, 0)$ and a minimum of -4 at $(2, -4)$ and $(-1, -4)$.

7. $f'(x) = 5x^4 - 4x^3 = x^3(5x - 4) = 0$ at $x = 0, \frac{4}{5}$. Testing the candidates:
$$f(0) = 0, f(\tfrac{4}{5}) = -\tfrac{256}{3125}, \ f(-1) = -2, \ f(1) = 0.$$
So we have a maximum of 0 at $(0, 0)$ and $(1, 0)$ and a minimum of -2 at $(-1, -2)$.

11. $f'(u) = 2 \sin u \cos u - \sin u = \sin u(2 \cos u - 1) = 0$ at $u = 0, \frac{\pi}{3}$, on $[0, 2]$. Testing the candidates:
$$f(0) = 1, \ f(\tfrac{\pi}{3}) = \tfrac{5}{4}, \ f(2) \approx 0.4107.$$
So we have a maximum of $\frac{5}{4}$ at $(\frac{\pi}{3}, \frac{5}{4})$ and a minimum of 0.4107 at $(2, 0.4107)$.

15. $f'(u) = -\frac{2}{3} x^{-\frac{1}{3}} = 0$ nowhere, but is undefined at $x = 0$.
Testing the candidates:
$$f(0) = 1, \ f(-1) = 0, \ f(1) = 0.$$
So we have a maximum of 1 at $(0, 1)$ and a minimum of 0 at $(-1, 0)$ and $(1, 0)$.

19. $f'(x) = 7560\pi \cos(378\pi x) = 0$ when $378\pi x = \frac{\pi}{2} + k\pi$ so $x = \frac{1}{756} + \frac{k}{378}$. Testing the candidates on $[-1, 1]$:

$f(-1) = 0, f(1) = 0$, and since this is a sine curve with period of $\frac{1}{189}$, $\frac{1}{756} + \frac{k}{378}$ will have maximum values when k is even and minimum values when k is odd. The first positive maximum value will be at $x = \frac{1}{756}$ ≈ 0.00132, and the first minimum for a positive x value will be at $x = \frac{3}{756}$ ≈ 0.00397. There will be 378 points with a maximum of 20, and 378 points with a minimum of -20. Your graphing calculator will most likely not be able to show this on $[-1, 1]$. A change of the horizontal scale gives the following graph:

19 (con't.)

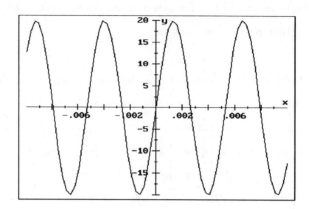

23.   Notice that whenever $f(x)$ has a common factor, $f'(x)$ will have that factor
as well. If we are looking for values for which $f'(x) = 0$, we can ignore
the factor. $f'(x) = 3x^2 - 12x + 9 = 3(x^2 - 4x + 3)$
$= 3(x - 3)(x - 1) = 0$ at $x = 1, 3$. Testing our candidates for extrema:
$f(0) = \frac{1}{6}$, $f(2) = \frac{1}{2}$, $f(1) = \frac{5}{6}$.
We have a minimum at $(0, \frac{1}{6})$ and a maximum at $(1, \frac{5}{6})$.

SURVIVAL HINT

If your function has a discontinuity on the specified interval you may have additional
"endpoints" to consider.

27. $h'(x) = \sec^2 x + \sec x \tan x = \sec x(\sec x + \tan x) = 0$ when $\sec x = 0$,
(never), or when $\sec x + \tan x = 0$
$$\frac{1 + \sin x}{\cos x} = 0$$
$$\sin x = -1, \quad x = \frac{3\pi}{2} \text{ on } [0, 2\pi].$$
We also need to consider points where $f'(x)$ is undefined: $\frac{\pi}{2} + k\pi$.
But $f(x)$ is also undefined at these points. Our only candidates are the
endpoints. Testing our candidates:
$$h(0) = 1, \quad h(2\pi) = 1.$$
Since the function is not continuous on the given interval, we are not guaranteed
any extrema, and in this case the range of $f(x)$ is $(-\infty, \infty)$.

27. (con't.)

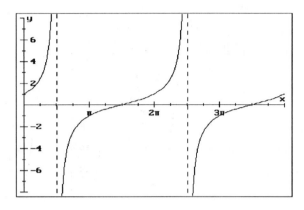

31. $f'(x) = 2x = 0$ at $x = 0$.

Testing our candidates: $f(0) = 0$, $f(-1) = 1$, $f(1) = 1$, so the smallest value on the interval is 0, at $(0, 0)$.

35. Since $\lim\limits_{t \to 1} (-t^2 - t + 2) = 0 \neq f(t) = 2$, the function is discontinuous at $t = 1$, and we will need to look at $[-2, 1)$ and $[1, 3]$ separately.

On $[-2, 1)$ $f'(t) = -2t - 1 = 0$ at $t = -\frac{1}{2}$. $f(-2) = 0$, $f(-\frac{1}{2}) = \frac{9}{4}$,

and as we have noted $\lim\limits_{t \to 1} (-t^2 - t + 2) = 0$.

On $[1, 3]$ we have a straight line with a slope of -1, so $f(1) = 2$ and $f(3) = 0$.

The largest value on the interval occurs at $(-\frac{1}{2}, \frac{9}{4})$.

43. $v(t) = s'(t) = 4t^3 - 6t^2 - 24t + 60$.

$v'(t) = 12t^2 - 12t - 24 = 12(t - 2)(t + 1) = 0$ when $t = -1, 2$.

Since this is a continuous function on a closed interval the extreme value theorem guarantees a maximum and a minimum value. Testing the candidates:

$$v(0) = 60, \quad v(2) = 20, \quad v(3) = 42.$$

The maximum velocity is 60 when $t = 0$.

47. Given that $3x + y = 126$, $xy = x(126 - 3x)$, maximize

$f(x) = -3x^2 + 126x$. $f'(x) = -6x + 126 = 0$ when $x = 21$.

The requirement that our value be in the first quadrant gives us an interval of $[0, \frac{126}{3}]$. Testing our candidates: $f(0) = 0$, $f(21) = 1323$, $f(42) = 0$.

The maximum value of 1323 occurs when $x = 21$ and $y = 63$.

51.   The average cost, $\overline{C}$, is the cost per item $= \frac{C}{x} = 0.125x + \frac{20\ 000}{x}$.
$\overline{C}'(x) = 0.125 - \frac{20\ 000}{x^2} = 0$ when $x^2 = 160\ 000$, $x = 400$.
Now consider $\overline{C}(x) = C'(x)$:

$$0.125x + \frac{20\ 000}{x} = 0.25x$$
$$0.125x^2 = 20\ 000$$
$$x = 400.$$

53.   Analysis: An excellent problem solving technique is to make a complex problem into a simpler one. When does $S(x) = (a_1 - x)^2$ have a minimum? When does $S(x) = (a_1 - x)^2 + (a_2 - x)^2$ have a minimum?

When does $S(x) = (a_1 - x)^2 + (a_2 - x)^2 + (a_3 - x)^2$ have a minimum?

Now use a little inductive reasoning and answer the question for the nth sum.

55.   Analysis: On the given interval one function is monotonic decreasing from ∞ to 8 and the other is monotonic increasing from 27 to ∞. Since they are both continuous on intermediate values, their sum is continuous and must have a finite minimum. To establish the relationship of x and θ at the minimum, set $f'(x) = 0$. You should find that $\tan x = \frac{2}{3}$. The graph of the component functions and their sum may be helpful:

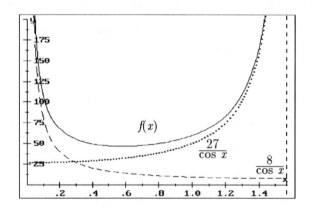

3.2 The Mean Value Theorem

SURVIVAL HINT

Never use a theorem until you have verified that the hypotheses are met.

3.   Polynomials are everywhere continuous and differentiable, so the hypotheses

of MVT are met. $f'(x) = 4x$, so there exists a c on the interval $[0, 2]$ such that

$f'(c) = \dfrac{f(2) - f(0)}{2 - 0}$. $\quad 4c = \dfrac{9 - 1}{2}$, $\quad 8c = 8$, $\quad c = 1$.

7.   Polynomials are everywhere continuous and differentiable, so the hypotheses

of MVT are met. $f'(x) = 4x^3$, so there exists a c on the interval $[-1, 2]$

such that $f'(c) = \dfrac{f(2) - f(-1)}{2 - (-1)}$. $\quad 4c^3 = \dfrac{18 - 3}{3}$, $\quad 12c^3 = 15$, $\quad c = \sqrt[3]{\dfrac{5}{4}}$.

11. $f(x)$ is continuous and differentiable everywhere except at $x = -1$, so on $[0, 2]$

the hypotheses are met. $f'(x) = -\dfrac{1}{(x + 1)^2}$, so there is at least one c on the

interval where $f'(c) = \dfrac{f(2) - f(0)}{2 - 0}$. $\quad -\dfrac{1}{(c + 1)^2} = \dfrac{\frac{1}{3} - 1}{2}$,

$\dfrac{2}{3}(c + 1)^2 = 2$, $(c + 1)^2 = 3$, $c + 1 = \pm\sqrt{3}$, $c = -1 \pm \sqrt{3}$.

Note that only $c = -1 + \sqrt{3}$ lies in the specified interval.

15. $f(x)$ is continuous on the closed interval, but is not differentiable on the open

interval as there is a cusp at $(2, 0)$. The theorem will not apply.

19. $f'(x) = \frac{1}{3}x^{-\frac{2}{3}}$, which is not differentiable at $x = 0$.

The theorem does not apply.

23. $f(x)$ is continuous everywhere, and $f'(x) = 2\sin x \cos x = \sin 2x$ so $f(x)$

is differentiable everywhere. $f(a) = f(b) = 1$. Rolle's theorem applies.

27. $g(x) = |x|$ is continuous everywhere, but has a cusp at $x = 0$,

so is not differentiable there. The theorem does not apply.

31.   **a.**   Let $f(x) = \cos x - 1$ on $[0, x]$. The hypotheses of the MVT apply,

so there exists a w on the interval such that

$$f'(w) = \dfrac{f(0) - f(x)}{0 - x}$$
$$\sin w = \dfrac{0 - \cos x + 1}{-x}$$

$(-\sin w)\,x = \cos x - 1$.

b.   Since w is on the interval $[0, x]$, as x approaches 0, w must approach 0 also.

So $\lim\limits_{x \to 0} \dfrac{\cos x - 1}{x} = \lim\limits_{w \to 0} \dfrac{\cos w - 1}{w} = \lim\limits_{w \to 0} (-\sin x) = 0$.

35.   A straight line will be continuous and differential, and $v(t_1) = v(t_2)$, so the

hypotheses of Rolle's theorem are met. This guarantees that there exists

some value c in the time interval for which $v'(c) = 0$. Since the acceleration is

$v'(t)$, this is what was to have been shown.

39.   If $x > 15$ then $f(x) = \sqrt{1 + x}$ is continuous and differentiable on $[15, x]$ and the

hypotheses of the MVT are met. Therefore there exists a c on $[15, x]$ such that

$$f'(c) = \frac{1}{2\sqrt{1 + c}} = \frac{f(x) - f(15)}{x - 15}$$

$$\frac{1}{2\sqrt{1 + c}} = \frac{\sqrt{1 + x} - 4}{x - 15}$$

$$\sqrt{1 + x} = 4 + \frac{x - 15}{2\sqrt{1 + c}},$$ but c is on $[15, x]$ and so $c > 15$,

making the denominator of the fraction greater then 8,

and $4 + \dfrac{x - 15}{2\sqrt{1 + c}} < 4 + \dfrac{x - 15}{8}$, so by transitivity

$$\sqrt{1 + x} < 4 + \frac{x - 15}{8}.$$

41.   Analysis: $f'(x) = \sec^2 x$, which does not have a horizontal tangent on $(0, \pi)$.

Therefore, part of the hypotheses of the MVT must not be met. Which part?

43.   Analysis: Here we will use the contrapositive of Rolle's theorem: if there is not a

horizontal tangent, then the hypotheses are not met. But the function is

continuous and differentiable, so we must fail to have $f(a) = f(b)$.

Show that $f'(x) \neq 0$.

45.   Analysis: It would seem that we could apply the MVT to each of the functions

and then substitute. The difficulty is that the c for $f(x)$ is not necessarily the

same c for $g(x)$. This proof is similar to the proof of the MVT. Define a new

function: $h(x) = [f(b) - f(a)][g(x) - g(a)] - [g(b) - g(a)][f(x) - f(a)]$.

47.   Analysis: Start with $f(x)$, find $f'(x)$, then let $2a = A$ and $b = B$.

3.3   First Derivative Test

SURVIVAL HINT

The first derivative gives us useful information about the behavior of a function, but

a better test for extrema will be presented in the next section.

3.   Notice that each function has a value of 0 when the other has a horizontal tangent. So this does not tell us which is which. Next consider that when f is increasing f' must be positive. This identifies which graph is f and which is f'.

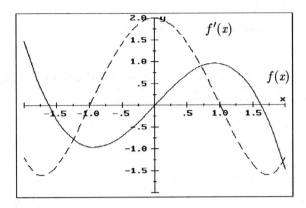

7.   For all values less than 2, $f'(x) = -1$. For all values greater than 2, $f'(x) = 1$. At 2, $f'(x)$ is not defined.

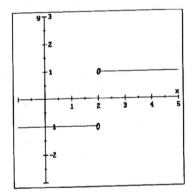

11.   We are given the graph of f' and asked to sketch f. Where f' is 0, f must have a horizontal tangent. Where f' is positive, f must be increasing. Where f' is negative, f must be decreasing.

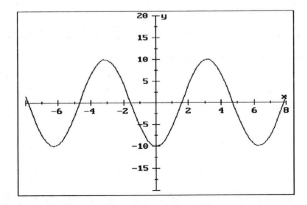

15. **a.** $f'(x) = 3x^2 + 70x - 125 = (x + 25)(3x - 5) = 0$ when $x = -25, \frac{5}{3}$.

b. If $x < -25$ both factors are negative and $f'(x) > 0$, so $f(x)$ is increasing.

If $-25 < x < \frac{5}{3}$ the first factor is positive and the second is negative so $f'(x) < 0$, and $f(x)$ is decreasing.

If $x > \frac{5}{3}$ both factors are positive and $f'(x) > 0$, so $f(x)$ is increasing.

$3x - 5$:

$x + 25$:

$f(x)$:

c. $f(-25) = 0$, $f(\frac{5}{3}) \approx -9481.48$

d. Note that the magnitude of the coefficients will require a change of scale.

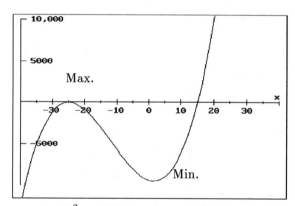

19. **a.** $f'(x) = \dfrac{(x^2 + 3)(1) - (x - 1)(2x)}{(x^2 + 3)^2}$

$= \dfrac{-x^2 + 2x + 3}{(x^2 + 3)^2}$

$= \dfrac{(-1)(x - 3)(x + 1)}{(x^2 + 3)^2} = 0$ when $x = -1, 3$.

19. (con't.)

b.   If $x < -1$ we have three negative factors and $f'(x) < 0$, and $f(x)$ is decreasing. (We are ignoring the denominator because it is always positive.)

If $-1 < x < 3$ we have factors that are $(-)(-)(+)$, so $f'(x) > 0$ and $f(x)$ is increasing.

If $x > 3$ our factors are $(-)(+)(+)$, so $f'(x) < 0$ and $f(x)$ is decreasing.

$x + 1$:

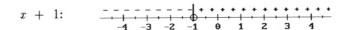

$x - 3$:

$f(x)$:

c. $f(-1) = -\frac{1}{2}$, $f(3) = \frac{1}{6}$

d.   Due to the small values a change of the vertical scale is necessary.

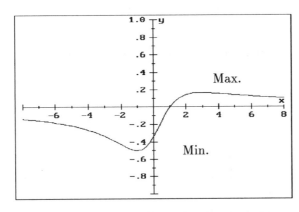

23.   **a.**   Write this as a polynomial before finding $f'(x)$.

$f(x) = 2x^3 - 19x^2 + 32x + 21$

$f'(x) = 6x^2 - 38x + 32 = 2(3x^2 - 19x + 16) = 2(3x - 16)(x - 1)$

$f'(x) = 0$ when $x = 1, \frac{16}{3}$.

b.   If $x < 1$ both factors are negative, $f'(x) > 0$, and $f(x)$ is increasing.

If $1 < x < \frac{16}{3}$, the factors are $(+)(-)$, $f'(x) < 0$, and $f(x)$ is decreasing.

If $x > \frac{16}{3}$, both factors are positive, $f'(x) > 0$, and $f(x)$ is increasing.

c. $f(1) = 36$, $f(\frac{16}{3}) = -\frac{1225}{27} \approx -45.4$

23. (con't.)

    d.   Change the vertical scale to accommodate the larger numbers.

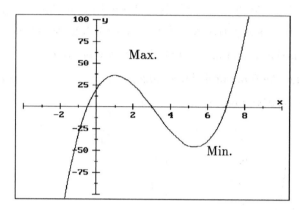

27. **a.**   Finding the derivative is probably easier if you multiply first.

$g(x) = 2x^{\frac{5}{3}} - 5x^{\frac{2}{3}}$, $g'(x) = \frac{10}{3}x^{\frac{2}{3}} - \frac{10}{3}x^{-\frac{1}{3}} = \frac{10}{3}x^{-\frac{1}{3}}(x-1)$

which is undefined at $x = 0$ and equal to 0 at $x = 1$.

    b.   If $x < 0$, $g'(x) > 0$ and $g(x)$ is increasing.

If $0 < x < 1$, $g'(x) < 0$ and $g(x)$ is decreasing.

If $x > 1$, $g'(x) > 0$ and $g(x)$ is increasing.

    c. $g(0) = 0$. Notice that the function is defined at $x = 0$, but it's derivative is not defined. This indicates a cusp, corner, or vertical tangent.

$g(1) = -3$.

    d.

31. **a.** $t'(x) = 2\tan x \sec^2 x = \dfrac{2\sin x}{\cos^3 x} = 0$ when $\sin x = 0$, $x = k\pi$

The function is undefined when $\cos x = 0$, $x = \frac{\pi}{2} + k\pi$.

So on the given interval, $[-\frac{\pi}{4}, \frac{\pi}{4}]$, the only critical point is $x = 0$.

31. (con't.)

    b.   The original function has a period of π, so we only need to examine

        $-\frac{\pi}{4} < x < 0$ and $0 < x < \frac{\pi}{4}$. On the first interval $\sin x$ is

        positive, but $\cos x$ is negative so $t'(x) < 0$ and the function is

        decreasing. On the second interval all factors are positive so

        $t'(x) > 0$ and the function is increasing.

    c. $t(0) = 0$, a relative minimum.

    d.

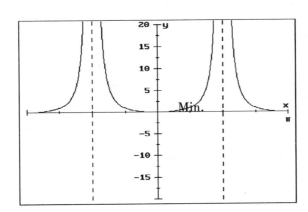

37. $f'(x) = 5(x^4 - 4x + 2)^4(4x^3 - 4) = 5(x^4 - 4x + 2)4(x^3 - 1)$

The last factor is 0 at the value of $x = 1$ so we have a critical point there.

To determine if it is a maximum, minimum, or neither we will look at the sign of

$f'(x)$ on either side of the critical point. $f'(1^-) = 5(+)(4)(-) < 0$.

$f'(1^+) = 5(+)(4)(+) > 0$. The slope goes from negative to zero

to positive, so we have a relative minimum at $(1, -1)$.

41. $f'(x)$ is undefined at $x = 1$, and equals 0 at $x = -3$ and $\frac{1}{2}$. So we need to look

at $f'(x)$ on four intervals: (We may ignore the donominator as it is always

positive.)

$x < -3, f'(x) = (-)(-) > 0$

$-3 < x < \frac{1}{2}, f'(x) = (-)(+) < 0$

$\frac{1}{2} < x < 1, f'(x) = (+)(+) > 0$

$x > 1, f'(x) = (+)(+) > 0$

At $x = -3$ the slope goes from positive to zero to negative. There is a relative

maximum at $x = -3$. At $x = \frac{1}{2}$ the slope goes from negative to zero to positive.

There is a relative minimum at $x = \frac{1}{2}$. At $x = 1$ the slope goes from positive to

undefined to positive. There is a vertical tangent or an asymptote at $x = 1$.

45.   This is a system of three variables for which we have three pieces of information. $(5, 12)$ is on the curve so: $12 = 25a + 5b + c$. $(0, 3)$ is on the curve so: $3 = c$. $f'(x) = 2ax + b = 0$ when x is 5, so: $0 = 10a + b$. Substituting the second and third equations into the first: $12 = 25a - 50a + 3$, $a = -\frac{9}{25}$, $b = \frac{18}{5}$, $c = 3$. $f(x) = -\frac{9}{25}x^2 + \frac{18}{5}x + 3$.

49. $v'(T) = v_0 \frac{1}{2}\left(1 + \frac{1}{273}T\right)^{-\frac{1}{2}}(\frac{1}{273})$. For $T > 0$ this is always positive, so the function is monotonic increasing. $v(0) = v_0 = 1$, so if we scale the vertical axis in terms of v_0. The fraction under the radical causes the function to increase very slowly, so the horizontal axis will need large values for the function not to look too horizontal.

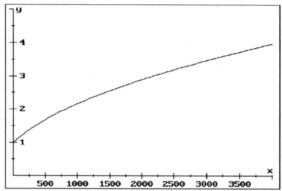

53.   The derivative will be easier to find if we write $f(x) = \left(Ax^{\frac{5}{3}} + Bx^{\frac{2}{3}}\right)^{\frac{1}{2}}$.

$$\text{Now } f'(x) = \frac{1}{2}\left(Ax^{\frac{5}{3}} + Bx^{\frac{2}{3}}\right)^{-\frac{1}{2}}\left(\frac{5}{3}Ax^{\frac{2}{3}} + \frac{2}{3}Bx^{-\frac{1}{3}}\right)$$

$$= \frac{1}{6}\left(Ax^{\frac{5}{3}} + Bx^{\frac{2}{3}}\right)^{-\frac{1}{2}}\left(5Ax^{\frac{2}{3}} + 2Bx^{-\frac{1}{3}}\right)$$

$$= \frac{1}{6}x^{-\frac{1}{3}}\left(Ax^{\frac{5}{3}} + Bx^{\frac{2}{3}}\right)^{-\frac{1}{2}}\left(5Ax + 2B\right)$$

Since the first two factors are in the denominator, $f'(x)$ will be zero when $5Ax + 2B = 0$, $x = -\frac{2B}{5A}$. We also have a critical value when $x = 0$, as the denominator will be 0. The square root is also 0 when $x = -\frac{B}{A}$. So there are two points at which $f'(x)$ is undefined: $x = -\frac{B}{A}$ and $x = 0$.

55.   Analysis: To show that $f'(x) \geq 0$ for all x in an interval I, show that it is true for any c in the interval. Use the definition of the derivative:
$$f'(c) = \lim_{h \to 0} \frac{f(c + h) - f(c)}{h}.$$
If the function is monotonic increasing on I, then show the numerator is positive.

3.4 Second-Derivative Test

5. $f'(x) = 4(x + 20) - 8 = 0$ at $x = -18$, our only critical point.

   $f'(x)$ is negative for $x < -18$ and so $f(x)$ is decreasing, $f'(x)$ is positive for

   $x > -18$ so $f(x)$ is increasing there.

   $f''(x) = 4$, which is always positive. The function is always concave up and has

   a minimum at $(-18, -1)$. You should recognize this as a translated parabola.

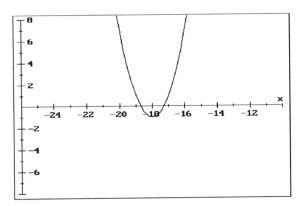

9. $f'(x) = 2 - \dfrac{18}{x^2} = 0$ when $x^2 = 9$, $x = \pm 3$. $f'(x)$ is undefined for $x = 0$.

   $f(x)$ is not defined at $x = 0$, meaning there is a cusp or vertical asymptote there.

   $f''(x) = \dfrac{36}{x^3}$ which is negative for $x = -3$, meaning that there is a relative

   maximum at $(-3, -11)$. $f''(x)$ is positive at $x = 3$, meaning that there

   is a relative minimum at $(3, 13)$. $f''(x)$ is never zero, so there are no

   candidates for inflection points.

   $f'(x)$ is positive on $(-\infty, -3)$ and $(3, \infty)$ and so $f(x)$ is increasing on those

   intervals. $f'(x)$ is negative on $(-3, 0)$ and $(0, 3)$ and so $f(x)$ is decreasing there.

   $f''(x)$ is negative on $(-\infty, 0)$ so the function is concave down there. $f''(x)$ is

   positive on $(0, \infty)$ and so $f(x)$ is concave up on that interval.

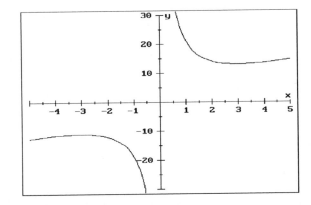

13. $f'(x) = \dfrac{(x + 1)(2x - 3) - (x^2 - 3x)(1)}{(x + 1)^2} = \dfrac{(x + 3)(x - 1)}{(x + 1)^2} = 0$

at $x = -3, 1$.

$f'(x)$ is undefined at $x = -1$, but so is $f(x)$. There is a vertical asymptote

at $x = -1$.

$f''(x) = \dfrac{(x + 1)^2(2x + 2) - (x^2 + 2x - 3)(2x + 2)}{(x + 1)^4} = \dfrac{8}{(x + 1)^3}.$

$f''(-3) < 0$, so we have a relative maximum at $(-3, -9)$.

$f''(1) > 0$, so we also have a relative minimum at $(1, -1)$.

$f'(x) > 0$ on $(-\infty, -3)$ and $(1, \infty)$ so $f(x)$ is increasing there.

$f'(x) < 0$ on $(-3, -1)$ and $(-1, 1)$ so $f(x)$ is decreasing there.

$f''(x) < 0$ on $(-\infty, -1)$ so $f(x)$ is concave down there.

$f''(x) > 0$ on $(-1, \infty)$ so $f(x)$ is concave up there.

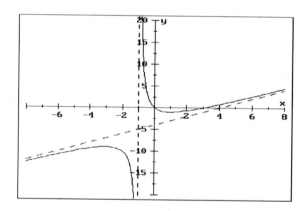

SURVIVAL HINT

$f''(x) = 0$ gives *candidates* for inflection points. Even though the second derivative
is zero, there may not be a change in concavity. They must be tested.

17. $f'(t) = 3t^2 - 12t^3 = 3t^2(1 - 4t) = 0$ at $t = 0, \frac{1}{4}$.

$f''(t) = 6t - 36t^2 = 6t(1 - 6t)$

$f''(0) = 0$, so the second derivative test fails. Use the first derivative test:

$f'(0^-) > 0$, $f'(0^+) > 0$, so we have an inflection point at $(0, 0)$.

$f''(\frac{1}{4}) < 0$, so the second derivative test tells us we have a relative maximum at $(\frac{1}{4}, \frac{1}{256})$. $f''(x)$ is also $= 0$ at $t = \frac{1}{6}$, so we have a candidate for an inflection point. Testing the concavity on either side of this point: $f''(\frac{1}{6}^-) > 0$, and $f''(\frac{1}{6}^+) < 0$ verifies an inflection point.

$f'(x) \geq 0$ on $(-\infty, \frac{1}{4})$ so $f(x)$ is increasing there and decreasing elsewhere.

$f''(x) \geq 0$ on $(0, \frac{1}{4})$ so $f(x)$ is concave up there and concave down elsewhere.

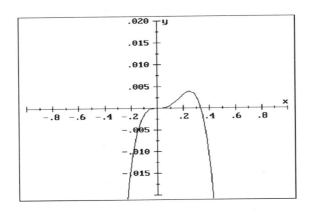

21. $f'(x) = \dfrac{(x^2 + 1)(1) - x(2x)}{(x^2 + 1)^2} = \dfrac{1 - x^2}{(x^2 + 1)^2} = 0$ when $x = -1, 1$.

$f''(x) = \dfrac{(x^2 + 1)^2(-2x) - (1 - x^2)(2)(x^2 + 1)(2x)}{(x^2 + 1)^4} = \dfrac{2x(x^2 - 3)}{(x^2 + 1)^3}$.

$f''(x) = 0$ at $x = 0, \pm\sqrt{3}$. Using the second derivative test on our critical points: $f''(-1) > 0$, $f''(1) < 0$. We have a relative minimum at $(-1, -\frac{1}{2})$ and a relative maximum at $(1, \frac{1}{2})$.

Testing our three candidates for inflection points:

$f''\left(-\sqrt{3}^-\right) < 0$, $f''\left(-\sqrt{3}^+\right) > 0$, so there is an inflection point at

$\left(-\sqrt{3}, \frac{\sqrt{3}}{4}\right)$. $f''\left(\sqrt{3}^-\right) < 0$, $f''\left(\sqrt{3}^+\right) > 0$, so there is also an inflection

point at $\left(\sqrt{3}, \frac{\sqrt{3}}{4}\right)$. Finally, testing the third candidate: $f''(0^-) > 0$, $f''(0^+) < 0$, so this is also an inflection point.

21. (con't.)

These inflection points establish the changes in concavity: down on $\left(-\infty, \ -\sqrt{3}\right)$ and $\left(0, \sqrt{3}\right)$ and up on $\left(-\sqrt{3}, 0\right)$ and $\left(\sqrt{3}, \infty\right)$. Since $f'(x)$ is positive on $\left(-1, 1\right)$ we know that $f(x)$ is increasing on that interval and decreasing elsewhere.

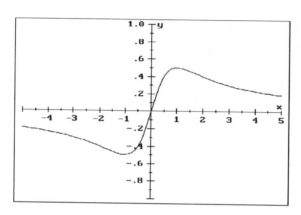

25. $t'(\theta) = 1 - 2\sin 2\theta = 0$ when $\sin 2\theta = \frac{1}{2}$, $2\theta = \frac{\pi}{6}, \frac{5\pi}{6}$, $\theta = \frac{\pi}{12}, \frac{5\pi}{12}$.

$t''(\theta) = -4\cos 2\theta$. Testing our critical points with the second derivative:

$t''(\frac{\pi}{12}) < 0$, so there is a relative maximum at $(\frac{\pi}{12}, \frac{\pi}{12} + \frac{\sqrt{3}}{2})$.

$t''(\frac{5\pi}{12}) > 0$, so there is a relative minimum at $(\frac{5\pi}{12}, \frac{5\pi}{12} - \frac{\sqrt{3}}{2})$.

Remember that the endpoints are also candidates for extrema.

$f(0) = 1$, $f(\pi) = \pi + 1$. There is an absolute maximum of $\pi + 1$ at $(\pi, \pi + 1)$.

To find our candidates for inflection points we set $t''(\theta) = -4\cos 2\theta = 0$.

$\cos 2\theta = 0$, $2\theta = \frac{\pi}{2}, \frac{3\pi}{2}$; $\theta = \frac{\pi}{4}, \frac{3\pi}{42}$. $t''(\frac{\pi}{4}^-) < 0$, $t''(\frac{\pi}{4}^+) > 0$, which verifies an inflection point at $(\frac{\pi}{4}, \frac{\pi}{4})$. $t''(\frac{3\pi}{4}^-) > 0$, $t''(\frac{3\pi}{4}^+) < 0$, which verifies an inflection point at $(\frac{3\pi}{4}, \frac{3\pi}{4})$. We see that the function is concave up on $(\frac{\pi}{4}, \frac{3\pi}{4})$ and concave down elsewhere. Looking at the sign of $f'(x)$ we see that it is positive on $(0, \frac{\pi}{12})$ and $(\frac{5\pi}{12}, \pi)$ and therefore increasing on these intervals and decreasing elsewhere.

25. (con't.)

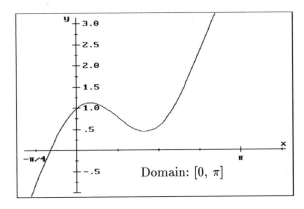

Domain: $[0, \pi]$

SURVIVAL HINT

Think Tank problems 28-30 make excellent test questions to determine if you are understanding the concepts. You should be able to visualize the graph of any of the 27 combinations of $f(x)$, $f'(x)$, $f''(x)$ positive, negative or zero.

33. $y' = 2Ax + B = 0$ when $x = -\dfrac{B}{2A}$.

Using the second derivative test: $y'' = 2A$, so our parabola is concave up if $A > 0$, and concave down when $A < 0$.

37. $D'(x) = 9x^3 - 21\ell x^2 + 10\ell^2 x = x(9x^2 - 21\ell x + 10\ell^2)$

$= x(3x - 2\ell)(3x - 5\ell) = 0$ when $x = 0, \dfrac{2\ell}{3}, \dfrac{5\ell}{3}$. Since the only candidate on the interval is $\dfrac{2\ell}{3}$, the maximum deflection occurs there and is $D(\dfrac{2\ell}{3}) = \dfrac{76\ell^4}{27}$.

39.   The marginal cost is given by: $C'(x) = 8x^3 - 18x^2 - 24x - 2$.

To find extrema we look at its derivative: $C''(x) = 24x^2 - 36x - 24$

$= 12(2x^2 - 3x - 2) = 12(2x + 1)(x - 2) = 0$ when $x = -\dfrac{1}{2}, 2$.

There is only one candidate on the given interval. To determine if it is a maximum or a minimum use the second derivative of C'. $C'''(x) = 24x - 36$.

$C'''(2) > 0$, so we have a relative minimum at $(2, -58)$. Do not forget that the endpoints are also candidates for extrema. $C'(0) = -2$, $C'(3) = -20$. Since this is a continuous function on a closed interval the extreme value theorem guarantees us an absolute minimum and an absolute maximum. These are:

Absolute maximum: $(0, -2)$, absolute minimum: $(2, -58)$.

39. (con't.)

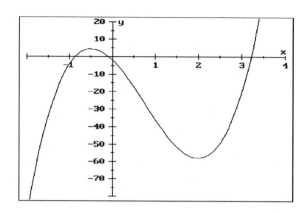

43.   Analysis: If $f(x)$ is concave up on I, then $f''(x) > 0$. If the second derivative is defined on I, then so is the first derivative; and if $f(x)$ is differentiable, then it is also continuous. Therefore, the hypotheses of the mean value theorem are met and we can use it. Solve the inequality in the hint for $f'(c)$ and compare to the results of the mean value theorem. Note that the slope-intercept form of a line, $y = mx + b$, can be written as $f(x) = f'(x)(x) + f(0)$, and translate.

3.5 Infinite Limits and Asymptotes

SURVIVAL HINT

Do not define an asymptote as a line that is approached, but not reached. f may cross the asymptote. For instance $y = \frac{\sin x}{x}$ has $y = 0$ as a horizontal asymptote, and it is crossed an infinite number of times.

3.   Divide numerator and denominator by x: $\displaystyle\lim_{x \to \infty} \frac{3 + \frac{5}{x}}{1 - \frac{2}{x}} = \frac{3 + 0}{1 - 0} = 3$

7.   Divide numerator and denominator by x^2: $\displaystyle\lim_{x \to \infty} \frac{3 - \frac{7}{x} + \frac{5}{x^2}}{-2 + \frac{1}{x} - \frac{9}{x^2}}$

$\displaystyle = \frac{3 - 0 + 0}{-2 + 0 - 0} = -\frac{3}{2}$

11.   Divide numerator and denominator by x^2: $\displaystyle\lim_{x \to \infty} \frac{8x - \frac{9}{x} + \frac{5}{x^2}}{1 + \frac{300}{x}}$

$\displaystyle = \frac{\infty - 0 + 0}{1 + 0} = \infty$

15.   The numerator is third degree and the denominator is fourth degree,

so $\displaystyle\lim_{x \to -\infty} f(x) = 0$

19.   Using the power rule: $\lim\limits_{t\to\infty}\left(\dfrac{8t+5}{3-2t}\right)^3 = \left(\lim\limits_{t\to\infty}\dfrac{8+\frac{5}{t}}{\frac{3}{t}-2}\right)^3 = (-4)^3 = -64$

23. $\sqrt{35} > 5.916$, so the denominator is growing faster than the numerator, and

$\lim\limits_{x\to\infty} f(x) = 0$

27. $\lim\limits_{x\to-2^-}\dfrac{x^2-3x+4}{x+2} = \lim\limits_{x\to-2^-}\dfrac{4+6+4}{0^-} = -\infty$

Note that as x approaches -2 through values less than -2 the denominator will always be negative, and the numerator positive, thus the infinite quotient will be negative.

31.   For small values of x, the $\sin x < x$, so the denominator will approach 0 through positive values. Meanwhile the numerator is approaching 1. The quotient $\frac{1}{0} = \infty$.

35.   As the expression stands the limit is $\frac{0}{0}$, which could have any value. We need to utilize some identities. Multiply numerator and denominator by $(1+\cos x)$

giving: $\dfrac{x^2\csc x\,(1+\cos x)}{1-\cos^2 x} = \dfrac{x^2(1+\cos x)}{\sin^3 x} = \dfrac{1+\cos x}{x\,\dfrac{\sin^3 x}{x^3}} = \dfrac{1+\cos x}{x\left(\dfrac{\sin x}{x}\right)^3}.$

Now take the limit, remembering that $\lim\limits_{x\to0}\dfrac{\sin x}{x} = 1$.

$\lim\limits_{x\to0}\dfrac{1+\cos x}{x\left(\dfrac{\sin x}{x}\right)^3} = \dfrac{1+1}{0(1)^3} = \dfrac{2}{0} = \infty$

SURVIVAL HINT

Values for which the denominator of a rational expression are zero are *candidates* for vertical asymptotes. If the numerator is also zero for that value, then you have a hole in the graph, and not an asymptote.

SURVIVAL HINT

When testing for vertical asymptotes it is useful to put a sign on 0. As we approach a from the right or left an expression in the denominator may approach zero; but it never *is zero* as we are *approaching* a. It will be approaching zero through either positive or negative values, so a sign is appropriate.

39.   Vertical asymptotes: only one candidate, $t = 3$. Testing: $\lim_{t \to 3^-} \left(4 + \frac{2t}{t-3} \right)$

$= 4 + \frac{6}{0^-} = -\infty.$ $\lim_{t \to 3^+} \left(4 + \frac{2t}{t-3} \right) = 4 + \frac{6}{0^+} = +\infty$

Horizontal asymptotes: $\lim_{x \to \infty} f(x) = 6^+$, $\lim_{x \to -\infty} f(x) = 6^-$

$f'(x) = \frac{-6}{(t-3)^2}$ is always negative, so $f(x)$ is always decreasing.

$f''(x) = \frac{12}{(t-3)^3}$ is positive for $t > 3$, and negative for $t < 3$, so

$f(x)$ is concave up for $t > 3$, and concave down for $t < 3$.

$f'(x)$ is never equal to 0, or undefined at a defined value of $f(x)$, and there are

no endpoints, so there are no candidates for relative extrema.

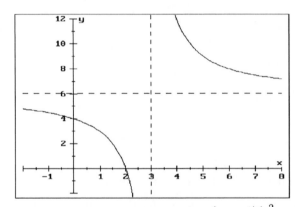

43.   Vertical asymptotes: factoring; $f(x) = \frac{(x+1)(x^2-x+1)}{(x-2)(x^2+2x+4)}$.

There is one candidate for a vertical asymptote: $x = 2$.

Testing: $\lim_{x \to 2^-} f(x) = \frac{3(3)}{0^-(12)} = -\infty$

$\lim_{x \to 2^+} f(x) = \frac{3(3)}{0^+(12)} = +\infty.$

Horizontal asymptotes: $\lim_{x \to +\infty} f(x) = 1^+$ (The numerator is always greater

than the denominator, so $f(x)$ approaches 1 from above.) $\lim_{x \to -\infty} f(x) = 1^-$

(The numerator is always less than the denominator in absolute value, so $f(x)$

approaches 1 from below).

$f'(x) = -\frac{27x^2}{(x^3-8)^2}$ is always negative, so $f(x)$ is always decreasing, except at

$x = 0$, $y = -\frac{1}{8}$, where there is a horizontal tangent.

43. (con't.)

$f''(x) = \dfrac{108x(x^3 + 4)}{(x^3 - 8)^3} = 0$ at $x = 0$. We have a candidate for an inflection

point at $x = 0$. Testing: $f''(0^-) > 0$. $f''(0^+) < 0$. So we have

an inflection point changing from concave up to concave down at $(0, -\frac{1}{8})$.

The third degree factor in the numerator also has one real root, $x = -\sqrt[3]{4}$.

Testing: $f''(-\sqrt[3]{4}^-) < 0$, $f''(-\sqrt[3]{4}^+) > 0$. So we have an inflection point

at $(-\sqrt[3]{4}, \frac{1}{4})$, going from concave down to concave up.

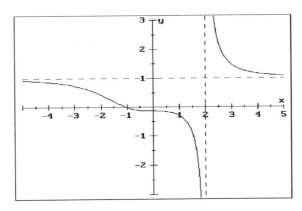

47.   For $x \le 1$ this is a piece of the parabola with x-intercepts of 0 and 3.

The endpoint, $(1, -2)$ is included. For $x > 1$ it is part of a half-parabola

opening to the right. The endpoint $(1, 1)$ is not included. Notice that since

$\lim\limits_{x \to 1^-} f(x) = -2$ is not equal to $\lim\limits_{x \to 1^+} f(x) = 1$, there is a discontinuity

at $x = 1$.

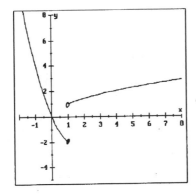

59.   **a.**   Divide the numerator and denominator of $f(x)$ by x^2 find the $\lim\limits_{x \to \pm\infty} f(x)$.

$$\lim_{x \to \pm\infty} \frac{a + \frac{b}{x} + \frac{c}{x^2}}{r + \frac{s}{x} + \frac{t}{x^2}} = \frac{a}{r} .$$ This, by definition, is the horizontal asymptote.

To find the value of x at which $f(x)$ will cross the asymptote,

solve $f(x) = \frac{a}{r}$. $\dfrac{ax^2 + bx + c}{rx^2 + sx + t} = \dfrac{a}{r}$, cross multiply:

$arx^2 + brx + cr = arx^2 + asx + at,$

$(br - as)x = at - cr, \quad x = \dfrac{at - cr}{br - as}.$

b.   As the denominator at the point of intersection becomes smaller, x becomes larger. If the denominator is 0, then the point of intersection goes to ∞, and the function will not cross the asymptote.

c.   Factoring: $g(x) = \dfrac{(x - 5)(x + 1)}{(2x + 5)(x - 2)}$, which will have vertical asymptotes at $x = -\frac{5}{2}$ and $x = 2$. $\lim\limits_{x \to \infty} g(x) = \frac{1}{2}$, so there is a horizontal asymptote of $y = \frac{1}{2}$.

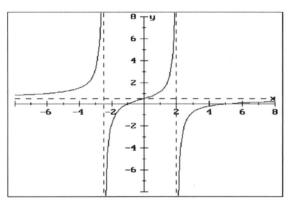

Factoring: $h(x) = -\dfrac{3x^2 - x - 7}{4(3x + 2)(x - 1)}$, which will have vertical asymptotes at $x = -\frac{2}{3}$ and $x = 1$. $\lim\limits_{x \to \pm\infty} h(x) = -\frac{1}{4}$, so there is a horizontal asymptote of $y = -\frac{1}{4}$.

59. (con't.)

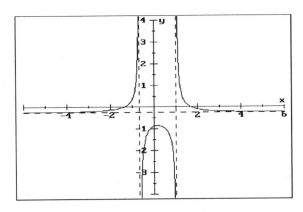

61.   Analysis:

    a.   If $m > n$ what will be $\lim\limits_{x \to \infty} f(x)$?

    b.   In the case of $m = n$, dividing numerator and denominator by x^n

        and taking $\lim\limits_{x \to \infty} f(x) = \dfrac{a_n}{b_n}$.

    c.   If $m > n$, $\lim\limits_{x \to \infty} f(x)$ will not have a finite limit L.

65.   Analysis:

$$\lim_{x \to c} \frac{f(x)}{g(x)} = \frac{\lim\limits_{x \to c} f(x)}{\lim\limits_{x \to c} g(x)} \ . \ \text{What happens to the quotient if the numerator}$$

is positive and the denominator is always negative?

3.6 Summary of Curve Sketching

SURVIVAL HINT

The steps listed on p.252 constitute a powerful set of analytic tools. But you need not use every tool in your toolbox on each problem. Skill in using the most efficient tool comes with practice. Often you have some idea of the general shape of the graph. Ask yourself what you would most like to know about the graph, then use the tool that will give you that information. In addition to the items listed, also make use of symmetry and graphing by composition.

7.   We know the general shape of a cubic equation, so what we need to determine
       are the extrema, inflection points, intercepts, and possibly a few other points.
       $f'(x) = 6x^2 + 12x + 6 = 6(x^2 + 2x + 1) = 6(x + 1)^2 = 0$ when
       $x = -1$. $f''(x) = 12(x + 1) = 0$ at $x = -1$. It is easy to see that the
       second derivative changes signs as we move across -1, so we have an inflection
       point at $(-1, 3)$ and the slope of the inflectional tangent is 0.

       When $x = 0$, $f(x) = 5$, so there is a y-intercept at $(0, 5)$. The x-intercepts
       are not easy to find, but we can find some nearby points, and the slopes there
       to aid in the graphing. $f(-2) = 1$ and the slope there is 6. $f(-3) = -13$,
       and the slope there is 24. $f(1) = 19$, and the slope there is 24.

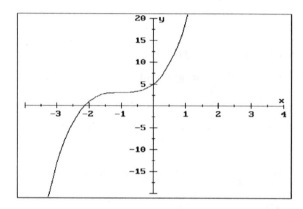

11.   We know the general shape of a cubic equation, so what we need to determine
       are the extrema, inflection points, intercepts, and possibly a few other points.
       $g'(x) = 6x^2 - 6x - 12 = 6(x^2 - x - 2) = 6(x + 1)(x - 2)$
              $= 0$ when $x = -1, 2$.
       $g''(x) = 12(2x - 1) = 0$ at $x = \frac{1}{2}$. Testing the candidates for extrema:
       $g''(-1) < 0$, so there is a relative maximum at $(-1, 21)$.
       $g''(2) > 0$, so there is a relative minimum at $(2, -6)$.

       The y-intercept is $(0, 14)$. It is easy to see that the second derivative changes
       sign as x passes through $\frac{1}{2}$. Therefore there is an inflection point at $(\frac{1}{2}, \frac{15}{2})$ and
       the slope of the inflectional tangent there is $-\frac{27}{2}$.

11. (con't.)

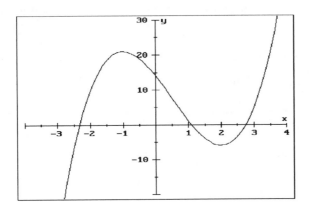

15. $f(x) = 9x^4 - 40x^3 + 72x^2 - 36x - 16$

$f'(x) = 36x^3 - 120x^2 + 144x - 36 = 12(3x^3 - 10x^2 - 12x - 3) = 0$

at $x = -\frac{1}{3}$ and the other two roots are complex. This root can be found with the rational roots theorem. So there is only one candidate for extrema. Checking the second derivative to test it: $f''(x) = 9x^2 - 20x + 12$ which is always positive. The discriminant is negative so there are no real roots. Our curve is always concave up and we have a minimum at approximately $(\frac{1}{3}, -21.37)$. The y-intercept is $(0, -16)$.

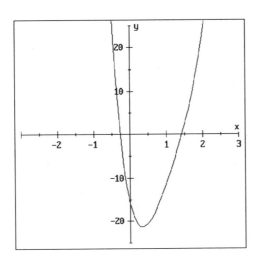

19.   The denominator of $f(x)$ is never 0, so there are no candidates for vertical

asymptotes. $\lim\limits_{x \to \infty} f(x) = 0$, so the x-axis is a horizontal asymptote. There are

no x-intercepts. There is a y-intercept at $(0, \frac{1}{12})$. $f'(x) = -\dfrac{2x}{(x^2 + 12)^2} = 0$

at $x = 0$. Find the second derivative to test this candidate for extrema:

$f''(x) = \dfrac{6(x^2 - 4)}{(x^2 + 12)^3} = 0$ at $x = \pm 2$. $f''(0) = -24$, so there is a relative

maximum at $(0, \frac{1}{12})$. There are inflection points at $(-2, \frac{1}{16})$ and $(2, \frac{1}{16})$.

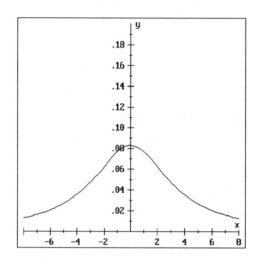

23. $f'(x) = 3x^2 + 6x = 3x(x + 2) = 0$ at $x = 0, -2$.

$f''(x) = 6x + 6 = 6(x + 1) = 0$ at $x = -1$.

Testing the candidates for extrema: $f''(0) = 6$, so there is a relative minimum

at $(0, 1)$. $f''(-2) = -6$, so there is a relative maximum at $(-2, 5)$.

It is easy to see that the sign of $f''(x)$ changes as x passes through $x = -1$, so

there is an inflection point at $(-1, 3)$. Notice that the relative maximum is one

unit left and two above the inflection point, and the relative minimum is one unit

right and two below the inflection point. Cubic polynomials are symmetric about

their inflection points $-$ a fact you can use in future graphing.

23. (con't.)

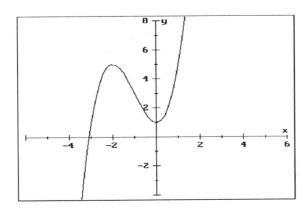

27. $f'(x) = 2 - \dfrac{18}{x^2} = 0$ when $x = \pm 3$. $f''(x) = \dfrac{36}{x^3} = 0$ for no x.

Testing our two candidates for extrema: $f''(-3) < 0$, so there is a relative

maximum at $(-3, -11)$. $f''(3) > 0$, so there is a relative minimum

at $(3, 13)$.

As x goes to ∞ the function "behaves like" $2x + 1$, so that is the equation

of the oblique asymptote.

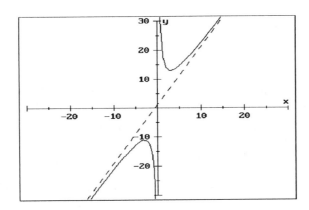

31. $f'(x) = 12x^3 - 6x^2 - 24x + 18 = 6(2x^3 - x^2 - 4x + 3)$.

Notice that the sum of the coefficients is 0, which means $x = 1$ is a root.

Using synthetic division we now have $f'(x) = 6(x - 1)(2x^2 + x - 3)$.

Factoring the trinomial: $f'(x) = 6(x - 1)(2x + 3)(x - 1) = 0$ at

$x = 1, -\dfrac{3}{2}$. $f''(x) = 6(6x^2 - 2x - 4) = 12(3x + 2)(x - 1) = 0$ at

$x = 1, -\dfrac{2}{3}$. Using the second derivative test on our candidates for

extrema: $f''(1) = 0$. The test fails. Using the first derivative test we see that

the slope is positive on both sides of the candidate, so we have an inflection point

at $(1, 7)$. Testing the other candidate: $f''(-\dfrac{3}{2}) > 0$, so there is a relative

minimum at $(-\dfrac{3}{2}, -\dfrac{513}{16})$.

31. (con't.)

Testing the other candidate for inflection point we can see that the concavity changes as we pass through $-\frac{2}{3}$, so there is an inflection point at $(-\frac{2}{3}, -\frac{436}{27})$.

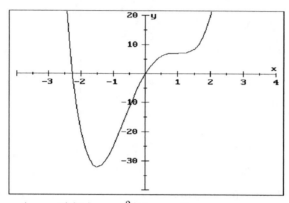

35. $f'(x) = \dfrac{(x-2)(2x) - x^2(1)}{(x-2)^2} = \dfrac{x^2 - 4x}{(x-2)^2} = 0$ at $x = 0, 4$ and is undefined

at $x = 2$. Testing the $\lim\limits_{x \to 2} f(x)$ tells us that there is a vertical asymptote at

$x = 2$. $f''(x) = \dfrac{(x-2)^2(2x-4) - (x^2 - 4x)2(x-2)}{(x-2)^4}$

$= \dfrac{8}{(x-2)^3}$ is nowhere $= 0$ so there are no inflection points.

Testing the candidates for extrema: $f''(0) < 0$, so there is a relative maximum at $(0, 0)$. $f''(4) > 0$, so there is a relative minimum at $(4, 8)$.

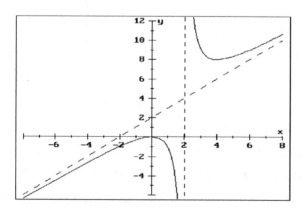

39. $f'(x) = 2 \sin x \cos x - 2 \cos x = 2 \cos x (\sin x - 1) = 0$ when

$x = \frac{\pi}{2}$ on $[0, \pi]$.

$f''(x) = (2 \cos x)(\cos x) + (\sin x - 1)(-2 \sin x)$

$\qquad = 2 \cos^2 x - 2 \sin^2 x + 2 \sin x$

$\qquad = 2(1 - \sin^2 x) - 2 \sin^2 x + 2 \sin x$

$\qquad = -2(2 \sin^2 x - \sin x - 1) = -2(2 \sin \theta + 1)(\sin \theta - 1) = 0$

at $\frac{\pi}{2}$ on $[0, \pi]$.   So there is either an extrema or an inflection point at $x = \frac{\pi}{2}$.

Checking the slopes on either side: $f'(\frac{\pi}{2}^-) = (+)(-) = -$,

$f'(\frac{\pi}{2}^+) = (-)(-) = +$. There is a relative minimum at $(\frac{\pi}{2}, 0)$.

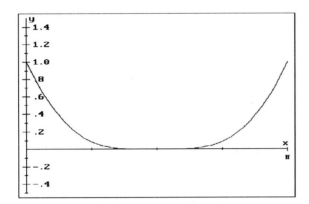

43. $f'(x) = \dfrac{(1 + \cos x)(\cos x) - (\sin x)(- \sin x)}{(1 + \cos x)^2} = \dfrac{\cos x + \cos^2 x + \sin^2 x}{(1 + \cos x)^2}$

$\qquad = \dfrac{\cos x + 1}{(1 + \cos x)^2} = \dfrac{1}{1 + \cos x}$, which is never 0, and is undefined only

at values for which $f(x)$ is also undefined.

$f''(x) = \dfrac{\sin x}{(1 + \cos x)^2} = 0$ at $x = -2\pi, -\pi, 0, \pi, 2\pi$ on $[-2\pi, 2\pi]$.

However, $f(x)$ is not defined at $\pm \pi$, and $\pm 2\pi$ are endpoints. All we have left is one candidate for an inflection point. It is easy to see by looking at $f''(x)$ that the concavity changes signs as x passes through π. There is, therefore, an inflection point at $(0, 0)$.

43. (con't.)

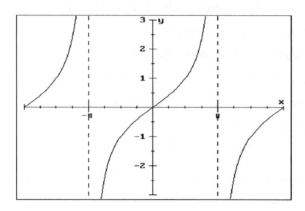

3.7 Optimization in the Physical Sciences and Engineering

SURVIVAL HINT

Much confusion will be avoided if you will begin each problem by writing

what it is you wish to do: maximize_____ given that _____ . Then, carefully

substitute the variables and constants into the stated relationship.

5.   **a.**   The problem is to maximize $A = \ell w$ given that $2\ell + 2w = 320$.

Substituting $w = 160 - \ell$ into the area formula: $A = \ell(160 - \ell)$.

$A' = 160 - 2\ell = 0$ when $\ell = 80$. Since this is a continuous function

on a closed interval, [0, 160], we are guaranteed an absolute maximum and

an absolute minimum by the extreme value theorem. The endpoints are

obvious minimums for the area, so our one remaining candidate must be

the maximum. The use of the extreme value theorem often eliminates the

need to use a second derivative test. Be certain that the hypotheses are

met in each case. So there is a maximum area of 6400 sq. ft. when $\ell = 80$

and $w = 80$, i.e. a square pasture.

   b.   Again the problem is to maximize $A = \ell w$, but this time $\ell + 2w = 320$.

Substituting: $A = w(320 - 2w) = 320w - 2w^2$. $A' = 320 - 4w$

$= 0$ when $w = 80$. $\ell = 320 - 2(80) = 160$. The extreme value

theorem guarantees that this is a maximum. The maximum area is

12,800 sq. ft. when the side parallel to the wall is 160 ft. and the other two

sides are 80 ft.

9.   The volume of the box will be given by $V = x(24 - 2x)(45 - 2x)$

$= 4x^3 - 138x^2 + 1080x$. Find V' to identify candidates for extrema.

$V' = 12x^2 - 276x + 1080 = 12(x - 5)(x - 18) = 0$ at $x = 0, 5, 18$.

0 is obviously a minimum, 18 is not in the domain, so the maximum volume

occurs when squares of 5 in are cut from the corners. The dimensions of the box

will be 5 by 14 by 35 in. and the volume will be 2450 cu in.

SURVIVAL HINT

If the hypotheses of the extreme value theorem apply to a particular problem we can

often avoid testing a candidate to see if it is a maximum or minimum. Suppose we have

three candidates and two are obviously minimums - the third *must* be a maximum.

13.

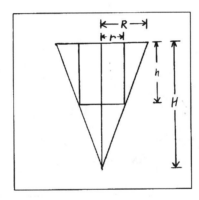

Draw the figure and label the dimensions of the cylinder r and h. We are asked

to maximize the volume of the cylinder: $V = \pi r^2 h$. By similar triangles:

$\frac{h}{R - r} = \frac{H}{R}$. Substituting: $V = \pi r^2 \left(\frac{H}{R}(R - r)\right) = \pi\left(Hr^2 - \frac{H}{R}r^3\right)$.

$V' = \pi\left(2Hr - \frac{3H}{R}r^2\right) = 0$ when $2Hr - \frac{3H}{R}r^2 = 0$, $\frac{H}{R}r(2R - 3r) = 0$,

$r = 0$ or $\frac{2}{3}R$. The extreme value theorem applies, so the maximum volume of the

cylinder occurs when its radius is $\frac{2}{3}$ of the radius of the cone. Its height will be

$\frac{H}{R}(R - \frac{2}{3}R) = \frac{1}{3}H$, i.e. $\frac{1}{3}$ the height of the cone.

17.   We need to maximize $V = \ell wh$, given that $\ell = w$,

and $5(\ell w) + 1(\ell w + 2\ell h + 2wh) = 72$. Solving the cost relationship for h:

$6\ell w + h(2\ell + 2w) = 72$, $h = \dfrac{72 - 6\ell w}{2\ell + 2w} = \dfrac{36 - 3\ell w}{\ell + w}$.

So $V = \ell(\ell)\Big(\dfrac{36 - 3\ell(\ell)}{\ell + \ell}\Big) = \dfrac{36\ell^2 - 3\ell^4}{2\ell} = \dfrac{3}{2}(12\ell - \ell^3)$.

$V' = 18 - \dfrac{9}{2}\ell^2 = 0$ when $\ell = \pm 2$. The extreme value theorem applies,

the end points and $\ell = -2$ are not reasonable, so the maximum volume with the

given restrictions occurs when $\ell = w = 2$ ft. and $h = 6$ ft.

21.

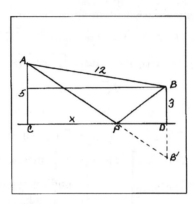

We want to minimize the $L = \overline{AP} + \overline{PB}$. Drawing a line parallel and equal to

$\overline{CD}$ through B, we have a right triangle with hypotenuse 12 and leg 2, so

$\overline{CD} = \sqrt{140}$. Let $\overline{CP} = x$. Now imagine that B is on the opposite side of the

road at B'. The shortest distance will be $\overline{AB'}$, which will be equal to $\overline{AP} + \overline{PB}$.

$\overline{AB'} = \sqrt{\sqrt{140}^2 + 8^2} = 2\sqrt{51} \approx 14.28$ mi. To find the desired point P on the

highway, use similar triangles: $\dfrac{x}{5} = \dfrac{\sqrt{140} - x}{3}$, $x \approx 7.395$ mi. from C.

25.   We are asked to minimize the cost of material to construct the cylinder.

$C = .5(\pi r^2) + .3(2\pi rh)$. The volume is not specified, but is a constant,

$V = \pi r^2 h$. To have the cost as a function of a single variable, solve for $h = \dfrac{V}{\pi r^2}$.

$C = .5\pi r^2 + .6\pi r\Big(\dfrac{V}{\pi r^2}\Big) = .5\pi r^2 + \dfrac{.6V}{r}$. $C' = \pi r - \dfrac{.6V}{r^2} = 0$ when

$r^3 = \dfrac{.6V}{\pi}$, $r = \sqrt[3]{\dfrac{.6V}{\pi}}$. Since the extreme value theorem applies, and the

endpoints are obviously maximums, this must be the radius that gives the

minimum cost.

29. $E' = -Mg + \dfrac{2mgx}{\sqrt{x^2 + d^2}} = 0$ when $Mg\sqrt{x^2 + d^2} = 2mgx$,

$$\sqrt{x^2 + d^2} = \frac{2mx}{M}, \quad x^2 + d^2 = \frac{4m^2x^2}{M^2}, \quad \left(\frac{4m^2}{M^2} - 1\right)x^2 = d^2,$$

$$x^2 = \frac{d^2M^2}{4m^2 - M^2}, \quad x = \frac{dM}{\sqrt{4m^2 - M^2}}.$$

33. If you used the trigonometric functions in problem 32, let us do the problem without them this time. $I(h) = \dfrac{k\sin\phi}{d^2}$ where $\sin\phi = \dfrac{h}{d}$ and $d = \sqrt{h^2 + 16}$.

$I(h) = \dfrac{kh}{d^3}$. $I'(h) = \dfrac{kd^3 - 3khd^2d'}{d^6}$ (Remember that d is a function of h, and we are using implicit differentiation.) $I'(h) = \dfrac{kd - 3khd'}{d^4} = 0$ when

$k(d - 3hd') = 0$, or substituting, when $\sqrt{h^2 + 16} = 3h\left(\dfrac{h}{\sqrt{h^2 + 16}}\right)$, $h^2 + 16 = 3h^2$, $h^2 = 8$, $h = 2\sqrt{2} \approx 3$ ft.

And, of course, this is the same answer as the previous problem because sine and cosine are co-functions. That is, $\theta + \phi = 90°$, and $\cos\theta = \dfrac{h}{d} = \sin\phi$.

35. Analysis: Consider the pipe as composed of two pieces, $L = x + y$. Parallel lines cut by a transversal form equal corresponding angles, θ. In one triangle $\sin\theta = \dfrac{2\sqrt{2}}{x}$, and in the other triangle $\cos\theta = \dfrac{2\sqrt{2}}{y}$. Solve for x and y and substitute into L.

37. Analysis: When you find $V'(T)$ you will have a quadratic equation with very messy coefficients. Its solution will give you two candidates for extrema, one near freezing, $0°$, and the other closer to boiling, $100°$. Note that the maximum density occurs when you have the minimum volume. The graph of this volume function is interesting:

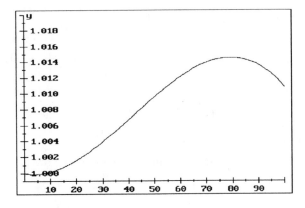

The maximum value occurs when $T \approx 80°$.

The minimum value occurs when $T \approx 4°$.

3.8 Optimization in Business, Economics, and the Life Sciences

3.   The profit is the revenue less the cost. $P(x) = x\,p(x) - C(x)$

$= \dfrac{70x - x^2}{x + 30} - \frac{1}{5}(x + 30)$. The profit is maximized when the marginal

revenue equals the marginal cost: $\dfrac{(x + 30)(70 - 2x) - (70x - x^2)}{(x + 30)^2} = \frac{1}{5}$.

$$(x + 30)(70 - 2x) - (70x - x^2) = \tfrac{1}{5}(x + 30)^2,$$

$$5(-x^2 - 60x + 2100) = x^2 + 60x + 900$$

$$6x^2 + 360x - 9600 = 0$$

$$x^2 + 60x - 1600 = 0$$

$$(x + 80)(x - 20) = 0$$

$$x = 20 \ \ (-80 \text{ is not in the domain}).$$

To maximize profit produce 20 items.

7.   **a.**   Total cost, $C(x)$, is the average cost, $A(x)$, times the number of items
produced, x. $C(x) = 5x + \dfrac{x^2}{50}$. Revenue, $R(x)$, is the price per item,
p, times the number of items sold, x. $R(x) = x\left(\dfrac{380 - x}{20}\right)$.

The profit, $P(x) = R(x) - C(x) = \dfrac{380x - x^2}{20} - \dfrac{250x + x^2}{50}$

$$P(x) = \dfrac{-7x^2 + 1400x}{100} = -0.07x^2 + 14x$$

b.   The maximum profit occurs when $R'(x) = C'(x)$.

$19 - \frac{x}{10} = 5 + \frac{x}{25}$, $14 = \frac{7}{50}x$, $x = 100$ items.

This gives a price per item, $p = \dfrac{380 - x}{20} = \$140$ per item.

The maximum profit will be $P(x) = \dfrac{(-7x + 1400)x}{100}$

$$= \dfrac{(-700 + 1400)100}{100} = \$700.$$

11.   **a.**   Average revenue per unit is total revenue divided by number of units:

$A(x) = -2x + 68 - \dfrac{128}{x}$. Marginal revenue is $R'(x) = -4x + 68$.

They are equal when: $-2x + 68 - \dfrac{128}{x} = -4x + 68$, $2x = \dfrac{128}{x}$,

$x^2 = 64$, $x = 8$. (-8 is not in the domain.)

11. (con't.)

b. $A'(x) = -2 + \dfrac{128}{x^2}$. $A'(8) = 0$. $A'(8^-) > 0$, $A'(8^+) < 0$.

So there is a relative maximum at $(8, 36)$. The function is increasing on

$(0, 8)$ and decreasing on $(8, \infty)$.

c.

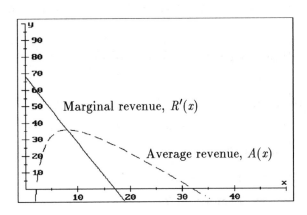

15. $P(x) = R(x) - C(x)$. Let $x =$ the increase in price, $40 + x =$ cost

per board, $50 - 3x =$ number of boards sold.

$P(x) = (40 + x)(50 - 3x) - 25(50 - 3x) = -3x^2 + 5x - 750$.

$P'(x) = -6x + 5 = 0$ when $x = \frac{5}{6}$. Since this is not an integer, and our

function has discreet rather than continuous values, we will test the nearest integer

values. $P(0) = 750$, $P(1) = 752$. Therefore, raise the price \$1, sell the boards

at a price of \$41, sell 47 per month, and have the maximum profit of \$752.

19. We want to maximize the yield, $Y(x)$. Let x be the additional number of trees to

be planted. Number of trees $= 60 + x$, average yield $= 400 - 4x$.

$Y(x) = (60 + x)(400 - 4x) = -4x^2 + 160x + 24\,000$.

$Y'(x) = -8x + 160 = 0$ when $x = 20$. So plant 80 total trees, have an

average yield per tree of 320 oranges, and a maximum total crop of

25,600 oranges.

23. The demand function is the number of items, x, that can be sold at a given

price, p.

a. The revenue function, $R(x)$, is the number of items sold times the price

per item. $R(x) = (120 - 0.1x^2)x = -0.1x^3 + 120x$

$R'(x) = -0.3x^2 + 120 = 0$ when $x^2 = 400$, $x = 20$ (-20 is not in

the domain unless we pay people to take away our product!)

$R''(x) = -0.6x$, which is negative for all values of x in the domain, so

our candidate is a maximum. Revenue is increasing for prices less than 20,

and decreasing for prices greater than 20. $p(20) = 80$

23. (con't.)

    b.   The demand function: $p(x) = 120 - 0.1x^2$

            The revenue function: $R(x) = -0.1x^3 + 120x$, on $[0, 1200]$

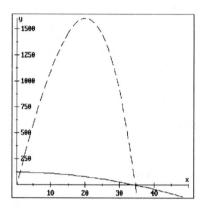

27.   We wish to minimize the surface given a fixed volume: $v_0 = \pi r^2 h$.

$$S(r) = 2\pi r(h + r) = 2\pi r\left(\frac{v_0}{\pi r^2} + r\right) = \frac{2\,v_0}{r} + 2\pi r^2.$$

$$S'(r) = -\frac{2\,v_0}{r^2} + 4\pi r = 0 \text{ when } 4\pi r^3 = 2\,v_0, \quad r = \sqrt[3]{\frac{v_0}{2\pi}}.$$

$$S''(r) = \frac{4\,v_0}{r^3} + 4\pi > 0, \text{ so the previous value of } r \text{ gives the minimum.}$$

31.   The total cost of production is the cost of setting up the machines plus the

    cost of supervising machine operation. Let $x =$ the number of machines used.

    The number of hours the machines must run is 8000 divided by 50 times the

    number of machines working. Supervision is \$35 for each of these hours.

    a. $C(x) = 800x + \dfrac{8000}{50x}(35) = 800x + \dfrac{5600}{x}.$

          $C'(x) = 800 - \dfrac{5600}{x^2} = 0 \text{ when } x^2 = 7, \; x \approx 2.65 .$

          Since this is a discreet rather than a continuous function, we need to test

           the nearest integer values. $C(2) = 4400, \; C(3) \approx 4267 .$

          So if we set up and run 3 machines, the number of operational hours will

          be $\dfrac{8000}{3(50)} \approx 53.3$ hrs.

    b.   The supervisor will earn $\approx (53.3)(35) \approx \$1866.67$

3.9 l'Hopital's Rule

SURVIVAL HINT

Do not attempt to apply l'Hôpital's rule unless you have verified that the limit has form $\frac{0}{0}$ or $\frac{\infty}{\infty}$. It is good practice to take the limit and indicate which form you have.

3. $\lim\limits_{x \to 2} \dfrac{x^4 - 16}{x^2 - 4}$ is of form $\frac{0}{0}$ so we may apply l'Hôpital's rule.

$$= \lim\limits_{x \to 2} \dfrac{4x^3}{2x} = \lim\limits_{x \to 2} 2x^2 = 8$$

7. $\lim\limits_{x \to 1} \dfrac{x^{10} - 1}{x - 1} = \lim\limits_{x \to 1} \dfrac{10x^9}{1} = 10$

11. $\lim\limits_{x \to \pi} \dfrac{\cos \frac{x}{2}}{\pi - x}$ is of form $\frac{0}{0}$ so $\lim\limits_{x \to \pi} \dfrac{-\frac{1}{2}\sin \frac{x}{2}}{-1} = \frac{1}{2}$

15. $\lim\limits_{x \to 0} \dfrac{x - \sin x}{\tan x - x}$ is of form $\frac{0}{0}$ so $\lim\limits_{x \to 0} \dfrac{1 - \cos x}{\sec^2 x - 1}$ is still of form $\frac{0}{0}$ so apply

l'Hôpital's rule again: $\lim\limits_{x \to 0} \dfrac{\sin x}{2 \sec^2 x \tan x} = \lim\limits_{x \to 0} \dfrac{1}{2 \sec^3 x} = \frac{1}{2}$

19. $\lim\limits_{x \to 0} \dfrac{3x + \sin^3 x}{x \cos x}$ is of form $\frac{0}{0}$ so $\lim\limits_{x \to 0} \dfrac{3 + 3 \sin^2 x \cos x}{\cos x - x \sin x} = \frac{3}{1} = 3$

23. $\lim\limits_{x \to \infty} x^{3/2} \sin \frac{1}{x}$ can be written as $\lim\limits_{x \to \infty} \dfrac{\sin \frac{1}{x}}{\frac{1}{x}} x^{1/2} = \lim\limits_{x \to \infty} \dfrac{\sin \frac{1}{x}}{\frac{1}{x}} \lim\limits_{x \to \infty} x^{1/2}$
$= 1(\infty) = \infty$

27. $\lim\limits_{x \to 0} \dfrac{x + \sin(x^2 + x)}{3x + \sin x} = \lim\limits_{x \to 0} \dfrac{1 + (2x + 1) \cos(x^2 + x)}{3 + \cos x} = \dfrac{1+1}{3+1} = \frac{1}{2}$

31. This is of form $\infty - \infty$, so do the subtraction: $\lim\limits_{x \to 0} \dfrac{x - \sin^2 x}{x \sin^2 x}$ which is now of

form $\frac{0}{0}$. So by l'Hôpital's rule, $\lim\limits_{x \to 0} \dfrac{1 - 2 \sin x \cos x}{2 x \sin x \cos x + \sin^2 x} = \dfrac{1}{0} = \infty$

35. Doing the subtraction:
$$\lim\limits_{x \to +\infty} \left(\dfrac{x^3}{x^2 - x + 1} - \dfrac{x^3}{x^2 + x - 1} \right) = \lim\limits_{x \to +\infty} \dfrac{2(x^4 - x^3)}{x^4 - x^2 + 2x - 1}.$$

Dividing by x^4, or several applications of l'Hôpital's rule, gives a limit $= 2$

39. Analysis: Change the difference of the sine functions into a product, using the appropriate sum to product identity:

$$\sin \alpha t - \sin \beta t = 2 \sin\left(\frac{\alpha t - \beta t}{2}\right)\cos\left(\frac{\alpha t + \beta t}{2}\right)$$ and note that $\lim_{\alpha \to \beta} f(t)$ has form $\frac{0}{0}$.

3.10 Antiderivatives

SURVIVAL HINT

Antiderivatives constitute a parametric family of curves. There is an infinite set of "parallel" curves that all have the same slope at any given value of x. Any curve that is an antiderivative can be translated vertically by C units and still be a solution. **Always** remember the $+C$.

1. Consider 2 as $2x^0$, then $\frac{x^{n+1}}{n+1} = x$. So $\int 2\,dx = 2x + C$

5. $\int (4t^3 + 3t^2)\,dt = 4\int t^3\,dt + 3\int t^2\,dt = 4\frac{t^4}{4} + 3\frac{t^3}{3} + C$
$= t^4 + t^3 + C$

9. $\int \sec^2\theta\,d\theta = \tan\theta + C$

13. Use the sum rule and the power rule: $\int (u^{3/2} - u^{1/2} + u^{-10})\,du$
$= \frac{u^{5/2}}{\frac{5}{2}} - \frac{u^{3/2}}{\frac{3}{2}} + \frac{u^{-9}}{-9} + C = \frac{2}{5}u^{5/2} - \frac{2}{3}u^{3/2} - \frac{1}{9}u^{-9} + C$

17. Negative exponents and the power rule work best on variables in the denominator.
$\int (t^{-2} - t^{-3} + t^{-4})\,dt = \frac{t^{-1}}{-1} - \frac{t^{-2}}{-2} + \frac{t^{-3}}{-3} + C$
$= -\frac{1}{t} + \frac{1}{2t^2} - \frac{1}{3t^3} + C$ or $-t^{-1} + \frac{1}{2}t^{-2} - \frac{1}{3}t^{-3} + C$

21. Do the division and use negative exponents and the power rule.
$\int (x^{-2} + 3x^{-3} - x^{-4})\,dx = -\frac{1}{x} - \frac{3}{2x^2} + \frac{1}{3\,x^3} + C$
or $-x^{-1} - \frac{3}{2}x^{-2} + \frac{1}{3}x^{-3} + C$

25. $\int (2x - 1)^2\,dx = \int (4x^2 - 4x + 1)\,dx = \frac{4}{3}x^3 - 2x^2 + x + C$, so
the equation of the line is $y = \frac{4}{3}x^3 - 2x^2 + x + C$. But we are given that
$(1, 3)$ is on the line, so we can determine which of the family of curves
is represented by the C. $3 = \frac{4}{3}(1)^3 - 2(1)^2 + 1 + C$, $C = \frac{8}{3}$.
So the equation of the line is $y = \frac{4}{3}x^3 - 2x^2 + x + \frac{8}{3}$

29.　$C(x) = \int C'(x)\, dx = \int (6x^2 - 2x + 5)\, dx = 2x^3 - x^2 + 5x + C.$

Now it is given that $C(1) = 5$, so we can find the constant of integration.

$5 = 2(1)^3 - (1)^2 + 5(1) + C,\ C = -1.$ So we now have:

$C(x) = 2x^3 - x^2 + 5x - 1,$ and $C(5) = 250 - 25 + 25 - 1 = \$249.$

33.　$a(t) = k,\ v(t) = \int a(t)\, dt = \int k\, dt = kx + C.$ But $v(0) = 0$, so $C = 0$

$s(t) = \int v(t)\, dt = \int kx\, dt = \dfrac{kx^2}{2} + C.$ But $s(0) = 0$, so again $C = 0.$

Now it is given that $s(6) = 360$, so $\dfrac{k(6)^2}{2} = 360,\ k = 20$

37.　Analysis: The linearity rule applies to the sum of two terms. To extend it to three terms, use the associative property to group two terms, and then apply the rule twice.

Chapter 3 Review

14.　This limit has the form $\frac{0}{0}$, so l'Hôpital's rule is applicable:

$$\lim_{x \to \frac{\pi}{2}} \frac{2\cos 2x}{-\sin x} = \frac{-2}{-1} = 2$$

15.　Using l'Hopital's rule: $\displaystyle\lim_{x \to 1} \frac{-\frac{1}{2} x^{-1/2}}{1} = -\frac{1}{2}$

16.　Do the subtraction: $\displaystyle\lim_{x \to \infty} \frac{1 - \sqrt{x}}{x}$ now has form $\frac{\infty}{\infty}$. Using l'Hôpital's rule:

$$\lim_{x \to \infty} \frac{-\frac{1}{2} x^{-1/2}}{1} = \frac{1}{\infty} = 0$$

17.　Do the subtraction: $\displaystyle\lim_{x \to \infty} \frac{4x^3}{x^4 - 4}$, dividing by x^3: $\displaystyle\lim_{x \to \infty} \frac{4}{x - \frac{4}{x^3}} = 0$

18.　$f(x) = x^3 + 3x^2 - 9x + 2,\quad f'(x) = 3x^2 + 6x - 9$

$$= 3(x + 3)(x - 1) = 0 \text{ when}$$

$x = -3, 1.\ f''(x) = 6x + 6 = 0$ when $x = -1.$ Using the second derivative test on our two candidates for extrema: $f''(-3) < 0$, and $f''(1) > 0.$ There is a relative maximum at $(-3, 29)$, and a relative minimum at $(1, -3).$ Testing our candidate for inflection point, it is easy to see that $f''(x)$ changes from negative to zero to positive as x passes through $-1.$ There is an inflection point at $(-1, 13).$

18. (con't.)

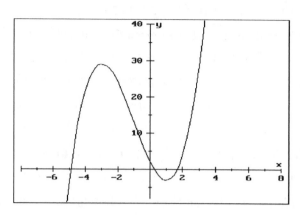

19. $f(x) = 27x^{1/3} - x^{4/3}$. $f'(x) = 9x^{-2/3} - \frac{4}{3}x^{1/3}$. $f'(x) = 0$ when

$\frac{9}{x^{2/3}} = \frac{4x^{1/3}}{3}$, $x = \frac{27}{4}$. $f''(x) = -6\,x^{-5/3} - \frac{4}{9}\,x^{-2/3} = \frac{-54 - 4x}{9x^{5/3}}$.

$f''(x) = 0$ when $x = -\frac{27}{2}$. $f''(-\frac{27}{2}^-) > 0$, $f''(-\frac{27}{2}^+) < 0$, so there is

an inflection point at approximately $(-\frac{27}{2}, -96.43)$. Points at which the second

denominator is zero are also candidates for inflection points. Looking at the sign

of $f''(x)$ as values pass through the origin we see that $x^{5/3}$ is negative when x is

negative, and positive when x is positive. So the concavity changes and there is an

inflection point at $(0, 0)$. Testing our one candidate for extrema with the second

derivative test: $f''(\frac{27}{4}) < 0$, so there is a relative maximum at approximately

$(\frac{27}{4}, 38.27)$. Note also that $f'(x)$ and $f''(x)$ are undefined when $x = 0$.

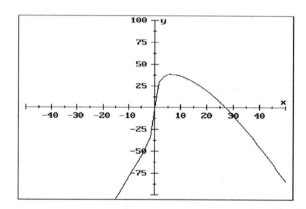

20. $f'(x) = 2 \sin x \cos x + 2 \sin x = 2 \sin x(\cos x + 1) = 0$ when

$x = 0, \pi, 2\pi$ on $[0, 2\pi]$. 0 and 2π are endpoints, $f(0) = -2$, $f(2\pi) = -2$.

All other values of $f(x)$ are greater than -2, so these are absolute minimums.

Testing our remaining candidate for extrema with the second derivative test:

$f''(x) = -2\sin^2 x + 2\cos^2 x + 2\cos x.$ $f''(\pi) = 0 + 2 - 2 = 0$, so

the test fails. Use the first derivative test to see if there is a change of slope:

$f'(\pi^-) > 0$, $f'(\pi^+) < 0$, so there is a relative maximum at $(\pi, 2)$.

To search for inflection points change $f''(x)$ to a function of cosines:

$$f''(x) = -2(1 - \cos^2 x) + 2\cos^2 x + 2\cos x$$

$$= 4\cos^2 x + 2\cos x - 2$$

$$= 2(2\cos^2 x + \cos x - 1)$$

$$= 2(2\cos x - 1)(\cos x + 1) = 0 \text{ when } \cos x = \tfrac{1}{2}, -1,$$

$x = \frac{\pi}{3}, \pi, \frac{5\pi}{3}$. Checking $f''(x)$ for change of concavity at these points:

$f''(\frac{\pi}{3}^-) > 0, f''(\frac{\pi}{3}^+) < 0$. So there is an inflection point at $(\frac{\pi}{3}, -\frac{1}{4})$.

$f''(\frac{5\pi}{3}^-) < 0, f''(\frac{5\pi}{3}^+) > 0$. So there is an inflection point at $(\frac{5\pi}{3}, -\frac{1}{4})$.

$f''(\pi^-) < 0$, $f''(\pi^+) < 0$. So there is no inflection point at $x = \pi$

and the function is concave down on both sides of $(\pi, 2)$.

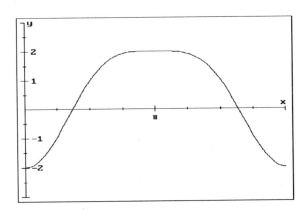

21. Checking for horizontal asymptotes: $\lim\limits_{x\to\infty} f(x) = 1$. $\lim\limits_{x\to-\infty} f(x) = 1$. So $y = 1$ is a horizontal asymptote. Factoring the denominator we see that there are two candidates for vertical asynmptotes; 2 and -2. Testing these: $\lim\limits_{x\to-2^-} f(x) = +\infty$. $\lim\limits_{x\to-2^+} f(x) = -\infty$. $\lim\limits_{x\to2^-} f(x) = -\infty$. $\lim\limits_{x\to2^+} f(x) = +\infty$. So there are two

vertical asymptotes. $f'(x) = \dfrac{(x^2 - 4)2x - (x^2 - 1)2x}{(x^2 - 4)^2} = \dfrac{-6x}{(x^2 - 4)^2}$.

This is equal to 0 when $x = 0$, and it is easy to see that the slope is positive to the left of 0, and negative to the right of 0. So there is a relative maximum at $(0, \frac{1}{4})$. This should be sufficient information to sketch the graph:

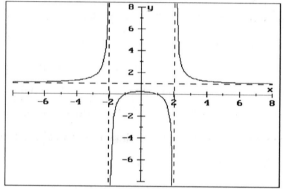

22. $f(x) = \displaystyle\int (3x^2 + 1)\, dx = x^3 + x + C$. If $f(0) = 10$ then $C = 10$ and $f(x) = x^3 + x + 10$

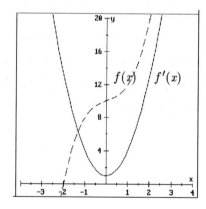

Note that $f'(x)$ designates the slope of $f(x)$.

23. This is a continuous function on a closed interval, so the extreme value theorem guarantees an absolute maximum and an absolute minimum. The candidates are the critical points and the endpoints. $f'(x) = 4x^3 - 10x^4 = x^3(4 - 10x)$ which is 0 at $x = 0, \frac{2}{5}$. Testing our candidates: $f(0) = 5$, $f(\frac{2}{5}) = 5\frac{16}{3125}$, $f(1) = 4$. The absolute maximum is at $(\frac{2}{5}, 5.00512)$ and the absolute minimum at $(1, 4)$.

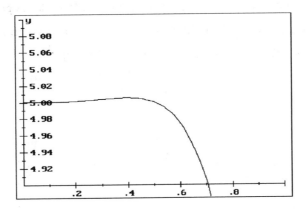

The vertical scale had to be changed to show the critical maximum, so the minimum is off the bottom of the graph at $(1, 4)$.

24. We are asked to minimize the amount of material, $S = x^2 + 4xh$, with the restriction that $V = 2$. We can use the volume formula to find a substitute value for h, and then have S as a function of x. $lwh = 2$, $x^2h = 2$, $h = \frac{2}{x^2}$.
$S = x^2 + 4x\left(\frac{2}{x^2}\right) = x^2 + \frac{8}{x}$. $S' = 2x - \frac{8}{x^2} = 0$ when $x^3 = 4$,
$x = \sqrt[3]{4}$. $h = \frac{2}{x^2} = \frac{2}{\sqrt[3]{4}^2}$. The dimensions of the box are
≈ 1.587 by 1.587 by $.794$ ft, or to the nearest inch: $19 \times 19 \times 10$ in

25. $P(.25) - P(0) = \left(20 - \frac{6}{.25 + 1}\right) - \left(20 - \frac{6}{0 + 1}\right) = 15.2 - 14$
$= 1,200$ additional people.

Chapter 4
Integration

4.1 Area as the Limit of a Sum; Summation Notation

3. $\displaystyle\sum_{1}^{n} k = \frac{n(n+1)}{2}$, so $\displaystyle\sum_{1}^{15} k = \frac{15(16)}{2} = 120$

7. $\displaystyle\sum_{1}^{100}(2k - 3) = 2\sum_{1}^{100} k - 3\sum_{1}^{100} 1 = 2\left(\frac{100(101)}{2}\right) - 3(100)$

$$= 10{,}100 - 300 = 9{,}800$$

11. $\displaystyle\lim_{n\to\infty}\sum_{1}^{n}\left(\frac{2}{n} + \frac{2k}{n^2}\right) = \lim_{n\to\infty}\left\{\frac{2}{n}(n) + \frac{2}{n^2}\left(\frac{n(n+1)}{2}\right)\right\} = \lim_{n\to\infty}\left\{2 + \left(1 + \frac{1}{n}\right)\right\}$

$$= 3$$

15. **a.**

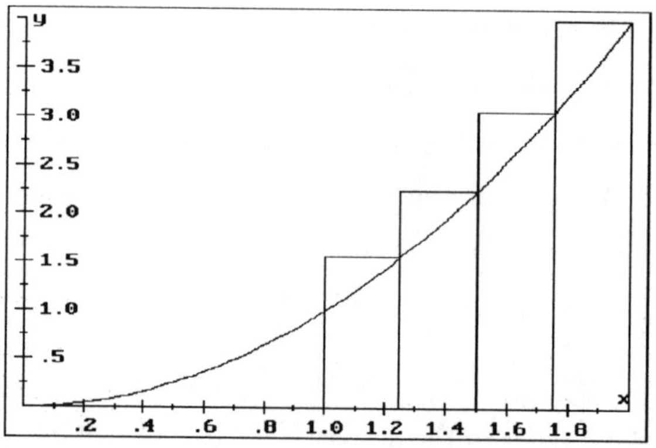

$$A \approx [f(1.25) + f(1.5) + f(1.75) + f(2)](.25)$$
$$= [(1.25)^2 + (1.5)^2 + (1.75)^2 + (2)^2](.25)$$
$$= [1.5625 + 2.25 + 3.0625 + 4](.25)$$
$$= [10.875](.25)$$
$$= 2.71875$$

b.

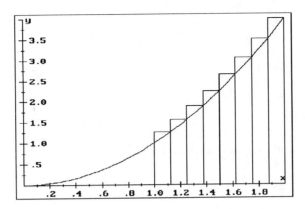

$$A \approx [\, f(1.125) \;+\; f(1.25) \;+\; f(1.375) \;+\; f(1.5) \;+ f(1.625)$$
$$+ \; f(1.75) \;+\; f(1.875) \;+\; f(2)](0.125)$$
$$= [\, (1.125)^2 \;+\; (1.25)^2 \;+\; (1.375)^2 \;+\; (1.5)^2 \;+ (1.625)^2$$
$$+ \; (1.75)^2 \;+\; (1.875)^2 \;+\; (2)^2](0.125)$$
$$= (20.1875)(0.125) \;=\; 2.5234375$$

19.

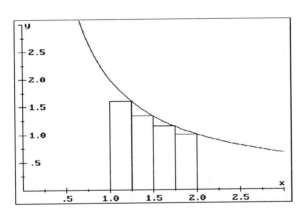

$$A \approx f(1.25)(.25) \;+\; f(1.5)(.25) \;+\; f(1.75)(.25) \;+\; f(2)(.25)$$
$$= [(\tfrac{2}{1.25}) \;+\; (\tfrac{2}{1.5}) \;+\; (\tfrac{2}{1.75}) \;+\; (\tfrac{2}{2})](.25) \;=\; 1.269$$

23. The beginning point of our rectangles, a, is 1, the ending point, b, is 2.

$$\Delta x = \frac{b - a}{n} = \frac{2 - 1}{n} = \frac{1}{n}.$$

$$A = \lim_{n \to \infty} \sum_{1}^{n} f(1 + k\Delta x)\, \Delta x = \lim_{n \to \infty} \sum_{1}^{n} f\left(1 + \frac{k}{n}\right) \Delta x$$

Evaluating the function at $\left(1 + \frac{k}{n}\right)$:

$$= \lim_{n \to \infty} \sum_{1}^{n} \left\{ 4\left(1 + \frac{k}{n}\right)^3 + 2\left(1 + \frac{k}{n}\right) \right\} \Delta x$$

Expand the algebraic expressions:

$$= \lim_{n \to \infty} \sum_{1}^{n} \left\{ 4\left(1 + \frac{3k}{n} + \frac{3k^2}{n^2} + \frac{k^3}{n^3}\right) + 2 + \frac{2k}{n} \right\} \Delta x$$

Combine terms:

$$= \lim_{n \to \infty} \sum_{1}^{n} \left\{ 6 + \frac{14k}{n} + \frac{12k^2}{n^2} + \frac{4k^3}{n^3} \right\} \Delta x$$

Apply the sigma formulas to the k expressions:

$$= \lim_{n \to \infty} \left\{ 6n + \frac{14n(n + 1)}{2n} + \frac{12n(n + 1)(2n + 1)}{6n^2} + \frac{4n^2(n + 1)^2}{4n^3} \right\} \frac{1}{n}$$

Multiply by the $\frac{1}{n}$:

$$= \lim_{n \to \infty} \left\{ 6 + \frac{14n^2 + 14n}{2n^2} + \frac{24n^3 + 36n^2 + 12n}{6n^3} + \frac{4n^4 + 8n^3 + 4n^2}{4n^4} \right\}$$

Take the limits of each term, remembering that $\lim_{n \to \infty} \frac{1}{n} = 0$:

$$= (6 + 7 + 4 + 1) = 18$$

27. The beginning point of our rectangles, a, is 0, the ending point, b, is 1.

$$\Delta x = \frac{b - a}{n} = \frac{1 - 0}{n} = \frac{1}{n}.$$

$$A = \lim_{n \to \infty} \sum_{1}^{n} f(k\Delta x)\, \Delta x = \lim_{n \to \infty} \sum_{1}^{n} f\left(\frac{k}{n}\right) \Delta x$$

Evaluating the function at $\left(\frac{k}{n}\right)$:

$$= \lim_{n \to \infty} \sum_{1}^{n} \left\{ 4\left(\frac{k}{n}\right)^3 + 3\left(\frac{k}{n}\right)^2 \right\} \Delta x$$

Apply the sigma formulas to the k expressions:

$$= \lim_{n \to \infty} \left\{ \frac{3n(n + 1)(2n + 1)}{6n^2} + \frac{4n^2(n + 1)^2}{4n^3} \right\} \frac{1}{n}$$

Multiply by the $\frac{1}{n}$:

$$= \lim_{n \to \infty} \left\{ \frac{6n^3 + 9n^2 + 3n}{6n^3} + \frac{4n^4 + 8n^3 + 4n^2}{4n^4} \right\}$$

Take the limits of each term, remembering that $\lim_{n \to \infty} \frac{1}{n} = 0$:

$$= (1 + 1) = 2$$

SURVIVAL HINT

After you have done several limits of the expanded sigma expressions you should realize that all but the first term of each sigma is going to vanish. Do not take the time to expand the polynomials.

31. The given curve is the half-circle in the first and second quadrants. Apply the
formula for the area of a circle. It is true.

35. $A = \lim\limits_{n \to \infty} \sum\limits_{1}^{n} f(a + k\Delta x)\Delta x$ where $\Delta x = \dfrac{b - a}{n}$.

For our rectangle $a = 0$, $b = \ell$, so $\Delta x = \dfrac{\ell}{n}$. $f(x)$ is the horizontal line $y = w$

$A = \lim\limits_{n \to \infty} \sum\limits_{1}^{n} f(0 + k\frac{\ell}{n})\,\frac{\ell}{n}$ where $f(0 + k\frac{\ell}{n}) = w$

$A = \lim\limits_{n \to \infty} \sum\limits_{1}^{n} w\left(\frac{\ell}{n}\right) = \lim\limits_{n \to \infty} nw\left(\frac{\ell}{n}\right) = \lim\limits_{n \to \infty} w\ell = w\ell$

39. $A = \lim\limits_{n \to \infty} \sum\limits_{1}^{n} f(a + k\Delta x)\Delta x$ where $\Delta x = \dfrac{b - a}{n}$

$a = 0$, $b = 4$, $f(x) = x^2$. $\Delta x = \dfrac{4}{n}$.

$A = \lim\limits_{n \to \infty} \sum\limits_{1}^{n} f(0 + \frac{4k}{n})\frac{4}{n} = \lim\limits_{n \to \infty} \sum\limits_{1}^{n} \frac{16k^2}{n^2}\frac{4}{n} = \lim\limits_{n \to \infty} \frac{64}{n^3}\frac{n(n + 1)(2n + 1)}{6}$

$= \lim\limits_{n \to \infty} \frac{64}{6}\left(2 + \frac{3}{n} + \frac{1}{n^2}\right) = \frac{64}{3}$. However, if instead of taking a limit, we
actually compute the sum for increasing values of n, we can see the convergence
of the sum toward the value of the area. First let $n = 4$, $\Delta x = 1$, then $n = 8$,
$\Delta x = \frac{1}{2}$, then $n = 16$, $\Delta x = \frac{1}{4}$, $n = 32$, $\Delta x = \frac{1}{8}$, etc.

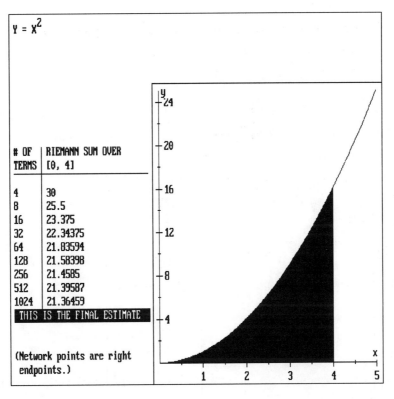

The actual area is $\dfrac{64}{3}$.

43.   Analysis: This is mostly algebraic manipulation: $k^2 - (k - 1)^2$
$= 2k - 1$, or $2k = [k^2 - (k - 1)^2] + 1$ or
$$k = \tfrac{1}{2}[k^2 - (k - 1)^2] + \tfrac{1}{2}.$$ Now use the summation
formulas. Part **b** is a "telescoping sum", as suggested, writing a few terms will
show the pattern.

45.   Analysis: Using the left endpoint, the height of the first rectangle is given by
$f(0)$. The height of the second rectangle is computed at $f(\Delta x)$, the third at
$f(2\Delta x)$, the fourth at $f(3\Delta x)$, and in general the kth at $f([k - 1]\Delta x)$. This will
give the limit expression in the problem.

47.   Analysis: If these are finite sums, that is, to some finite value n, the properties
of the reals can be applied. We can associate, commute, and distribute. Writing
out the sigma on the left of the equation, with a $+ \ldots +$ in the middle, and using
these properties, will give the sigma expressions on the right of the equation.

4.2 Riemann Sums and the Definite Integral

SURVIVAL HINT

Thoroughly understand the significance of each symbol in the limit of the Riemann sum
as given on p. 315. This concept of an infinite sum of infinitely small parts will be
used repeatedly throughout the text. Any successful calculus student should be able
to state and explain the definitions of limit, derivative, and definite integral.

3. $a = 1, b = 3, \Delta x = \dfrac{b - a}{n} = \dfrac{2}{4} = \dfrac{1}{2}$
$S_n \approx [f(1.5) + f(2) + f(2.5) + f(3)](\tfrac{1}{2}) = (2.25 + 4 + 6.25 + 9)(\tfrac{1}{2})$
$= \dfrac{21.5}{2} = 10.75$

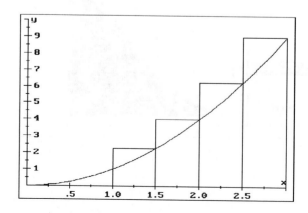

7. $a = -\frac{\pi}{2}, \ b = 0, \ \Delta x = \frac{b - a}{n} = \frac{\pi}{8}$

  $S_n \approx [f(-\frac{3\pi}{8}) + f(-\frac{\pi}{4}) + f(-\frac{\pi}{8}) + f(0)](\frac{\pi}{8})$

  $\approx [0.382683 + 0.707107 + 0.923880 + 1](0.392699)$

  ≈ 1.183465

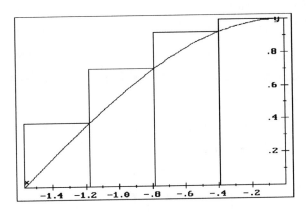

11. $a = 0, \ b = \pi, \ \Delta x = \frac{b - a}{n} = \frac{\pi}{4}$

  $D \approx [f(0) + f(\frac{\pi}{4}) + f(\frac{\pi}{2}) + f(\frac{3\pi}{4})](\frac{\pi}{4})$

  $= \left(0 + \frac{\sqrt{2}}{2} + 1 + \frac{\sqrt{2}}{2}\right)(\frac{\pi}{4})$

  $= (1 + \sqrt{2})(\frac{\pi}{4})$

  ≈ 1.896119

15. $\displaystyle\int_{-1}^{2} (2x^2 - 3x)\,dx$ by linearity $= 2\displaystyle\int_{-1}^{2} x^2\,dx - 3\displaystyle\int_{-1}^{2} x\,dx = 2(3) - 3(\frac{3}{2}) = \frac{3}{2}$

19.   On $[0, 1]$ $x^3 \le x$, so $\displaystyle\int_{0}^{1} x^3\,dx \le \displaystyle\int_{0}^{1} x\,dx = \frac{1}{2}$

23.   By the subdivision and opposite properties, $\displaystyle\int_{-1}^{2} f(x)\,dx$

  $= \displaystyle\int_{-1}^{1} f(x)\,dx + \displaystyle\int_{1}^{3} f(x)\,dx + \displaystyle\int_{3}^{2} f(x)\,dx$

  $= \displaystyle\int_{-1}^{1} f(x)\,dx + \displaystyle\int_{1}^{3} f(x)\,dx - \displaystyle\int_{2}^{3} f(x)\,dx = 3 + 5 - (-2) = 10$

27. Analysis: The area designated by the definite integral will be either a trapezoid or, if $Cx + D$ has a root on the given interval, two triangles. For a trapezoid, $A = \frac{h}{2}(B_1 + B_2)$, $h = b - a$, $B_1 = f(a) = Ca + D$, $B_2 = f(b) = Cb + D$. Substitute these values into the area formula.

29. Analysis: Find the five differences, Δx_k, the largest of the set is the norm, which is 1.1

31. Analysis: $R_5 = \sum\limits_{1}^{5} f(x_k) \, \Delta x_k.$

35. Analysis: **a.** Show that if $f(x) < 0$ on $[a, b]$, $-f(x) > 0$ on $[a, b]$, and represents the area, A.

    b. Use the subdivision rule to divide $\int\limits_{a}^{b} f(x) \, dx$ into two integrals where $f(x) > 0$ and $f(x) < 0$.

    c. $\int\limits_{-2}^{1} 2x \, dx = -\int\limits_{-2}^{0} -2x \, dx + \int\limits_{0}^{1} f(x) \, dx$

4.3 The Fundamental Theorem of Calculus; Integration by Substitution

SURVIVAL HINT

The *Fundamental Theorem* of Calculus sounds really important - and it is. However, most calculus students seem not to appreciate its significance upon first exposure. It will become more meaningful as you encounter it in more advanced contexts, when you have a greater perspective. For now, see that it relates the two major concepts of elementary calculus — the derivative and the Riemann sum. Try to see that the increase in the area under a curve, as x is incremented, depends upon the slope of the curve.

3. $\int\limits_{-3}^{5} (2x + a) \, dx = x^2 + ax \Big|_{-3}^{5} = (5)^2 + 5a - [(-3)^2 + (-3)a]$

         $= 16 + 8a$

7. Write the function with a negative exponent and use the power rule:

$\int\limits_{1}^{2} x^{-3} \, dx = \frac{x^{-2}}{-2} \Big|_{1}^{2} = -\frac{1}{2}[\frac{1}{4} - 1] = \frac{3}{8}$

11. Multiply and use rational exponents: $\displaystyle\int_0^1 (t^{3/2} - t)\, dt = \frac{2t^{5/2}}{5} - \frac{t^2}{2}\ \Big|_0^1$

$= \frac{2}{5} - \frac{1}{2} = -\frac{1}{10}$

SURVIVAL HINT

It might be a good idea to take a few minutes to review the derivatives of the trigonometric functions. You need to be competent with the derivatives in order to do antiderivatives.

15. Make use of the Pythagorean identity $\tan^2 x = \sec^2 x - 1$:

$$\int_0^{\frac{\pi}{4}} [\sec^2 x - (\sec^2 x - 1)]\, dx = \int_0^{\frac{\pi}{4}} dx = x\ \Big|_0^{\frac{\pi}{4}} = \frac{\pi}{4}$$

19. a. $\displaystyle\int_0^4 x^{\frac{1}{2}}\, dx = \frac{2x^{3/2}}{3}\ \Big|_0^4 = \frac{16}{3}$

b. $\displaystyle\int_{-4}^0 \sqrt{-x}\, dx$ Note that since $-4 \le x \le 0$, $-x \ge 0$ and the radical is defined.

If $u = -x$, then $du = -dx$

$$\int_{-4}^0 \sqrt{-x}\, dx = -\int_{-4}^0 \sqrt{-x}\, (-dx) = -\frac{2(-x)^{3/2}}{3}\ \Big|_{-4}^0 = 0 + \frac{2(8)}{3} = \frac{16}{3}$$

On **a** and **b** notice that $y = \sqrt{x}$ and $y = \sqrt{-x}$ represent half-parabolas which are symmetric about the y axis. So they have the same areas.

SURVIVAL HINT

Do not try to do any but the simplest of the u substitutions in your head. You will have fewer errors, and save time in the long run, if for each problem you write down:

$u =$, and $du =$. Most errors in u substitutions come from a failure to properly introduce constants to "set-up the du".

23. Let $u = 2x + 3$, then $du = 2\, dx$

$$\int (2x + 3)^4 dx = \frac{1}{2}\int u^4\, du = \frac{u^5}{10} + C = \frac{(2x + 3)^5}{10} + C$$

27. All the first term needs is the power rule. For the second term let $u = 3x$,

$du = 3\, dx$. $\displaystyle\int x^2\, dx - \frac{1}{3}\int \cos u\, du = \frac{x^3}{3} - \frac{1}{3}\sin u + C$

$= \frac{1}{3}(x^3 - \sin 3x) + C$

31. Let $u = t^{3/2} + 5$, $du = \frac{3}{2} t^{1/2} dt$

$$\frac{2}{3} \int (t^{3/2} + 5)^3 \frac{3}{2} t^{1/2} dt = \frac{2}{3} \int u^3 du = \frac{2}{3} \frac{u^4}{4} + C = \frac{1}{6}(t^{3/2} + 5)^4 + C$$

35. **a.** Let $u = 2x^2 + 1$, $du = 4x\,dx$. Then $\frac{1}{4} \int (2x^2 + 1)^{\frac{1}{2}} 4x\,dx$

$$= \frac{1}{4} \int u^{1/2} du = \frac{1}{4} \frac{u^{3/2}}{\frac{3}{2}} + C = \frac{1}{6}(2x^2 + 1)^{3/2} + C$$

b. Again let $u = 2x + 1$, $du = 2\,dx$. This time there is a "left over" x.

But from our substitution, $x = \frac{u - 1}{2}$.

$$\int x\sqrt{2x + 1}\,dx = \frac{1}{2} \int \frac{u - 1}{2} u^{1/2} du = \frac{1}{4} \int (u^{3/2} - u^{1/2})\,du$$

$$= \frac{1}{4}\left(\frac{u^{5/2}}{\frac{5}{2}} - \frac{u^{3/2}}{\frac{3}{2}}\right) + C = \frac{1}{2} u^{3/2}\left(\frac{u}{5} - \frac{1}{3}\right) + C$$

$$= \frac{(2x + 1)^{3/2}}{2}\left(\frac{2x + 1}{5} - \frac{1}{3}\right) + C$$

$$= \frac{(2x + 1)^{3/2}}{2}\left(\frac{3(2x + 1) - 5}{15}\right) + C$$

$$= \frac{(2x + 1)^{3/2}}{2}\left(\frac{6x - 2}{15}\right) + C = \frac{(2x + 1)^{3/2}(3x - 1)}{15} + C$$

39. Let $u = \sin t$, $du = \cos t\,dt$. Then $\int \sin^3 t \cos t\,dt = \int u^3 du = \frac{u^4}{4} + C$
$= \frac{1}{4}\sin^4 t + C$

43. $\displaystyle\int_0^{\frac{\pi}{6}} (\sin x + \cos x)\,dx = -\cos x + \sin x \Big|_0^{\frac{\pi}{6}}$

$$= -\frac{\sqrt{3}}{2} + \frac{1}{2} - (-1 + 0) = \frac{3 - \sqrt{3}}{2}$$

47. For $2 \le x \le 3$, $|x| = x$. $\displaystyle\int_2^3 x\,dx = \frac{x^2}{2}\Big|_2^3 = \frac{9}{2} - \frac{4}{2} = \frac{5}{2}$

53. The multiply-defined function requires two integrals:

$$\int_0^{\frac{\pi}{2}} \cos x\,dx + \int_{\frac{\pi}{2}}^{\pi} x\,dx, \quad \text{each of which is continuous on its domain.}$$

$$= \sin x \Big|_0^{\frac{\pi}{2}} + \frac{x^2}{2}\Big|_{\frac{\pi}{2}}^{\pi} = (1 - 0) + \left(\frac{\pi^2}{2} - \frac{\pi^2}{8}\right) = 1 + \frac{3\pi^2}{8}$$

57. $f(-x) = -f(x)$. By the symmetry about the origin, $\displaystyle\int_{-a}^{a}$ (an odd function) $= 0$.

61. **a.** At the instant when the particle changes from positive velocity to negative velocity its velocity will be zero. $v(t) = t^2(t^3 - 8)^{1/3} = 0$ when $t = 0$ or 2.

b. $v(t) = s'(t)$, so $s(t) = \int v(t)\, dt = \int t^2(t^3 - 8)^{1/3}\, dt$.

Let $u = (t^3 - 8)$, $du = 3t^2\, dt$

$$\int t^2(t^3 - 8)^{1/3}\, dt = \tfrac{1}{3}\int (t^3 - 8)^{1/3}\, 3\, t^2\, dt = \tfrac{1}{3}\int u^{1/3}\, du$$

$$= \tfrac{1}{3}\frac{u^{4/3}}{\frac{4}{3}} + C = \tfrac{1}{4}(t^3 - 8)^{4/3} + C$$

Now when $t = 0$, $s(0) = 1$. $s(0) = \dfrac{(0 - 8)^{4/3}}{4} + C = 1$,

$4 + C = 1$, $C = -3$. $s(t) = \tfrac{1}{4}(t^3 - 8)^{4/3} - 3$.

In part **a** we found that the particle turns when $t = 2$, so

$s(2) = \tfrac{1}{4}(2^3 - 8)^{4/3} - 3 = -3$. It turns at $s = -3$.

65. Analysis: **a.** Substitute $f'(x)$ for $f(x)$ before integrating.

b. Note that $f^2(x) = f(x)f'(x)$, let $u = f(x)$, $du = f'(x)dx$.

Now integrate.

4.4 Introduction to Differential Equations

SURVIVAL HINT

The solution to a differential equation, like the process of antidifferentiation, has an infinite set of vertically translated functions. Unless you are given initial values that will specify one particular function from this set, you must remember to include $+\ C$ in the general solution.

3. By implicit differentiation: $x\dfrac{dy}{dx} + y = 0$, $\dfrac{dy}{dx} = -\dfrac{y}{x}$.

7. Separating variables: $y\, dy = -x\, dx$, $\int y\, dy = -\int x\, dx$,

$\dfrac{y^2}{2} = -\dfrac{x^2}{2} + C_1$, $x^2 + y^2 = C$

(Since C_1 is any constant, let $2C_1 = C$.)

11. Separating variables: $y\, dy = x\sqrt{1 - x^2}\, dx$, $\int y\, dy = \int x\sqrt{1 - x^2}\, dx$,

On the right let $u = (1 - x^2)$, $du = -2x\, dx$, $\int u^{1/2} du$ form.

Integrating: $\dfrac{y^2}{2} = -\dfrac{1}{2}\dfrac{(1 - x^2)^{3/2}}{\frac{3}{2}} + C$, $\dfrac{(1 - x^2)^{3/2}}{3} + \dfrac{y^2}{2} = C$,

Multiplying by 6: $2(1 - x^2)^{3/2} + 3y^2 = C$.

15. $\int \cos y\, dy = \int \sin x\, dx$, $\sin y = -\cos x + C$, $\cos x + \sin y = C$

19. $y\, dx = x\, dy$, $y\, dx - x\, dy = 0$, $y - x\dfrac{dy}{dx}$ suggests we set up the quotient rule:

Dividing by y^2, $\dfrac{y - x\dfrac{dy}{dx}}{y^2} = 0$, which is the derivative of $\dfrac{x}{y} = 0$.

So $\dfrac{x}{y} = 0 + C$ or $x = Cy$.

23. Let T be temperature, t be time, T_n be temperature of the surrounding medium, and c be the constant of proportionality. Then: $\dfrac{dT}{dt} = c(T - T_n)$.

27. Let $P(x, y)$, be a point where a curve of the family $2x - 3y = C$ intersects the orthogonal trajectory curve. Orthogonal implies: $\dfrac{dx}{dy} = -\dfrac{1}{\dfrac{dy}{dx}}$

$= -\dfrac{1}{\dfrac{2}{3}} = -\dfrac{3}{2}$. The family of curves with slope of $-\dfrac{3}{2}$ is $y = -\dfrac{3}{2}x + C$,

or $3x + 2y = C$.

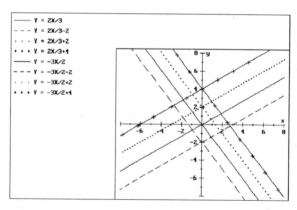

31. Let P, (x, y), be a point where a curve of the family $xy^2 = C$ intersects the orthogonal trajectory curve. By implicit differentiation of the given function:

$x(2y\dfrac{dy}{dx}) + y^2 = 0$, $\dfrac{dy}{dx} = -\dfrac{y^2}{2xy} = -\dfrac{y}{2x}$.

Orthogonal implies that the new family of curves has slopes

$= -\dfrac{1}{\dfrac{dy}{dx}} = -\dfrac{1}{-\dfrac{y}{2x}} = \dfrac{2x}{y}$. For this family $\dfrac{dy}{dx} = \dfrac{2x}{y}$.

Separating variables: $\int y\, dy = \int 2x\, dx$, $\dfrac{y^2}{2} = x^2 + C$, $2x^2 - y^2 = C$

31. (con't.)

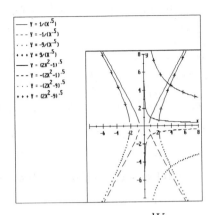

35. By Toricelli's Law (see example 4), $\dfrac{dV}{dt} = -4.8\, A_0\, \sqrt{h}$. Given a square hole

with $s = 1.5$ in. $= \frac{1}{8}$ ft, $A_0 = (\frac{1}{8})^2 = \frac{1}{64}$. So $\dfrac{dV}{dt} = -4.8\,\frac{1}{64}\,\sqrt{h}$.

Since $V = \pi\, 3^2\, h$, we can differentiate to get $\dfrac{dV}{dt} = 9\,\pi\,\dfrac{dh}{dt}$. Substituting

these two formulas we have: $-4.8\,\dfrac{\sqrt{h}}{64} = 9\,\pi\,\dfrac{dh}{dt}$. Separating variables:

$\int -\dfrac{4.8}{9\pi(64)}\, dt = \int h^{-1/2}\, dh,\; -\dfrac{1}{120\pi}\, t = 2\, h^{1/2} + C,\; t = -240\pi\,\sqrt{h} + C.$

When the tank is full, $t = 0$ and $h = 5$. $\;0 = -240\pi\,\sqrt{5} + C,\; C = 240\pi\sqrt{5}$.

We have the formula relating time and the height of the water:

$t = -240\pi\sqrt{h} + 240\pi\,\sqrt{5}$. The tank will be empty when $h = 0$. The time

required for this is: $\; t = 240\pi\,\sqrt{5}$ min ≈ 1686 sec $= 28.2$ min

39. The rate of change in population, P, is given by: $\dfrac{dP}{dt} = 1500\, t^{-1/2}$.

Separating variables: $\int dP = \int 1500\, t^{-1/2} dt,\; P = 3000\,\sqrt{t} + C.$

1994 is four years after 1990, so $P(4) = 39,000$, and $39,000 = 3000\sqrt{4} + C,$

$C = 33\,000.$ Our formula is now: $P = 3000\,\sqrt{t} + 33,000.$

  a. $P(0) = 33,000$

  b. $P(9) = 3000\,\sqrt{9} + 33,000 = 42,000$

43. It is given that: $\dfrac{dP}{dt} = k\,\sqrt{P}$. Separating variables, $\int P^{-1/2}\, dP = \int k\, dt.$

$2\,\sqrt{P} = kt + C.$ We were given two pieces of information which we can use

to find k and C. $P(0) = 9000$, so $C = 60\,\sqrt{10}$, and $P(-10) = 4000$, so

$2\,\sqrt{4000} = -10\,k + 60\,\sqrt{10},\; k = 2\,\sqrt{10}$. We now have the equation:

$2\,\sqrt{P} = 2\,\sqrt{10}\, t + 60\,\sqrt{10}$, or $\sqrt{P} = \sqrt{10}\, t + 30\,\sqrt{10}.$

To find t for $P = 16\,000$, $\sqrt{16,000} = \sqrt{10}\, t + 30\,\sqrt{10}$, $t = 10$ years.

47. 50 cm on a side of a square means $A = 2500$, and $k = 0.0025$

$\frac{dQ}{dt} = -0.0025\,(2500)\,\frac{dT}{ds}$. Separate the variables, integrate, and solve for T, remembering that $\frac{dQ}{dt}$ is a constant. $\frac{dQ}{dt}\int ds = -6.25\int dT$,

$\frac{dQ}{dt}\,s = -6.25\,T + C$. We are given two pieces of information that will allow us to find C and $\frac{dQ}{dt}$. When $s = 0$, $T = 60$, $\frac{dQ}{dt}(0) = -6.25\,(60) + C$,

$C = 375$. When $s = 2$, $T = 5$, $\frac{dQ}{dt}(2) = -6.25\,(5) + 375$,

$\frac{dQ}{dt} = 171.875$ calories/s

51. Analysis: $F = \frac{mk}{s^2}$, $F = ma$, so $a = \frac{k}{s^2}$. We need to establish that $k = -gR^2$. Newton's law of gravitation says that the force of attraction between two bodies is inversely proportional to the square of the distance between their centers of mass. $F = \frac{k}{R^2}$. The force of attraction is called gravity.

4.5 The Mean Value Theorem for Integrals; Average Value

1. $f(x)$ is continuous on $[1, 2]$, so the MVT guarantees the existence of a c such that:

$$\int_1^2 4x^3\,dx = f(c)\,(2 - 1).\quad f(c) = x^4\,\Big|_1^2 = 16 - 1 = 15.$$

$$4c^3 = 15,\quad c^3 = \frac{15}{4},\quad c = \frac{\sqrt[3]{30}}{2}.$$

SURVIVAL HINT

Remember that this is a *theorem*, and as such has a hypothesis that must be verified before the theorem is valid. Know the hypotheses, and understand why they are necessary, for every theorem that you learn.

5. $y = \csc x$ is discontinuous at $x = 0$, so MVT does not apply.

9. $\frac{dF}{dx} = \frac{d}{dx}\int_4^x (t^{1/4} - 4)\,dt$, By second FTC, $F'(x) = x^{1/4} - 4$.

Verification: $\int_4^x (t^{1/4} - 4)\,dt = \frac{4\,t^{5/4}}{5} - 4t\,\Big|_4^x$

$$F(x) = \frac{4x^{5/4}}{5} - 4x - \frac{4^{5/4}}{5} + 16$$

$$\frac{d}{dx}\left(\frac{4x^{5/4}}{5} - 4x - \frac{4^{5/4}}{5} + 16\right) = x^{1/4} - 4 = F'(x)$$

13. $A = \displaystyle\int_0^{10} \frac{x}{2}\, dx = \frac{x^2}{4}\Big|_0^{10} = 25 = f(c)(b - a) = f(c)10.\ f(c) = 2.5$

So $\frac{c}{2} = 2.5$, $c = 5$

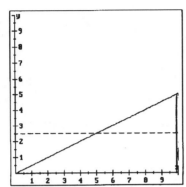

17. $A = \displaystyle\int_{-1}^{1.5} \cos x\, dx = \sin x \Big|_{-1}^{1.5} = \sin 1.5 - \sin(-1) \approx 1.83897$

$A = f(c)(b - a) = f(c)(2.5) \approx 1.83897,\ f(c) \approx 0.735586$

$\cos c \approx 0.735586$, $c \approx 0.744264$ or -0.744264 on $[-1, 1.5]$.

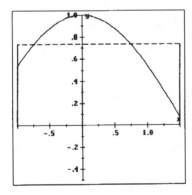

21. Average Value $= \dfrac{1}{\frac{\pi}{4}}\displaystyle\int_0^{\frac{\pi}{4}} \sin x\, dx = \frac{4}{\pi}(-\cos x)\Big|_0^{\frac{\pi}{4}} = \frac{4}{\pi}\left(-\frac{\sqrt{2}}{2} + 1\right) \approx 0.372923$

25. Average Value $= \dfrac{1}{1 - 0}\displaystyle\int_0^1 (2x - 3)^3\, dx.$ Let $u = 2x - 3,\ du = 2\, dx$

$= \frac{1}{2}\displaystyle\int_{-3}^{-1} u^3\, du = \frac{u^4}{8}\Big|_{-3}^{-1} = \frac{1}{8} - \frac{81}{8} = -10$

29. Average Value $= \dfrac{1}{3-(-3)} \displaystyle\int_{-3}^{3} \sqrt{9-x^2}\, dx = \dfrac{1}{6}$ (area of a semi-circle with $r = 3$)

$$= \dfrac{1}{6}\left(\dfrac{9\,\pi}{2}\right) = \dfrac{3\pi}{4}$$

33. By Leibniz's Rule if $F(x) = \displaystyle\int_{v(x)}^{u(x)} f(t)\, dt$, then $F'(x) = f(u)\dfrac{du}{dx} - f(v)\dfrac{dv}{dx}$

If $H(t) = \displaystyle\int_{3t}^{1-t} \dfrac{\sin x}{x+1}\, dx$, then $H'(t) = \dfrac{\sin (1-t)1}{(1-t)+1}(-1) - \dfrac{\sin 3t}{3t+1}(3)$

$$= -\dfrac{\sin (1-t)}{2-t} - \dfrac{3\sin 3t}{3t+1}$$

35. Average height $= \dfrac{1}{t_1 - t_0} \displaystyle\int_{t_0}^{t_1} \left(-\tfrac{1}{2}gt^2 + v_0 t\right) dt$

$$= \dfrac{1}{t_1 - t_0}\left(-\tfrac{1}{6}gt^3 + v_0\dfrac{t^2}{2}\right)\Bigg|_{t_0}^{t_1}$$

$$= \dfrac{1}{t_1 - t_0}\left\{-\tfrac{1}{6}gt_1{}^3 + v_0\dfrac{t_1{}^2}{2} - \left(-\tfrac{1}{6}gt_0{}^3 + v_0\dfrac{t_0{}^2}{2}\right)\right\}$$

$$= \dfrac{1}{t_1 - t_0}\left\{-\dfrac{g}{6}(t_1{}^3 - t_0{}^3) + \dfrac{v_0}{2}(t_1{}^2 - t_0{}^2)\right\}$$

Factoring the difference of 2 cubes and the difference of 2 squares:

$$= \dfrac{1}{t_1 - t_0}\left\{-\dfrac{g}{6}(t_1 - t_0)(t_1{}^2 + t_1 t_0 + t_0{}^2) + \dfrac{v_0}{2}(t_1 - t_0)(t_1 + t_0)\right\}$$

Remove and cancel the common factor of $(t_1 - t_0)$:

$$= -\tfrac{1}{6}g(t_1{}^2 + t_1 t_0 + t_0{}^2) + \tfrac{1}{2}v_0(t_1 + t_0)$$

37. Average temperature $= \dfrac{1}{12-9} \displaystyle\int_{9}^{12} (-0.3t^2 + 4t + 10)\, dt$

$$= \tfrac{1}{3}\left[-0.1t^3 + 2t^2 + 10t\right]\Big|_{9}^{12}$$
$$= \tfrac{1}{3}(235.2 - 179.1)$$
$$= 18.7^\circ\ C$$

39. Analysis:

    a.   Check the hypotheses and apply the Fundamental Theorem of Calculus.

    b.   By the Fundamental Theorem the derivative of the integral is $f(x)$, so

$$f(x) = \sec^2 x$$

41.   Analysis: The average value of $f(t)$ on $[x, x^2]$ is given by:

$$A(x) = \frac{1}{x} = \frac{1}{x^2 - x} \int_x^{x^2} f(t)\, dt$$

$$x - 1 = \int_x^{x^2} f(t)\, dt$$

Now differentiate both sides using Leibniz's rule, and evaluate at $x = 2$.

4.6 Numerical Integration: The Trapezoidal Rule and Simpson's Rule

SURVIVAL HINT

Successful use of the trapezoidal and Simpson's rule is highly dependent upon your careful organization of the data. To simultaneously compute and keep a running total of values in your calculator takes considerable skill. Make a table of x_n values, $f(x_n)$, and $f(x_n)$ times the appropriate multiplier. It will save you time in the long run.

1.   Trapezoidal rule: $\Delta x = \dfrac{2 - 1}{4} = \dfrac{1}{4}$

$$x_0 = 1 \qquad f(x_0) = 1$$
$$x_1 = 1.25 \quad f(x_1) = 1.5625$$
$$x_2 = 1.5 \quad f(x_2) = 2.25$$
$$x_3 = 1.75 \quad f(x_3) = 3.0625$$
$$x_4 = 2 \qquad f(x_4) = 4$$
$$A \approx \tfrac{1}{2}[1 + 2(1.5625) + 2(2.25) + 2(3.0625) + 4](\tfrac{1}{4})$$
$$= \tfrac{1}{8}(18.75) = 2.34375$$

Simpson's rule:

$$A \approx \tfrac{1}{3}[1 + 4(1.5625) + 2(2.25) + 4(3.0625) + 4](\tfrac{1}{4})$$
$$= \tfrac{1}{12}(28) = 2.33333$$

Exact:
$$A = \int_1^2 x^2\, dx = \frac{x^3}{3}\Big|_1^2 = \frac{8}{3} - \frac{1}{3} = 2\tfrac{1}{3}$$

5.   **a.**   Trapezoidal rule: $\Delta x = \dfrac{4 - 2}{4} = \dfrac{1}{2}$

$x_0 = 2 \qquad f(x_0) = 1.381773$

$x_1 = 2.5 \qquad f(x_1) = 1.264307$

$x_2 = 3 \qquad f(x_2) = 1.068232$

$x_3 = 3.5 \qquad f(x_3) = 0.805740$

$x_4 = 4 \qquad f(x_4) = 0.493151$

$A \approx \frac{1}{2}[1.381773 + 2(1.264307) + 2(1.068232) + 2(0.805740)$

$\qquad + .493151](\frac{1}{2}) = \frac{1}{4}(8.151482) = 2.037871$

b.   Simpson's rule:

$A \approx \frac{1}{3}[1.381773 + 4(1.264307) + 2(1.068232) + 4(0.805740)$

$\qquad + .493151](\frac{1}{2})$

$\qquad = \frac{1}{6}(12.291576) = 2.048596$

9.   For the trapezoidal rule, $|E_n| \le \dfrac{(b - a)^3}{12n^2} M$, where M is the maximum value of

$f''(x)$ on $[a, b]$. $f(x) = \dfrac{1}{x^2 + 1}$, $f'(x) = -2x(x^2 + 1)^{-2} = \dfrac{-2x}{(x^2 + 1)^2}$

$f''(x) = \dfrac{-2(x^2 + 1)^2 - (-2x)(2)(x^2 + 1)(2x)}{(x^2 + 1)^4} = \dfrac{6x^2 - 2}{(x^2 + 1)^3}$

Candidates for extrema are $\{0, 1\}$. $f''(0) = -2$, $f''(1) = \frac{1}{2}$

So we need $\dfrac{1}{12n^2}(2) \le 0.05$, $\dfrac{1}{n^2} \le 0.30$, $n^2 \ge \dfrac{10}{3}$, $n \ge 2$.

$A \approx \frac{1}{2}[f(0) + 2f(0.5) + f(1)](\frac{1}{2}) = \frac{1}{4}[1 + 2(0.8) + 0.5] = 0.775$

13.   For the trapezoidal rule, $|E_n| \le \dfrac{(b - a)^3}{12n^2} M$, where M is the maximum value of

$f''(x)$ on $[a, b]$. $f(x) = x(4 - x)^{1/2}$, $f'(x) = \dfrac{8 - 3x}{2(4 - x)^{1/2}}$,

$f''(x) = \dfrac{3x - 16}{4(4 - x)^{3/2}}$.   On $[0, 4]$ the candidates for extrema are the endpoints

and where $f' = 0$; $x = \frac{8}{3}$. $f''(4)$ is undefined, $f''(\frac{8}{3}) \approx -1.30$,

$f''(0) = -\frac{1}{2}$, so $M = 1.30$. $0.1 \le \dfrac{4^3}{12n^2}(1.30)$, $n^2 \ge \dfrac{640}{12}(1.30)$, $n \ge 8.32$.

Use of 9 terms gives 8.5

17.   **a.** $f(x) = x^{-1/2}, \; f'(x) = -\frac{1}{2}x^{-3/2}, \; f''(x) = \frac{3}{4}x^{-5/2}.$

The only candidates for extrema are the endpoints: $\{1, 4\}$.

$f''(1) = \frac{3}{4}, \; f''(4) = \frac{3}{128}.$ For the trapezoidal rule we are required

to have $|E_n| \leq \dfrac{(b-a)^3}{12n^2} M \leq 0.00005, \quad \dfrac{27}{12n^2}\left(\frac{3}{4}\right) \leq 0.00005,$

$n^2 \geq 33750, \; n \geq 184$ terms.

b. $f'''(x) = -\frac{15}{8}x^{-7/2}, \; f^{(4)}(x) = \frac{105}{16}x^{-9/2}.$ This is a decreasing

function, so the maximum occurs at the left endpoint: $f^{(4)}(1) = \frac{105}{16}.$

For Simpson's rule $|E_n| \leq \dfrac{(b-a)^5}{180n^4} K = \dfrac{3^5}{180n^4}\dfrac{105}{16} \leq 0.00005$

$n^4 \geq 177188, \; n \approx 20.52.$ Since n must be even use $n \geq 22.$

21. $f(x) = x^{-1}, \; f'(x) = -x^{-2}, \; f''(x) = 2x^{-3}.$ This is a decreasing

function, so the maximum occurs at the left endpoint: $f''(1) = 2.$

For the trapezoidal rule we are required to have:

$$|E_n| \leq \frac{(b-a)^3}{12n^2} M \leq 0.0000005$$

$$\frac{1}{12n^2}(2) \leq 0.0000005$$

$$n^2 \geq 333\,333$$

$$n \geq 578$$

25. $\Delta x = 0.5, \; A \approx \frac{1}{3}[10 + 4(9.75) + 2(10) + 4(10.75) + 2(12) + 4(13.75)$

$$+ \; 2(16) + 4(18.75) + 2(22) + 4(25.75) + 30]\left(\tfrac{1}{2}\right)$$

$$= \tfrac{1}{6}(475) = 79\tfrac{1}{6}$$

29.   Analysis: Apply the formula of Problem **28** with $a = -1, \; b = 2,$

$b - a = 3, \; \dfrac{b-a}{6} = \tfrac{1}{2},$ and $\dfrac{a+b}{2} = \tfrac{1}{2}.$

31.   Analysis: Using the error formula for Simpson's rule, $f^{(4)}(x)$ for a third degree

polynomial will be zero, so there has to be a number c in the interval for which

the error is zero and the value is exact.

33.   Analysis: Substitute the values for the points into the parabola formula to give

three equations. Solve this system of equations for A, B, and C. Now integrate

this function from $-h$ to h.

4.7 Area Between Two Curves

SURVIVAL HINT

Students generally enjoy these problems. But you must be neat, organized, and follow these steps:

- Draw a careful sketch, finding the coordinates at the points of intersection.
- Decide whether horizontal or vertical strips will be most efficient. Draw the strip.
- Be careful about which is the leading curve, and express it properly; y as a function of x, or x as a function of y.
- Sum the strips between the appropriate values; x values for vertical strips, and y values for horizontal strips.

3.

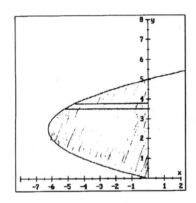

The curves intersect at $(0, 0)$ and $(0, 5)$. Using horizontal strips, the leading curve is $x = 0$, and the trailing curve is $x = y^2 - 5y$.

$$A = \int_0^5 [0 - (y^2 - 5y)]\, dy = \int_0^5 (5y - y^2)\, dy = \frac{5y^2}{2} - \frac{y^3}{3} \Big|_0^5$$

$$= \frac{125}{2} - \frac{125}{3} = 125(\tfrac{1}{2} - \tfrac{1}{3}) = \frac{125}{6}$$

7.

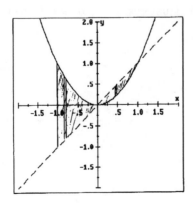

7. (con't.)

The curves intersect at $(0, 0)$ and $(1, 1)$. The parabola is above the line on $[-1, 0)$, and the line is above the parabola on $(0, 1]$.

$$A = \int_{-1}^{0} (x^2 - x)\, dx + \int_{0}^{1} (x - x^2)\, dx = \frac{x^3}{3} - \frac{x^2}{2}\Big|_{-1}^{0} + \frac{x^2}{2} - \frac{x^3}{3}\Big|_{0}^{1}$$

$$= \left(\tfrac{1}{3} + \tfrac{1}{2}\right) + \left(\tfrac{1}{2} - \tfrac{1}{3}\right) = \tfrac{5}{6} + \tfrac{1}{6} = 1$$

11.

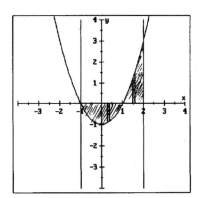

$$A = \int_{-1}^{1} [0 - (x^2 - 1)]\, dx + \int_{1}^{2} [(x^2 - 1) - 0]\, dx$$

$$= x - \frac{x^3}{3}\Big|_{-1}^{1} + \frac{x^3}{3} - x\Big|_{1}^{2}$$

$$= \left(1 - \tfrac{1}{3}\right) - \left(-1 + \tfrac{1}{3}\right) + \left(\tfrac{8}{3} - 2\right) - \left(\tfrac{1}{3} - 1\right) = \tfrac{8}{3}$$

15.

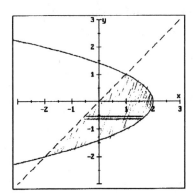

The curves intersect when $y = 2 - y^2$, $y^2 + y - 2 = 0$, $y = -2, 1$.

If we use horizontal strips the parabola will be the leading curve over the interval.

$$A = \int_{-2}^{1} [(2 - y^2) - y]\, dy = 2y - \frac{y^3}{3} - \frac{y^2}{2}\Big|_{-2}^{1}$$

$$= \left(2 - \tfrac{1}{3} - \tfrac{1}{2}\right) - \left(-4 + \tfrac{8}{3} - 2\right) = 4\tfrac{1}{2}$$

19.

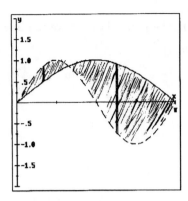

The curves intersect when $\sin x = \sin 2x$

$$\sin x = 2 \sin x \cos x$$

$$\sin x - 2 \sin x \cos x = 0$$

$$\sin x(1 - 2 \cos x) = 0$$

$$\sin x = 0 \text{ or } \cos x = \tfrac{1}{2}. \quad x = 0, \pi, \tfrac{\pi}{3}.$$

$\sin 2x \geq \sin x$ on $[0, \tfrac{\pi}{3}]$ and $\sin x \geq \sin 2x$ on $[\tfrac{\pi}{3}, \pi]$.

$$A = \int\limits_{0}^{\frac{\pi}{3}} (\sin 2x - \sin x)\, dx + \int\limits_{\frac{\pi}{3}}^{\pi} (\sin x - \sin 2x)\, dx$$

$$= -\tfrac{1}{2} \cos 2x + \cos x \Big|_{0}^{\frac{\pi}{3}} + \left(-\cos x + \tfrac{1}{2} \cos 2x \right)\Big|_{\frac{\pi}{3}}^{\pi}$$

$$= \tfrac{1}{4} + \tfrac{1}{2} - \left(-\tfrac{1}{2} + 1 \right) + \left(1 + \tfrac{1}{2} \right) - \left(-\tfrac{1}{2} - \tfrac{1}{4} \right) = 2\tfrac{1}{2}$$

23.

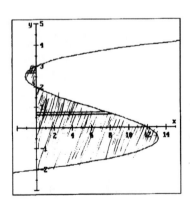

The curves intersect when $y^3 - 3y^2 - 4y + 12 = 0$,

$y^2(y - 3) - 4(y - 3) = 0$, $(y - 3)(y^2 - 4) = 0$, $y = 3, 2, -2$.

If we use horizontal strips, the cubic is the leading curve on $[-2, 2)$ and

$x = 0$ is the leading curve on $(2, 3]$.

23. (con't.)

$$A = \int_{-2}^{2} (y^3 - 3y^2 - 4y + 12)\, dy + \int_{2}^{3} [0 - (y^3 - 3y^2 - 4y + 12)]\, dy$$

$$= \frac{y^4}{4} - y^3 - 2y^2 + 12y \Big|_{-2}^{2} - \left(\frac{y^4}{4} - y^3 - 2y^2 + 12y\right)\Big|_{2}^{3}$$

$$= 4 - 8 - 8 + 24 - (4 + 8 - 8 - 24)$$

$$+ (-\tfrac{81}{4} + 27 + 18 - 36) - (-4 + 8 + 8 - 24)$$

$$= 12 + 20 - 11\tfrac{1}{4} + 12 = 32\tfrac{3}{4}$$

27.   Consumer surplus $= \int_{0}^{q_0} D(q)\, dq - p_0\, q_0, \quad p = D(q) = 150 - 6q,$

$p(5) = 120, \quad p(0) = 150.$

a.   Consumer surplus $= \int_{0}^{5} (150 - 6q)\, dq - (120)(5)$

$$= 150q - 3q^2 \Big|_{0}^{5} - (600)$$

$$= 150(5). - 75 - 600 = \$75$$

b. $p(12) = 150 - 6(12) - 78.$

Consumer surplus $= \int_{0}^{12} (150 - 6q)\, dq - (78)(12)$

$$= 150q - 3q^2 \Big|_{0}^{12} - (936)$$

$$= 150(12) - 432 - 936 = \$432$$

31.   Equilibrium occurs when $D(q) = S(q)$. $25 - q^2 = 5q^2 + 1,$

$6q^2 = 24, \quad q = 2.$ $(-2$ is meaningless here.$)$ $p = D(2) = 21$

Consumer surplus $= \int_{0}^{2} (25 - q^2)\, dq - 2(21) = 25q - \frac{q^3}{3}\Big|_{0}^{2} - 42$

$$= 50 - \tfrac{8}{3} - 42 = \$5.33$$

35.

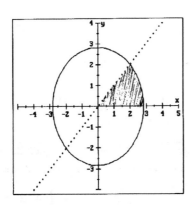

This is one-eighth of a circle with radius $\sqrt{8}$. $A = \tfrac{1}{8}\pi(\sqrt{8})^2 = \pi.$

39.   **a.**   Equating the two profit functions: $100 + x^2 = 220 + 2x$,

$x^2 - 2x - 120 = 0$, $(x - 12)(x + 10) = 0$, $x = 12$ years.

b.   Excess profit $= \displaystyle\int_0^{12} [(220 + 2x) - (100 + x^2)] \, dx$

$$= 120x + x^2 - \frac{x^3}{3}\Big|_0^{12} = 1440 + 144 - 576 = \$1008$$

43. $p(q) = \frac{1}{4}(10 - q)^2$. $\dfrac{dC}{dq} = \frac{3}{4}q^2 + 5$.

a.   Total revenue $= p(q)q$

$$= \frac{q}{4}(10 - q)^2$$

$$= \frac{1}{4}(q^3 - 20q^2 + 100q)$$

Marginal revenue $= \frac{1}{4}(3q^2 - 40q + 100)$

$$= \frac{3}{4}q^2 - 10q + 25$$

b.   Profit is maximized when marginal revenue equals marginal cost. $R' = C'$.

$\frac{3}{4}q^2 - 10q + 25 = \frac{3}{4}q^2 + 5$, $10q = 20$, $q = 2$.

c.   Consumer surplus $= \displaystyle\int_0^{q_0} D(q) \, dq - p_0 \, q_0$. $p(2) = 16$

$$= \int_0^2 \tfrac{1}{4}(10 - q)^2 \, dq - 2 \, p(2) = \int_0^2 \tfrac{1}{4}(q^2 - 20q + 100) \, dq - 32$$

$$= \tfrac{1}{4}\Big(\frac{q^3}{3} - 10q^2 + 100q\Big)\Big|_0^2 - 32 = \tfrac{1}{4}(\tfrac{8}{3} - 40 + 200) - 32$$

$$= \tfrac{1}{4}(162\,\tfrac{2}{3}) - 32 = \$8.67$$

Chapter 4 Review

PRACTICE PROBLEMS

19.   Using the linearity rule:

$$\int_0^1 (2x^4 - 3x^2) \, dx = 2\int_0^1 x^4 \, dx - 3\int_0^1 x^2 \, dx = 2(\tfrac{1}{5}) - 3(\tfrac{1}{3}) = -\tfrac{3}{5}$$

20.   Using Leibniz's rule: $F'(x) = \dfrac{d}{dx}\displaystyle\int_3^x t^5 \sqrt{\cos(2t + 1)} \, dt = x^5 \sqrt{\cos(2x + 1)}$

21. $\displaystyle\int_1^4 (x^{1/2} + x^{-3/2}) \, dx = \dfrac{2x^{3/2}}{3} - \dfrac{2x^{-1/2}}{1} \Big|_1^4 = \tfrac{16}{3} - 1 - \tfrac{2}{3} + 2 = \tfrac{17}{3}$

22. $\displaystyle\int_0^1 (2x^3 - 18x^2 + 40x - 12)\,dx = \frac{x^4}{2} - 6x^3 + 20x^2 - 12x\,\Big|_0^1$

$$= \tfrac{1}{2} - 6 + 20 - 12 = \tfrac{5}{2}$$

23. Let $u = 1 + \cos x$, $du = -\sin x\,dx$. For the new limits, when $x = 0$,

$u = 0$, and when $x = \frac{\pi}{2}$, $u = 1$.

$$\int_0^{\frac{\pi}{2}} (1 + \cos x)^{-2}\sin x\,dx = -\int_2^1 u^{-2}\,du = \frac{1}{u}\,\Big|_2^1 = 1 - \tfrac{1}{2} = \tfrac{1}{2}$$

24. Let $u = 2x^2 + 2x + 5$, $du = (4x + 2)\,dx$. For new limits, when $x = -2$,

$u = 9$, and when $x = 1$, $u = 9$.

$$\tfrac{1}{2}\int_9^9 u^{1/2}\,du = 0$$

25. **a.** $\displaystyle A = \int_{-1}^3 (3x^2 + 2)\,dx = x^3 + 2x\,\Big|_{-1}^3 = 27 + 6 + 1 + 2 = 36$

b. The curves intersect when $x^6 = x$

$$x(x^5 - 1) = 0$$

$$x(x - 1)(x^4 + x^3 + x^2 + x + 1) = 0$$

$$x = 0,1$$

They intersect at $(0, 0)$, $(1, 1)$. On $[0, 1]$ the curve $y^3 = x$ is above $y = x^2$.

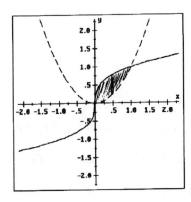

25. (con't.)

We can use either vertical or horizontal strips. Vertical:

$$A = \int_0^1 (x^{1/3} - x^2)\, dx = \frac{3x^{4/3}}{4} - \frac{x^3}{3}\bigg|_0^1 = \frac{3}{4} - \frac{1}{3} = \frac{5}{12}$$

Horizontal: (The parabola is the leading curve.)

$$A = \int_0^1 (y^{1/2} - y^3)\, dy = \frac{2y^{3/2}}{3} - \frac{y^4}{4}\bigg|_0^1 = \frac{2}{3} - \frac{1}{4} = \frac{5}{12}$$

26. Average value $= \dfrac{1}{b-a}\displaystyle\int_0^{\frac{\pi}{2}} \cos 2x\, dx = \dfrac{2}{\pi}\displaystyle\int_0^{\frac{\pi}{2}} \dfrac{1}{2}\cos 2x\,(2\,dx) = \dfrac{1}{\pi}(\sin 2x)\bigg|_0^{\frac{\pi}{2}}$

$$= \tfrac{1}{\pi}(0) = 0$$

27. $v'(t) = a(t),\ v(t) = \int a(t)\, dt = \int (2t + 1)\, dt = t^2 + t + C.$

But when $t = 0,\ v = 2$, so $C = 2$ and $v(t) = t^2 + t + 2.$

$s'(t) = v(t),\ s(t) = \int v(t)\, dt = \int (t^2 + t + 2)\, dt = \dfrac{t^3}{3} + \dfrac{t^2}{2} + 2t + C.$

But $s = 4$ when $t = 0$, so $C = 4$, and

$s(t) = \dfrac{t^3}{3} + \dfrac{t^2}{2} + 2t + 4$

28. **a.** The profit ends when $R' = C'.\ \ 1575 - 5x^2 = 1200 + 10x^2,$

$15x^2 = 375,\ x^2 = 25,\ x = 5$ years.

b. Earnings, the difference between the revenue and the cost at any given time,

$$= \int_0^5 [(1575 - 5x^2) - (1200 + 10x^2)]\, dx = \int_0^5 (-15x^2 + 375)\, dx$$

$$= -5x^3 - 375x\bigg|_0^5 = -625 + 1875 = \$1250.$$

29. Separating variables: $\dfrac{dy}{y^2} = \sin 3x\, dx,\ \ \int y^{-2}dy = \tfrac{1}{3}\int \sin 3x\,(3\,dx),$

$-\dfrac{1}{y} = -\dfrac{1}{3}\cos 3x + C,\ y = \dfrac{3}{\cos 3x + C}$

30. **a.** For the trapezoidal rule we need $\dfrac{(b-a)^3}{12n^2}\, M \le 0.0005$ where M is the

maximum of $|f''(x)|$ on $[a, b].\ \ f(x) = \cos x,\ f'(x) = -\sin x,$

$f''(x) = -\cos x$, so on $[0, \frac{\pi}{2}],\ M = |-1| = 1.$

$\dfrac{(\frac{\pi}{2})^3}{12n^2}(1) \le 0.0005,\ n^2 \ge 646,\ n \ge 26.$

30. (con't.)

    b.   For Simpson's rule we need $\dfrac{(b-a)^5}{180n^4} K \le 0.0005$ where K is the

maximum of $|f^{(4)}(x)|$ on $[0, \frac{\pi}{2}]$. $f'''(x) = \sin x$, $f^{(4)}(x) = \cos x$.

So on $[0, \frac{\pi}{2}]$, $M = 1$.

$\dfrac{(\frac{\pi}{2})^5}{180n^4} (1) \le 0.0005$, $\quad n^4 \ge 106.3$, $\quad n \ge 4$

Chapter 5
Exponential, Logarithmic, and Inverse Trigonometric Functions

5.1 Exponential Functions: The Number e

1. $y = 3^x$

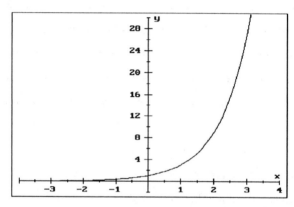

5. $32^{2/5} + 9^{3/2} = (2^5)^{2/5} + (3^2)^{3/2} = 2^2 + 3^3 = 4 + 27 = 31$

9. $(e^{1.3})^2 = e^{2.6} \approx 13.463\ 7$

13. $5{,}000(1 + 0.011\ 25)^{60} = 5{,}000(1.011\ 25)^{60} \approx 9{,}783.23$

17. $3^{x^2-x} = 3^2,\ x^2 - x = 2,\ x^2 - x - 2 = 0,\ (x - 2)(x + 1) = 0,$
$x = 2, -1$

21. $(2^{1/3})^{x+10} = 2^{x^2},\ \dfrac{x + 10}{3} = x^2,\ 3x^2 = x + 10,\ 3x^2 - x - 10 = 0,$
$(3x + 5)(x - 2) = 0,\ x = 2, -\frac{5}{3}$

25. **a.** If $b = 1,\ y = 1^x = 1,$ which is a horizontal line and an
algebraic function.

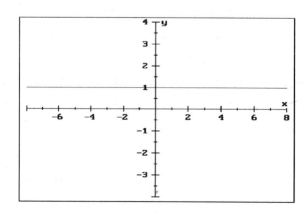

25. (con't.)

    b.   If $b = 0$, $y = 0^x = 0$, which is the x-axis and an algebraic function.

         However, if x is negative we would be dividing by a power of 0, so $x > 0$.

29. $A = P(1 + i)^n$ where $P = 9,400$, $n = 6(30) = 180$, $i = \frac{.14}{360}$

       $A = 9,600(1.000\ 388\ 889)^{180} \approx \$10,081.44$

33.   **a.** $p(t) = 100e^{-0.03t}$ where $t = 40$. $p(40) \approx 30.12\%$

      b.   Failure rate $= 1 - p(50) = 1 - 100e^{-0.03(50)} \approx 77.69\%$

      c. $p(40) - p(50) = 30.12 - 22.31 = 7.81\%$

37. $T = A + (B - A)e^{-kt}$ where $A = 75$, $B = 120$, $t = 30$, $k = 0.01$.

       $T = 75 + (120 - 75)e^{-.3} \approx 75 + (45)(.740\ 8) \approx 108°$

39.   Analysis: Treat the exponent as an instruction, specifying the number of factors

       of b. m factors and n more factors make a total of $m + n$ factors.

41.   Analysis: Treat the exponent as an instruction, specifying the number of factors

       of b. There are m factors in the numerator and n factors in the denominator.

       Now consider the cancellation if $m \geq n$ or if $m < n$.

43.   Analysis: Use the definition of symmetry about the y-axis: $f(x) = f(-x)$.

45.   Analysis: Since we have not officially established the inverse function for

       e^x yet, we cannot simply establish $e^x > N$ when $x > \ln N$. Since $e > 2$, use the

       inequality $e^x > 2^x > 2^N$, and show that letting N approach $+\infty$ causes e^x

       to approach $+\infty$.

5.2 Inverse Functions; Logarithms

SURVIVAL HINT

You need to be comfortable with logarithms as inverses of exponentials. $y = e^x$ has

inverse $x = e^y$, where y is described as $\ln x$, i.e. the exponent of e that gives x.

Logarithms are exponents. $\ln 5$ is the exponent of e that gives 5. $e^{\ln 3} = 3$ because

e to the exponent of e that gives 3 will, of course, be 3. If you are not really comfortable

with the concepts of 5.1 and 5.2, it might be a good idea to review the sections

on logarithms and inverses in a trigonometry or precalculus text. For applications

involving the way our world and universe is built the two most important numbers are

π and e.

3. $f[g(x)] = \frac{4}{5}(\frac{5}{4}x + 3) + 4 = x + \frac{12}{5} + 4 \neq x$. These are not inverse

       functions.

7. To find the inverse interchange the domain and range values:

$\{(5, 4), (3, 6), (1, 7), (4, 2)\}$.

11. Interchanging domain and range: $x = \dfrac{2y - 6}{3y + 3}$. Solving for y:

$3xy + 3x = 2y - 6$, $(3x - 2)y = -3x - 6$, $y = \dfrac{-3(x + 2)}{3x - 2}$.

So $h^{-1}(x) = -\dfrac{3(x + 2)}{3x - 2}$, $x \neq \frac{2}{3}$.

15. Write the problem as: $5 \log_3 3^2 - 2 \log_2 2^4$. Since the exponent of 3 that gives 3^2 is 2, and the exponent of 2 that gives 2^4 is 4, the expression becomes:

$5(2) - 2(4) = 10 - 8 = 2$.

19. The exponent of 3 that gives 3^4 is 4, and the exponent of e that gives $e^{0.5}$ is 0.5. The value of the expression is $4 - .5 = 3.5$

23. This can be rewritten as an exponential equation: $x^2 = 16$, $x = 4$

(The negative 4 is not in the domain of the log function.)

27. Consider this as $(\frac{1}{7})^x = 15$, and write as a logarithm: $\log_{1/7} 15 = x$.

Using the change of base formula: $x = \dfrac{\ln 15}{\ln (\frac{1}{7})} \approx \dfrac{2.708\,050\,201}{-1.945\,910\,149}$

$\approx -1.391\,662\,509$

31. Using the product rule this becomes: $\log_3 x(2x + 1) = \log_3 3$

(1 is the exponent of 3 that gives 3.) Taking the antilog base 3 of both sides:

$x(2x + 1) = 3$, $2x^2 + x - 3 = 0$, $(2x + 3)(x - 1) = 0$, $x = 1$.

(The $-\frac{3}{2}$ is not in the domain of the original function.)

35. $f(x) = \sin x$ is monotonic increasing on $[0, \frac{\pi}{2}]$, so there is an inverse which is also a function. $f^{-1}(x) = \arcsin x$. Using the derivative of an inverse function theorem:

$(f^{-1})'(x) = \dfrac{1}{f'[f^{-1}(x)]}$. Since $f'(x) = \cos x$, and since $f(\frac{\pi}{4}) = \dfrac{\sqrt{2}}{2}$,

$f^{-1}\left(\dfrac{\sqrt{2}}{2}\right) = \frac{\pi}{4}$. So $(f^{-1})'\left(\dfrac{\sqrt{2}}{2}\right) = \dfrac{1}{f'[f^{-1}(\frac{\sqrt{2}}{2})]} = \dfrac{1}{f'(\frac{\pi}{4})} = \dfrac{1}{\cos \frac{\pi}{4}}$

$= \dfrac{1}{\left(\frac{\sqrt{2}}{2}\right)} = \sqrt{2}$

39. $f(x) = x^2$ on $(-\infty, 0]$. $f^{-1}(x) = -\sqrt{x}$ on $y \le 0$

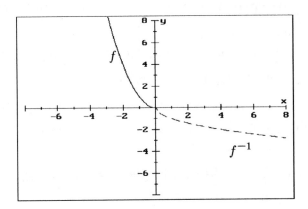

43. $f(x) = \cos x$ is monotonic decreasing on $[0. \pi]$ and therefore has an inverse which is also a function. $f^{-1}(x) = \arccos x$.

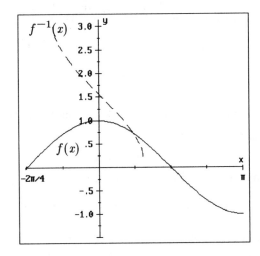

47. $A = Pe^{rt}$. A is to be $2P$, and $r = .06$.

$2P = Pe^{0.06t}$, $2 = e^{0.06t}$, $\ln 2 = 0.06t$, $t = \dfrac{\ln 2}{0.06} \approx 11.55$ years

51. $A = Pe^{rt}$. $A = 2.52 \times 10^{10}$, $P = 24$, $t = 364$

$2.52 \times 10^{10} = 24e^{364r}$, $e^{364r} = \dfrac{2.52 \times 10^{10}}{24}$, $364r = \ln \dfrac{2.52 \times 10^{10}}{24}$,

$364r \approx 20.772$, $r \approx 0.0571$ or 5.71%

55.   Analysis: Use the power rule on the exponent of b: $b^{\log_b x^x}$. Now read this as b to the power of b that gives x^x.

57.   Analysis: Substitute (u, v) into the first formula, then switch the domain and range to obtain the inverse function. Write that logarithmic function as an exponential function. Now verify that (v, u) is on that graph.

59. Analysis: If $\dfrac{df}{dx} > 0$, then by the derivative of an inverse function theorem the derivative of the inverse function is the reciprocal, and therefore, also positive.

61. Analysis: Set up the substitutions in the hint to show that

$$g_1(x) = g_1[f(g_2(x))] = f[g_1(g_2(x))] = g_2(x).$$

5.3 Derivatives Involving e^x and $\ln x$

3. If $y = \ln u$, then $\dfrac{dy}{dx} = \dfrac{1}{u}\dfrac{du}{dx}$. In this case $u = 3x^4 + 5x$.
$\dfrac{dy}{dx} = \dfrac{1}{3x^4 + 5x}(12x^3 + 5) = \dfrac{12x^3 + 5}{3x^4 + 5x}$

7. Note that π^e and e^π are constants and their derivatives are 0.
$\dfrac{dy}{dx} = e^x + ex^{e-1}$

11. If $y = e^u$, then $\dfrac{dy}{dx} = e^u\dfrac{du}{dx}$. In this case $\dfrac{dy}{dx} = e^{-4x}(-4) + e^{3-2x}(-2)$,

$\dfrac{dy}{dx} = -4e^{-4x} - 2e^{3-2x}$

15. $f'(u) = \dfrac{1}{2}\left[e^{2u}(2) - e^{-2u}(-2)\right] = e^{2u} + e^{-2u}$

19. Use the quotient rule on the first term, the second term is a constant.

$g'(t) = \dfrac{(3t+5)e^{2t}(2) - e^{2t}(3)}{(3t+5)^2} = \dfrac{e^{2t}(6t+7)}{(3t+5)^2}$

23. This is a function of a function, so we need the chain rule.

$f'(u) = \dfrac{1}{\ln u}\left(\dfrac{1}{u}\right) = \dfrac{1}{u\ln u}$

27. Product rule is needed on the left. $xe^{-x}(-1) + e^{-x} = \dfrac{dy}{dx}$

$\dfrac{dy}{dx} = e^{-x}(1 - x)$

31. In the first term when using $e^u\,du$, the du needs the product rule.

$e^{xy}(xy' + y) + \dfrac{1}{y^2}(2yy') = 1, \quad y'\left(xe^{xy} + \dfrac{2}{y}\right) = 1 - ye^{xy},$

$y' = \dfrac{1 - ye^{xy}}{xe^{xy} + \dfrac{2}{y}}$

39. $\ln y = \ln \dfrac{e^{3x^2}}{(x^3 + 1)^2(4x - 7)^{-2}}$

$\qquad = 3x^2 \ln e - 2 \ln(x^3 + 1) + 2 \ln(4x - 7)$

Differentiating both sides:

$\dfrac{1}{y} y' = 6x - 2 \dfrac{3x^2}{x^3 + 1} + 2 \dfrac{4}{4x - 7}$

$y' = y \left(6x - \dfrac{6x^2}{x^3 + 1} + \dfrac{8}{4x - 7}\right)$

43. $\ln y = \ln (\sin x)^{\sqrt{x}} = \sqrt{x} \ln (\sin x)$. Differentiating both sides:

$\dfrac{1}{y} y' = \sqrt{x} \dfrac{1}{\sin x} (\cos x) + \ln |\sin x| \dfrac{1}{2\sqrt{x}}$

$y' = y \sqrt{x} \left(\cot x + \dfrac{\ln |\sin x|}{2x}\right)$

47. $\ln y = \ln (\sin x)^x = x \ln |\sin x|$

$\dfrac{1}{y} y' = x \dfrac{\cos x}{\sin x} + \ln|\sin x|$

$y' = y[x \cot x + \ln|\sin x|]$

51. We will use the point-slope form of a line to find the equation of the tangent. To do so we need y' at the specified point. To find y' we will use logarithmic differentiation. $\ln y = \ln x^{x^x} = x^x \ln x$. To differentiate both sides we will need to differentiate x^x. Let $u = x^x$, $\ln u = \ln x^x = x \ln x$.

$\dfrac{1}{u} u' = \dfrac{x}{x} + \ln x$. $u' = u(1 + \ln x) = x^x (1 + \ln x)$. Now differentiate the original equation: $\dfrac{1}{y} y' = x^x \left(\dfrac{1}{x}\right) + (\ln x)x^x (1 + \ln x)$.

$y' = y\, x^x\left[\dfrac{1}{x} + (\ln x)(1 + \ln x)\right]$

$y'(1, 1) = 1$

So the equation of the tangent line is: $y - 1 = 1(x - 1)$ or $y = x$

55. Using implicit differentiation: $2x + 2y' = e^y y'$, $y' = \dfrac{2x}{e^y - 2}$

$y'(1, 0) = \dfrac{2}{1 - 2} = -2$. So the equation of the tangent line is:

$y - 0 = -2(x - 1)$ or $2x + y = 2$.

59. Before finding the derivative write $\ln \sqrt{x}$ as $\dfrac{1}{2} \ln x$. Now $y = \dfrac{\ln x}{2x}$.

$y' = \dfrac{2x(\frac{1}{x}) - 2 \ln x}{4x^2} = \dfrac{(1 - \ln x)}{2x^2}$. $y'(4, \dfrac{\ln 2}{4}) = \dfrac{1 - \ln 4}{32}$.

So the equation of the tangent line is: $y - \dfrac{\ln 2}{4} = \dfrac{1 - \ln 4}{32} (x - 4)$

63. Use the point-slope form of a line, but the slope of the normal is the negative reciprocal of the slope of the tangent. Using implicit differentiation:

$2 + y' = e^y y'$, $y' = \dfrac{2}{e^y - 1}$. $y'(1, 0) = \dfrac{2}{0} = \infty$. So the slope of the normal is 0. The equation of the line: $y - 0 = 0(x - 1)$ or $y = 0$.

65.   Analysis: **a.** Use $F'(x) = \frac{1}{u} \, du$.

   b. Use $F'(x) = \frac{1}{u} \, du$, factor and reduce.

67.   If $f(x) = e^x$ then by definition $f'(x) = \lim\limits_{h \to 0} \dfrac{e^{x+h} - e^x}{h} = \lim\limits_{h \to 0} \dfrac{e^x e^h - e^x}{h}$

$$= \lim_{h \to 0} \frac{e^x(e^h - 1)}{h} \, .$$

Now finish the limit using the results of Problem 66.

5.4 Applications Involving Derivatives of e^x and ln x

3. $f(0) = +\infty$, so it must be graph **b** or **e**. $f(1) = e$, so it is **b**.

7.   **a.**   The domain of ln x, and therefore this function is $(0, +\infty)$.

   b. $f'(x) = 1 - \frac{1}{x} = \frac{x-1}{x}$, which is undefined at 0 and equals 0 at 1.
   The critical values are: $\{0, 1\}$. By putting the critical values on a
   sign line we see that $f'(x) > 0$ on the intervals $(-\infty, 0)$ and $(1, \infty)$, and
   is rising there, and $f'(x) < 0$ on $(0, 1)$ and is falling on that interval.
   Since the function is only defined for positive values, it is falling on
   $(0, 1)$, and rising on $(1, \infty)$.

   c.   Since there are no endpoints, the only candidates for extrema are the critical
   points. $f(0)$ is undefined, and $f(1) = 1$. $(1, 1)$ is a relative minimum.
   $f(x) = 0$ when ln $x = x$, or when $x = e^x$, which is never.

   d. $f''(x) = \dfrac{x(1) - (x-1)(1)}{x^2} = \dfrac{1}{x^2}$. This is always positive, so the
   graph is concave upward everywhere. There are no inflection points.

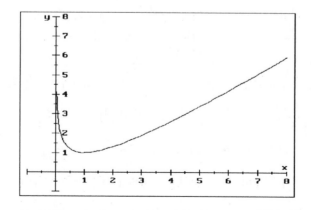

11. **a.** $\sqrt{x}$ must be ≥ 0, but $\ln x$ must be > 0. The domain is $(0, +\infty)$.

b. Use the power rule for exponents to write $f(x) = \dfrac{\ln x}{2\sqrt{x}}$, then

$$f'(x) = \frac{2\sqrt{x}\left(\frac{1}{x}\right) - (\ln x)\left(\frac{1}{\sqrt{x}}\right)}{4x} = \frac{2 - \ln x}{4x\sqrt{x}} = 0 \text{ when } \ln x = 2,$$

$x = e^2$. To the left of e^2, $f'(x) > 0$, and to the right of e^2, $f'(x) < 0$.

c. The change in sign of $f'(x)$ means that there is a relative maximum at $(e^2, \approx 0.3679)$. There is an x-intercept at $(1, 0)$.

d. $f''(x) = \dfrac{4x\sqrt{x}\left(-\frac{1}{x}\right) - (2 - \ln x)(6\sqrt{x})}{16x^3} = \dfrac{3\ln x - 8}{8x^{5/2}}.$

This is 0 when $\ln x = \frac{8}{3}$, or $x = e^{8/3}$. To the left of this point $f''(x) < 0$ and the function is concave down, to the right of this point $f''(x) > 0$ and the function is concave up. There is an inflection point at $(e^{8/3}, \frac{4}{3}e^{-4/3})$.

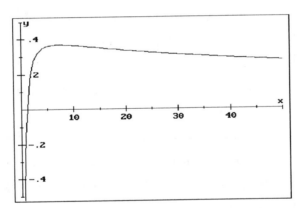

15. **a.** $\dfrac{t^2}{1 - t} > 0$. Since t^2 is positive, $1 - t > 0$, or $t < 1$. And since $t \neq 0$ the domain is $(-\infty, 0) \cup (0, 1)$.

b. $f'(t) = \dfrac{1 - t}{t^2}\left[\dfrac{(1 - t)2t - t^2(-1)}{(t - 1)^2}\right] = \dfrac{t - 2}{t(t - 1)}.$

This is undefined at $t = 0, 1$ and is 0 at $t = 2$. Putting $f'(x)$ on a sign line we see that it is positive and rising on $(0, 1), (2, \infty)$, and negative and decreasing on $(-\infty, 0), (1, 2)$. But keeping the domain in mind, we see that it is falling on $(-\infty, 0)$, undefined at 0, rising on $(0, 1)$, and undefined at 1.

15. (con't.)

c. There are no candidates for extrema.

It has intercepts at $\left(\dfrac{-1 \pm \sqrt{5}}{2},\ 0\right)$

d. $f''(t) = \dfrac{(t^2 - t)(1) - (t - 2)(2t - 1)}{(t^2 - t)^2} = \dfrac{-t^2 + 4t - 2}{(t^2 - t)^2}$.

$t^2 - 4t + 2 = 0$ when $t = 2 \pm \sqrt{2}$, but only $2 - \sqrt{2} \approx 0.5858$
is in the domain of the function. For $t < 0$, $f''(t) < 0$, and the function
is concave down. For t on $(0,\ 2 - \sqrt{2})$, $f''(t) < 0$, and the function is
concave down. For t on $(2 - \sqrt{2},\ 1)$, $f''(t) > 0$, and the function is
concave up. The change in concavity gives an inflection point
at approximately $(2 - \sqrt{2},\ -0.1882)$.

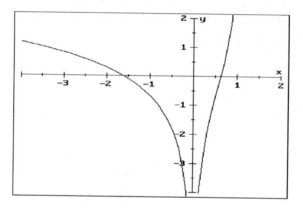

19. **a.** For $\ln x$ to be defined the domain is $(0,\ +\infty)$.

b. $f'(x) = 2\,(\ln x)\dfrac{1}{x} = \dfrac{2\ln x}{x}$. This is undefined at 0, and is 0 at $x = 1$.
On $(0,\ 1)$, $f'(x) < 0$, and the function is falling. On $(1,\ \infty)$, $f'(x) > 0$,
and the function is rising.

c. The change in slope at $(1,\ 0)$ tells us that there is a relative minimum there,
and also an intercept at $(1,\ 0)$.

19. (con't.)

    d. $f''(x) = \dfrac{x\left(\frac{2}{x}\right) - (2\ln x)(1)}{x^2} = \dfrac{2(1 - \ln x)}{x^2} = 0$ when $x = e$,

and changes signs there, so there is an inflection point at $(e, 1)$.

The function is concave up on $(0, e)$ and concave down on (e, ∞).

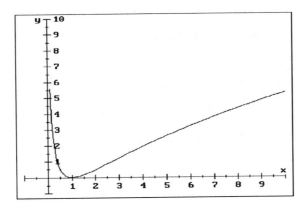

23.   **a.**   If $b > 1$, $\log_b x$ is monotonic increasing, as is its inverse $y = b^x$.

    $y = \ln x$ is an example $(b = e)$.

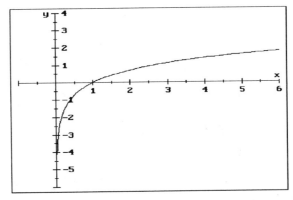

    b.   If $b < 1$, $\log_b x$ is monotonic decreasing, as is its inverse $y = b^x$.

    As an example let $b = \frac{1}{e}$.

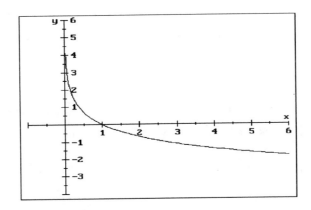

27. $f'(t) = t(e^{-t^2})(-2t) + e^{-t^2}(1) = e^{-t^2}(1 - 2t^2) = 0$ when

$t^2 = \frac{1}{2}$, $t = \pm\frac{\sqrt{2}}{2}$. The candidates for extrema are $\{\pm\frac{\sqrt{2}}{2}, -1, 1\}$.

$f''(t) = e^{-t^2}(-4t) + (1 - 2t^2)e^{-t^2}(-2t) = e^{-t^2}(4t^3 - 6t)$.

Using the second derivative test on the critical points:

$f''\left(\frac{\sqrt{2}}{2}\right) < 0$, so there is a relative maximum at $\left(\frac{\sqrt{2}}{2}, \frac{\sqrt{2}}{2\sqrt{e}}\right)$.

$f''\left(-\frac{\sqrt{2}}{2}\right) > 0$, so there is a relative minimum at $\left(-\frac{\sqrt{2}}{2}, -\frac{\sqrt{2}}{2\sqrt{e}}\right)$.

$f(-1) = -\frac{1}{e}$, which is greater than $-\frac{\sqrt{2}}{2\sqrt{e}}$, so this is not a minimum.

$f(1) = \frac{1}{e}$, which is less than $\frac{\sqrt{2}}{2\sqrt{e}}$, so this is not a maximum.

So $\left(\frac{\sqrt{2}}{2}, \frac{\sqrt{2}}{2\sqrt{e}}\right)$ and $\left(-\frac{\sqrt{2}}{2}, -\frac{\sqrt{2}}{2\sqrt{e}}\right)$ become the absolute maximum and minimum.

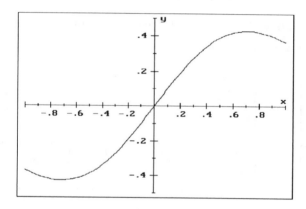

SURVIVAL HINT

When dealing with limits expressions having the form $\frac{\infty}{0}$ and $\frac{0}{\infty}$ are *not* indeterminate forms. The first is ∞, and the second is 0. The six most common indeterminate forms are $\frac{0}{0}$ and $\frac{\infty}{\infty}$, for which l'Hôpital's rule is applicable; $\infty - \infty$, where the functions yielding these values should be combined to a single fraction, usually one of the first two forms; 0^0, ∞^∞, and 1^∞, where the method outlined in example 5 should be used. Remember that $\lim\limits_{x \to \infty} \ln f(x) = \ln \lim\limits_{x \to \infty} f(x)$.

31.   This limit has the form $(0)(-\infty)$. Rewrite as $\lim\limits_{x \to \infty} \frac{\ln x}{\csc x}$, which now has the form $\frac{\infty}{\infty}$, which meets the hypothesis of l'Hôpital's rule.

$\lim\limits_{x \to \infty} \frac{\frac{1}{x}}{-\csc x \cot x} = \frac{0}{\infty} = 0$

35. Let $y = (e^x + x)^{1/x}$, $\ln y = \ln (e^x + x)^{1/x} = \frac{1}{x} \ln (e^x + x)$,

$\lim\limits_{x \to 0} \ln y = \lim\limits_{x \to 0} \dfrac{\ln (e^x + x)}{x}$, is of form $\frac{0}{0}$ so we can use l'Hôpital's rule:

$\lim\limits_{x \to 0} \ln y = \lim\limits_{x \to 0} \dfrac{e^x + 1}{e^x + x} = 2$. Now the limit of a log is the log of the limit:

$\ln \lim\limits_{x \to 0} y = 2$, $\lim\limits_{x \to 0} y = e^2$.

39. Recall that $\lim\limits_{x \to \infty} \left(1 + \frac{1}{x}\right)^x = e$. Now if $k > m$, let $m = k - n$.

$$\lim_{x \to \infty}\left(1 + \frac{1}{x^k}\right)^{x^m} = \lim_{x \to \infty}\left(1 + \frac{1}{x^k}\right)^{x^{k-n}}$$
$$= \lim_{x \to \infty}\left(1 + \frac{1}{x^k}\right)^{x^k x^{-n}} = \lim_{x \to \infty} (e)^{x^{-n}}$$
$$= \lim_{x \to \infty} (e)^{1/x^n} = e^0 = 1.$$

Now if $k < m$, let $m = k + n$.

$$\lim_{x \to \infty}\left(1 + \frac{1}{x^k}\right)^{x^m} = \lim_{x \to \infty}\left(1 + \frac{1}{x^k}\right)^{x^{k+n}}$$
$$= \lim_{x \to \infty}\left(1 + \frac{1}{x^k}\right)^{x^k x^n} = \lim_{x \to \infty} (e)^{x^n} = e^\infty = \infty.$$

And finally, if $k = m$,

$$\lim_{x \to \infty}\left(1 + \frac{1}{x^k}\right)^{x^m} = \lim_{x \to \infty}\left(1 + \frac{1}{x^k}\right)^{x^k} = e.$$

43. Our candidates for extrema are the endpoints and the critical points:

$g'(t) = t(\frac{1}{t}) + (\ln t)(1) = 1 + \ln t = 0$ when $\ln t = -1$, $t = \frac{1}{e}$.

Candidates: $\{0, \frac{1}{e}, 4\}$. Using the second derivative test on the critical point:

$g''(t) = \frac{1}{t}$, $g''(\frac{1}{e}) = e > 0$, so there is a relative minimum at $(\frac{1}{e}, 1 - \frac{1}{e})$.

$g(0) = 1$, $g(4) = 4 \ln 4 + 1 \approx 6.545$. So $(\frac{1}{e}, 1 - \frac{1}{e})$ is an absolute

minimum and $(4, 6.545)$ is an absolute maximum.

47. **a.** $f'(t) = Ae^{-kt}(-k) = -kf(t)$

   b.   The percentage rate of change $= 100\dfrac{f'(t)}{f(t)} = \dfrac{-100kf(t)}{f(t)} = -100k$

   which is a constant.

51. The problem is to maximize $p\,'(t)$.

$p(t) = 160(1 + 8e^{-0.01t})^{-1}$.

$p\,'(t) = -160(1 + 8e^{-0.01t})^{-2}(-0.01)(8e^{-0.01t})$

$p\,'(t) = 12.8e^{-0.01t}(1 + 8e^{-0.01t})^{-2}$. To maximize we need its derivative.

This will require the product rule and several applications of the chain rule:

$p\,''(t) = 12.8\Big(e^{-0.01t}(-2)(1 + 8e^{-0.01t})^{-3}(8e^{-0.01t})(-0.01) +$

$\qquad (1 + 8e^{-0.01t})^{-2}(e^{-0.01t})(-0.01)\Big)$

$\qquad = -0.128(1 + 8e^{-0.01t})^{-2}(e^{-0.01t})\left(-16e^{-0.01t}(1 + 8e^{-0.01t})^{-1} + 1\right)$

This expression $= 0$ when the last factor $= 0$. After some algebra:

$e^{0.01t} = 8,\; t = \dfrac{\ln 8}{0.01} \approx 208$ years

55. Analysis: The percentage rate of change of the square of the function

can be represented by $\dfrac{\{[f(x)]^2\}'}{[f(x)]^2} = k\cos x$, which is of the form $\dfrac{du}{u}$.

Integrate both sides and use the given initial value to find C.

57. Analysis: Find C' and C'' and note that the critical value t_c requires

$C' = 0$, which in turn requires that $ae^{-at_c} = be^{-bt_c}$. Substitute into C''

and show that the resulting expression is less than 0.

5.5 Integrals Involving e^x and $\ln x$

SURVIVAL HINT

The most common error in problems having the form $\displaystyle\int e^u\,du$ is the failure to properly

set up the du. It pays to write out for each problem: $u = \quad$, $du = \quad$, and introduce

the proper constants.

1. Let $u = 5x$, then $du = 5\,dx$. $\dfrac{1}{5}\displaystyle\int e^{5x}(5\,dx)$ is of form $\displaystyle\int e^u\,du = e^u + C$

$\dfrac{1}{5}\displaystyle\int e^{5x}(5\,dx) = \dfrac{1}{5}e^{5x} + C$

5. Recall that $\ln e^x = x$. $\displaystyle\int \ln e^x\,dx = \int x\,dx = \dfrac{x^2}{2} + C$

9. Let $u = 1 - x^2$. $du = -2x\,dx$

$\displaystyle\int x\,e^{1-x^2}\,dx = -\dfrac{1}{2}\int e^{1-x^2}(-2x\,dx) = -\dfrac{1}{2}e^{1-x^2} + C$

13. Let $u = 2x^2 + 3$. $du = 4x\,dx$.

$$\frac{1}{4}\int \frac{4x\,dx}{2x^2 + 3} \text{ is of form } \int \frac{du}{u} = \ln|u| + C$$

$$\frac{1}{4}\int \frac{4x\,dx}{2x^2 + 3} = \frac{1}{4}\ln|2x^2 + 3| + C = \frac{1}{4}\ln(2x^2 + 3) + C$$

17. Let $u = x\sqrt{x} = x^{3/2}$. $du = \frac{3}{2}\sqrt{x}\,dx$.

$$\frac{2}{3}\int e^{x^{3/2}}(\tfrac{3}{2}\sqrt{x}\,dx) \text{ is of form } e^u\,du, \text{ so it } = \frac{2}{3}e^{x^{3/2}} + C$$

21. Let $u = \ln x$. $du = \frac{dx}{x}$. $\int (\ln x)\frac{dx}{x}$ is of form $\int u\,du = \frac{u^2}{2} + C$

$$\int (\ln x)\frac{dx}{x} = \frac{(\ln x)^2}{2} + C. \text{ (Note that domain of original problem is } x > 0.)$$

25. This is of the form $\int \frac{du}{u} = \ln|u| + C$

$$\int \frac{e^t\,dt}{e^t + 1} = \ln|e^t + 1| + C$$

29. Let $u = \sqrt{x}$. $du = \frac{dx}{2\sqrt{x}}$. $2\int \frac{\cot\sqrt{x}}{2\sqrt{x}}\,dx = 2\int \cot u\,du = 2\ln|\sin\sqrt{x}| + C$

33. Let $u = 2x^3 + 1$. $du = 6x^2\,dx$. $\frac{5}{6}\int_0^1 \frac{6x^2\,dx}{2x^3 + 1}$ is of form $\int \frac{du}{u} = \ln|u| + C$

Changing limits on integration: when $x = 0$, $u = 1$, when $x = 1$, $u = 3$.

$$\frac{5}{6}\int_1^3 \frac{du}{u} = \frac{5}{6}(\ln 3 - \ln 1) = \frac{5}{6}\ln 3$$

37. Let $u = \frac{1}{x}$. $du = -\frac{dx}{x^2}$. $-\int_1^2 e^{1/x}\left(-\frac{dx}{x^2}\right)$ is now of form $e^u\,du$.

$$\int_1^2 \frac{e^{1/x}\,dx}{x^2} = -\int_1^2 e^{1/x}\left(-\frac{dx}{x^2}\right) = -e^{1/x}\Big|_1^2 = e - e^{1/2}$$

41. Let $u = -0.2x$. $du = -0.2\,dx$. Set up $\frac{du}{u}$ form:

Multiply numerator and denominator by $e^{0.2x}$:

$$\int_0^5 \frac{0.58\,e^{0.2x}}{e^{0.2x} + 1}\,dx = \frac{0.58}{0.2}\int_0^5 \frac{e^{0.2x}(0.2\,dx)}{e^{0.2x} + 1} = \frac{0.58}{0.2}\ln(e^{0.2x} + 1)\Big|_0^5$$

$$= 2.9[\ln(e + 1) - \ln 2] = 2.9\ln\frac{e + 1}{2} \approx 1.798\,33$$

45. To find the intersection of the curves: $2^x = 2^{1-x}$, $x = 1 - x$, $x = \frac{1}{2}$.

$A = \displaystyle\int_0^{1/2} (2^{1-x} - 2^x) \, dx$. Since $\frac{d}{dx} b^x = b^x \ln b$, $\displaystyle\int b^x \ln b \, dx = b^x + C$

$= \displaystyle\int_0^{1/2} 2^{1-x} dx - \int_0^{1/2} 2^x \, dx$

$= -\dfrac{1}{\ln 2} \displaystyle\int_0^{1/2} 2^{1-x}(\ln 2)(-dx) - \dfrac{1}{\ln 2} \int_0^{1/2} (2^x)(\ln 2) \, dx$

$= -\dfrac{1}{\ln 2} (2^{1-x}) \Big|_0^{1/2} - \dfrac{1}{\ln 2} 2^x \Big|_0^{1/2}$

$= -\dfrac{1}{\ln 2} (\sqrt{2} - 2) - \dfrac{1}{\ln 2} (\sqrt{2} - 1)$

$= -\dfrac{1}{\ln 2}(2\sqrt{2} - 3) = \dfrac{1}{\ln 2} (3 - 2\sqrt{2})$

49. Average value $= \dfrac{1}{\frac{\pi}{4}} \displaystyle\int_0^{\pi/4} \tan x \, dx = \dfrac{4}{\pi} (-\ln |\cos x|) \Big|_0^{\pi/4} = \dfrac{4}{\pi}(-\ln \dfrac{\sqrt{2}}{2} + \ln 1)$

$= \dfrac{4}{\pi} \ln \sqrt{2} = \dfrac{\ln 4}{\pi} \approx 0.44$

53. Since $v(t) = s'(t)$, $s(t) = \displaystyle\int v(t) \, dt = \int \left(\dfrac{1}{t} + t\right) dt = \ln t + \dfrac{t^2}{2} + C$

On the given time interval, $s = \displaystyle\int_1^{e^2} \left(\dfrac{1}{t} + t\right) dt = \left(\ln t + \dfrac{t^2}{2}\right)\Big|_1^{e^2} = 2 + \dfrac{e^4}{2} - \dfrac{1}{2}$

$= \dfrac{3 + e^4}{2} \approx 29 \text{ ft}$

57. Finding the intersections: $\dfrac{2}{x} = 3 - x$, $x^2 - 3x + 2 = 0$, $x = 1, 2$.

$A = \displaystyle\int_1^2 (3 - x - \dfrac{2}{x}) \, dx = 3x - \dfrac{x^2}{2} - 2 \ln x \Big|_1^2$

$= (6 - 2 - \ln 4) - (3 - \dfrac{1}{2}) = \dfrac{3}{2} - \ln 4$

61. Analysis:

a. If $1 < t$, $\dfrac{1}{t} < 1$, so the inequality holds for the integrals, which will give

$\ln \left(1 + \dfrac{1}{n}\right) \leq \dfrac{1}{n}$.

b. Integrate the given inequality and manipulate the result to

$\ln\left(1 + \dfrac{1}{n}\right) \geq \dfrac{1}{n + 1}$

c. Use the antilog of the inequalities in **a.** and **b.** to complete the proof.

d. Do the computation.

5.6 The Inverse Trigonometric Functions

SURVIVAL HINT

One of the easiest ways to remember the domains and ranges of the inverse trigonometric functions is to look at the graphs. Firmly sketch the graph of the trigonometric function. Hold the graph by the line $y = x$, and rotate it about that axis to see the back-side of your paper. This physically interchanges the domain and the range. The x-axis will have become the y-axis and the y-axis the x-axis. What you see is the inverse of your original trigonometric function. Restrict it to either $-\frac{\pi}{2}$ to $\frac{\pi}{2}$ or 0 to π (whichever gives a function), and you have the graph of the inverse function.

5. **a.** The angle that has a tangent of -1 in the range of $(-\frac{\pi}{2}, \frac{\pi}{2})$ is $-\frac{\pi}{4}$.

  b. The angle that has a cotangent of $-\sqrt{3}$ in the range of $(0, \pi)$ is $\frac{5\pi}{6}$.

9. $\cos$ (the angle which has a sine of $\frac{1}{2}$) $= \cos\left(\frac{\pi}{6}\right) = \frac{\sqrt{3}}{2}$

13. Using the addition identity: $\cos(\alpha + 2\beta) = \cos\alpha\cos 2\beta - \sin\alpha\sin 2\beta$.

  Now use the double angle identities:

  $= \cos\alpha(\cos^2\beta - \sin^2\beta) - \sin\alpha(2\sin\beta\cos\beta)$.

  Draw triangles showing $\sin\alpha = \frac{1}{5}$, and $\cos\beta = \frac{1}{5}$.

  You can use Pythagoras to find the other leg of these triangles, and then see that

  $\cos\alpha = \frac{2\sqrt{6}}{5}$ and $\sin\beta = \frac{2\sqrt{6}}{5}$. So $\cos\left(\sin^{-1}\frac{1}{5} + 2\cos^{-1}\frac{1}{5}\right)$

  $= \frac{2\sqrt{6}}{5}\left\{\left(\frac{1}{5}\right)^2 - \left(\frac{2\sqrt{6}}{5}\right)^2\right\} - \frac{1}{5}\left\{2\left(\frac{2\sqrt{6}}{5}\right)\left(\frac{1}{5}\right)\right\} = \frac{2\sqrt{6}}{5}\left\{\left(\frac{1-24}{25}\right) - \left(\frac{2}{25}\right)\right\}$

  $= \frac{2\sqrt{6}}{5}\left(\frac{-25}{25}\right) = -\frac{2\sqrt{6}}{5}$

17. $\frac{d}{dx}\tan^{-1}u = \frac{1}{1+u^2}\frac{du}{dx}$. $\frac{d}{dx}\tan^{-1}\sqrt{x^2+1} = \frac{1}{1+(x^2+1)}\frac{x}{\sqrt{x^2+1}}$

  $= \frac{x}{(x^2+2)\sqrt{x^2+1}}$

21. Use the power rule on $u^{1/2}$, and the chain rule twice on the radicand:

  $y' = \frac{1}{2\sqrt{\tan^{-1}(2x)}}\frac{1}{1+4x^2}(2) = \frac{1}{(4x^2+1)\sqrt{\tan^{-1}(2x)}}$

25. $y' = \frac{1}{1+\left(\frac{1}{x}\right)^2}\left(-\frac{1}{x^2}\right) = -\frac{1}{x^2+1}$

29. Use implicit differentiation and the product rule on each term:

$$x \frac{1}{\sqrt{1-y^2}} y' + \sin^{-1} y + y \frac{1}{1+x^2} + y' (\tan^{-1} x) = 1$$

$$y' \left(\frac{x}{\sqrt{1-y^2}} + \tan^{-1} x \right) = 1 - \sin^{-1} y - \frac{y}{1+x^2}$$

$$y' = \frac{1 - \sin^{-1} y - \dfrac{y}{1+x^2}}{\dfrac{x}{\sqrt{1-y^2}} + \tan^{-1} x}$$

SURVIVAL HINT

The three integration formulas on p. 433 probably need to be memorized. Knowing these will also give you the six differentiation formulas on p. 430. When using these formulas follow the pattern set by examples 8 through 11; write down the substitute value for u, du, and a.

33. To set up the form for the $\sin^{-1} u$, factor $\sqrt{5}$ from the denominator:

$$\frac{1}{\sqrt{5}} \int \frac{dx}{\sqrt{1 + \frac{2}{5} x^2}} . \text{ Now let } u^2 = \tfrac{2}{5} x^2, \quad u = \sqrt{\tfrac{2}{5}} x, \quad du = \sqrt{\tfrac{2}{5}}\, dx.$$

$$\frac{1}{\sqrt{2}} \int \frac{\sqrt{\tfrac{2}{5}}\, dx}{\sqrt{1 + \frac{2}{5} x^2}} = \frac{1}{\sqrt{2}} \sin^{-1} \sqrt{\tfrac{2}{5}}\, x + C$$

37. The x in the numerator suggests a $\frac{du}{u}$ form. But if $u = x^2 + x + 1$, then $du = 2x + 1$. To get the $+1$ we will have to add and subtract 1, and then use two integrals: $\displaystyle\int \frac{x\, dx}{x^2 + x + 1} = \tfrac{1}{2} \int \frac{2x + 1 - 1}{x^2 + x + 1}\, dx$

$$= \tfrac{1}{2} \int \frac{2x + 1}{x^2 + x + 1}\, dx - \tfrac{1}{2} \int \frac{1}{x^2 + x + 1}\, dx$$

$$= \tfrac{1}{2} \ln |x^2 + x + 1| - \tfrac{1}{2} \int \frac{dx}{(x + \tfrac{1}{2})^2 + \tfrac{3}{4}}$$

$$= \tfrac{1}{2} \ln |x^2 + x + 1| - \tfrac{2}{3} \int \frac{dx}{\tfrac{4}{3}(x + \tfrac{1}{2})^2 + 1}$$

$$= \tfrac{1}{2} \ln |x^2 + x + 1| - \tfrac{2}{3} \int \frac{dx}{\left(\tfrac{2}{\sqrt{3}}\right)^2 (x + \tfrac{1}{2})^2 + 1}$$

37. (con't.)

$$= \tfrac{1}{2} \ln |x^2 + x + 1| - \tfrac{2}{3} \tfrac{\sqrt{3}}{2} \int \frac{\tfrac{2}{\sqrt{3}} \, dx}{\left(\tfrac{2}{\sqrt{3}}\right)^2 (x + \tfrac{1}{2})^2 + 1}$$

$$= \tfrac{1}{2} \ln |x^2 + x + 1| - \tfrac{\sqrt{3}}{3} \tan^{-1}\!\left(\tfrac{2}{\sqrt{3}}\right)(x + \tfrac{1}{2}) + C$$

41. Completing the square gives the form for the inverse tangent:

$$\int \frac{dx}{(x+1)^2 + 1} = \tan^{-1}(x + 1) + C$$

45. If we divide the denominator by 4 we will have a $u^2 + 1$ form.

$$\tfrac{1}{4} \int_0^{\ln 2\sqrt{3}} \frac{e^x \, dx}{\frac{e^{2x}}{4} + 1}. \quad \text{Now if } u^2 = \frac{e^{2x}}{4}, \quad du = \tfrac{1}{2} e^x dx.$$

$$\tfrac{1}{2} \int_0^{\ln 2\sqrt{3}} \frac{\tfrac{1}{2} e^x \, dx}{\frac{e^{2x}}{4} + 1} = \tfrac{1}{2} \tan^{-1} \tfrac{1}{2} e^x \Big|_0^{\ln 2\sqrt{3}} = \tfrac{1}{2} \left(\tan^{-1}\sqrt{3} - \tan^{-1} \tfrac{1}{2}\right)$$

$$= \tfrac{1}{2}(\tfrac{\pi}{3} - \tan^{-1} \tfrac{1}{2}) = \tfrac{\pi}{6} - \tfrac{1}{2} \tan^{-1}\tfrac{1}{2}$$

49. Using the double angle identity: $\sin 2\theta = 2 \sin \theta \cos \theta$, so
$\sin (2 \tan^{-1} x) = 2 \sin \theta \cos \theta$. Drawing a triangle in which $\tan \theta = x$,
and using the Pythagorean theorem to find the hypotenuse, we have:

$$\sin (2 \tan^{-1} x) = 2 \left(\frac{x}{\sqrt{x^2 + 1}}\right)\!\left(\frac{1}{\sqrt{x^2 + 1}}\right) = \frac{2x}{x^2 + 1}$$

53. Using the addition identity: $\sin (\alpha + \beta) = \sin \alpha \cos \beta + \sin \beta \cos \alpha$,
$\sin (\sin^{-1} x + \cos^{-1} x) = \sin \alpha \cos \beta + \sin \beta \cos \alpha$, where $\alpha = \sin^{-1} x$
and $\beta = \cos^{-1} x$. Drawing triangles where $\sin \alpha = x$, and $\cos \beta = x$, we can
see that $\cos \alpha = \sqrt{1 - x^2}$, and $\sin \beta = \sqrt{1 - x^2}$. So now
$\sin (\sin^{-1} x + \cos^{-1} x) = x(x) + \sqrt{1 - x^2} \sqrt{1 - x^2} = 1$.
If we had been observant we would have noticed that if $\sin \alpha = x$ and $\cos \beta = x$
then $\alpha + \beta = \tfrac{\pi}{2}$ and $\sin \left(\tfrac{\pi}{2}\right) = 1$

57. $$A = \int_{\sqrt{2}}^{2} \frac{dx}{x \sqrt{x^2 - 1}} = \sec^{-1} x \Big|_{\sqrt{2}}^{2} = \frac{\pi}{3} - \frac{\pi}{4} = \frac{\pi}{12}$$

61. $f'(x) = 1 - \dfrac{1}{1 + (2x)^2}(2) = \dfrac{4x^2 - 1}{4x^2 + 1} = 0$ when $x = \pm\frac{1}{2}$.

Use the second derivative test to determine if these are maximums or minimums.

$$f''(x) = \frac{(4x^2 + 1)(8x) - (4x^2 - 1)(8x)}{(4x^2 + 1)^2} = \frac{16x}{(4x^2 + 1)^2}.$$

So there is a relative maximum at $(-\frac{1}{2}, -\frac{1}{2} + \frac{\pi}{4})$ and a relative minimum at $(\frac{1}{2}, \frac{1}{2} - \frac{\pi}{4})$. There are intercepts when $f(x) = 0$, $\tan^{-1}(2x) = x$, which is not easily solved except for the point $(0, 0)$.

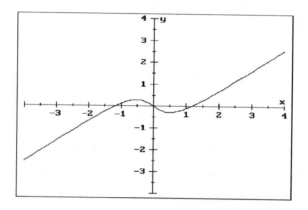

65. Let the horizontal distance from the camera to the wall be x, and name the resulting angles as indicated:

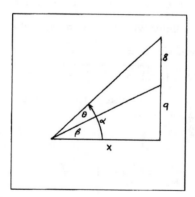

65. (con't.)

$$\theta = \alpha - \beta = \tan^{-1}\frac{17}{x} - \tan^{-1}\frac{9}{x}. \text{ Maximize with derivatives:}$$

$$\frac{d\theta}{dt} = \frac{1}{1 + (\frac{17}{x})^2}\left(-\frac{17}{x^2}\right) - \frac{1}{1 + (\frac{9}{x})^2}\left(-\frac{9}{x^2}\right) = -\frac{17}{x^2 + 289} + \frac{9}{x^2 + 81}$$

$$= \frac{1224 - 8x^2}{(x^2 + 289)(x^2 + 81)} = 0 \text{ when } 8(x^2 - 153) = 0, x = \sqrt{153}.$$

Since the endpoint $x = 0$ is obviously a minimum, we have the maximum angle when $x = \sqrt{153} \approx 12.4$ ft.

67. Analysis: Divide numerator and denominator by a^2 to set up the form $\dfrac{du}{1 + u^2}$, with $u = \frac{u}{a}$.

69. Analysis: Since the derivative of the constant $\frac{\pi}{2}$ is 0, this will give $-\dfrac{d}{dx}(\tan^{-1}x)$.

5.7 An Alternative Approach: The Logarithm as an Integral

SURVIVAL HINT

The material of this section may seem confusing because you already "know" too much about logarithms and exponents. In trigonometry you were introduced to $y = b^x$ without any proof that it was continuous for irrational values. The approach here is really better because the area function used *is* continuous to begin with, so the other functions derived from it, e^x, b^x, and $\log_b x$ will also be continuous. Try to read the section as if you were seeing $\ln x$ for the first time.

1. Analysis: Consider what happens to $\ln 2^{-N} = -N \ln 2$ as n approaches 0 through positive values. Follow steps listed in **a.** and **b.**

3. Analysis: $|E_n| \le \dfrac{(b - a)^5}{180 n^4} K$, where K is the minimum value of $f^{(4)}(x)$.

Let $E_n = 0.00005$ and solve for n.

5. Analysis: Follow the rather explicit instructions in **a.** and **b.**

7. Analysis:

a. $f(1y) = f(1) + f(y)$, so $f(1) = 0$

b. Let $x = y = -1$: $f(1) = f(-1) + f(-1)$, $0 = 2f(-1)$,

$f(-1) = 0$

7. (con't.)

c. Let $y = -1$: $f(-1x) = f(x) + f(-1)$, so if $f(-1) = 0$,

$f(x) = f(-x)$.

d. In $f(x, y) = f(x) + f(y)$ hold x fixed and differentiate with respect to y, remembering to use the chair rule on the left:

$x f'(xy) = 0 + f'(y)$

when $y = 1$, $x f'(x) = f'(1)$ so $f'(x) = \dfrac{f'(1)}{x}$

e. The result of **d.** is continuous on the closed interval $[a, b]$ and so we can apply the fundamental theorem of calculus:

$$f(x) - f(c) = \int_c^x f'(t)\, dt = f'(1)\int_c^x \frac{dt}{t} \text{ (switch limits if } x \text{ and } c < 0)$$

Since $f(1) = 0$ we can set $c = 1$ to get $f(x) = f'(1)\int_1^x \frac{dt}{t}$ if $x > 0$

If $x < 0$, then $-x > 0$ and since $f(x) = f(-x)$ we get

$$f(x) = f'(1)\int_1^{-x} \frac{dt}{t} \quad x < 0. \text{ Now combining the two formulas,}$$

$$f(x) = f'(1)\int_1^{|x|} \frac{dt}{t} \text{ if } x \neq 0. \text{ And now if } f'(1) \neq 0 \text{ we can divide}$$

and let $F(x) = \dfrac{f(x)}{f'(1)}\displaystyle\int_1^{|x|} \frac{dt}{t}$. It can be shown that if $f(xy) = f(x) + f(y)$

then $F(xy) = F(x) + F(y)$. All solutions of $f(xy) = f(x) + f(y)$ can be obtained as multiples of $F(x)$. More detailed work can be found in Apostle's calculus text, vol. I, 2nd edition pp. 228-229.

9. Analysis: If $\ln 2 < 1 < \ln 3$, then $e^{\ln 2} < e^1 < e^{\ln 3}$.

Chapter 5 Review

17. **a.** This is the product of two functions, so the product rule gives:

$y' = x^2\left(e^{-\sqrt{x}}\right)\left(-\dfrac{1}{2\sqrt{x}}\right) + e^{-\sqrt{x}}(2x) = xe^{-\sqrt{x}}\left(2 - \dfrac{\sqrt{x}}{2}\right)$

b. This is the quotient of two functions, so the quotient rule gives:

$y' = \dfrac{(\ln 3x)(\frac{1}{x}) - (\ln 2x)(\frac{1}{x})}{(\ln 3x)^2} = \dfrac{\ln 1.5}{x(\ln 3x)^2}$

18.   **a.** $y' = \dfrac{1}{\sqrt{1 - (3x + 2)^2}} (3) = \dfrac{3}{\sqrt{1 - (3x + 2)^2}}$

   b. $y' = \dfrac{1}{1 + (2x)^2} (2) = \dfrac{2}{1 + 4x^2}$

19. $\ln y = \ln \dfrac{\ln (x^2 - 1)}{\sqrt[3]{x} \, (1 - 3x)^3} = \ln \ln (x^2 - 1) - \tfrac{1}{3} \ln x - 3 \ln (1 - 3x)$

Taking the derivative of both sides:

$$\tfrac{1}{y} y' = \dfrac{1}{\ln (x^2 - 1)} \dfrac{1}{x^2 - 1} (2x) - \dfrac{1}{3x} - \dfrac{3}{1 - 3x} (-3)$$

$$= \dfrac{2x}{(x^2 - 1)\ln (x^2 - 1)} - \dfrac{1}{3x} - \dfrac{9}{3x - 1}$$

$$y' = y \left(\dfrac{2x}{(x^2 - 1)\ln (x^2 - 1)} - \dfrac{1}{3x} - \dfrac{9}{3x - 1} \right)$$

20. $y = (x^2 - 3)e^{-x} = 0$ when $x = \pm \sqrt{3}$, which are the x-intercepts.

When $x = 0$, $y = -3$, which is the y-intercept.

$y' = (x^2 - 3)e^{-x}(-1) + e^{-x}(2x) = -e^{-x}(x^2 - 2x - 3) = 0$ when

$(x - 3)(x + 1) = 0$, $x = -1, 3$, which are candidates for extrema.

$y'' = -e^{-x}(2x - 2) + (x^2 - 2x - 3)e^{-x} = e^{-x}(x^2 - 4x - 1)$

$= 0$ when $x = 2 \pm \sqrt{5}$. These are candidates for inflection points.

Using the second derivative to test the candidates of extrema we see that:

$y''(-1) = 4e > 0$, so there is a relative minimum at $(-1, -2e)$.

$y''(3) = -\dfrac{4}{e^x} < 0$, so there is a relative maximum at $(3, 6e^{-3})$.

$\lim\limits_{x \to \infty} f(x) = \lim\limits_{x \to \infty} \dfrac{x^2 - 3}{e^x}$ which is $\dfrac{\infty}{\infty}$ form so we can use l'Hôpital's rule:

$= \lim\limits_{x \to \infty} \dfrac{2x}{e^x}$, use the rule again, $= \lim\limits_{x \to \infty} \dfrac{2}{e^x} = 0$

$\lim\limits_{x \to -\infty} f(x) = \lim\limits_{x \to -\infty} \dfrac{x^2 - 3}{e^x} = \dfrac{\infty}{0} = \infty$

20. (con't.)

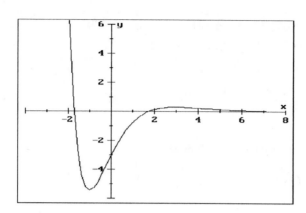

21. **a.** $\dfrac{1}{2}\displaystyle\int_0^{\ln 2} \left[e^{2x}(2) + e^{-2x}(-2) \right] dx = \dfrac{1}{2}\left(e^{2x} + e^{-2x} \right)\Big|_0^{\ln 2}$

$= \dfrac{1}{2}\left(e^{2\ln 2} + e^{-2\ln 2} - e^0 - e^0 \right) = \dfrac{1}{2}\left(e^{\ln 2^2} + e^{\ln 2^{-2}} - 2 \right)$

$= \dfrac{1}{2}\left(4 + \dfrac{1}{4} - 2 \right) = \dfrac{9}{8}$

b. $\displaystyle\int (\ln x)^1 \left(\dfrac{1}{x}\right) dx$ is of form $u^n\, du$, and $= \dfrac{(\ln x)^2}{2} + C$

c. $\displaystyle\int \tan 2x\, dx = \int \dfrac{\sin 2x}{\cos 2x}\, dx = -\dfrac{1}{2}\int \dfrac{(-\sin 2x)(2\, dx)}{\cos 2x}$, which is

now of form $\dfrac{du}{u}$, and $= -\dfrac{1}{2}\ln|\cos 2x| + C$

22. **a.** Recall and use the limit definition of e: $\displaystyle\lim_{x\to 0^+}\left(\left(1 + \dfrac{1}{x}\right)^x\right)^4$, and

$\lim f\Big(g(x)\Big) = f\Big(\lim g(x)\Big)$, so we have:

$\left(\displaystyle\lim_{x\to 0^+}(1 + \dfrac{1}{x})^{x4} \right) = e^4$

b. $\displaystyle\lim_{x\to 1^+}\left(\dfrac{1}{1-x}\right)^x = (-\infty)^x$ so the limit does not exist.

c. This is of form 0^0, so let $y = x^{\tan x}$, $\ln y = \ln x^{\tan x} = (\tan x)(\ln x)$.

$\displaystyle\lim_{x\to 0^+}\ln y = \lim_{x\to 0^+}(\tan x)(\ln x)$, which is of $(0)(\infty)$ form, so write it as:

$\displaystyle\lim_{x\to 0^+}\ln y = \lim_{x\to 0^+}\dfrac{\ln x}{\cot x}$, which has form $\dfrac{\infty}{\infty}$, now we can use l'Hôpital's rule:

$\displaystyle\lim_{x\to 0^+}\ln y = \lim_{x\to 0^+}\dfrac{\frac{1}{x}}{-\csc^2 x} = \lim_{x\to 0^+} -\dfrac{\sin^2 x}{x} = \lim_{x\to 0^+} -\dfrac{2\sin x\cos x}{1} = 0.$

So now $\displaystyle\lim_{x\to 0^+}\ln y = \ln\Big(\lim_{x\to 0^+} y\Big) = 0$, $\displaystyle\lim_{x\to 0^+} y = e^0 = 1$

23. $A = \int\limits_{0}^{1} (e^{2x} - e^{x})\, dx = \frac{1}{2} e^{2x} - e^{x} \Big|_{0}^{1} = \frac{1}{2} e^{2} - e - \frac{1}{2} + 1$

$= \frac{1}{2}(e^{2} - 2e + 1) \approx 1.476\ 24$

24. **a.** $A = P(1 + \frac{r}{t})^{nt}.\ 5 = 2\left(1 + \frac{0.08}{4}\right)^{4t},\ \ln 2.5 = 4t \ln 1.02,$

$t = \dfrac{\ln 2.5}{4 \ln 1.02} \approx 11.57$ years or 11 years 207 days

b. $5 = 2\left(1 + \frac{0.08}{12}\right)^{12t},\ \ln 2.5 = 12t \ln 1.006\ 666\ 7,$

$t = \dfrac{\ln 2.5}{12 \ln 1.006\ 666\ 67} \approx 11.49$ years or 11 years 180 days

c. $2.5 = e^{0.08t},\ 0.08t = \ln 2.5,\ t \approx 11.45$ years or 11 years 166 days

25. Profit = Revenue − Cost. $P = p\left(800 e^{-0.01p}\right) - 40\left(800 e^{-0.01p}\right).$

$P = \left(800 e^{-0.01p}\right)(p - 40).$ To find candidates for extrema:

$P' = 800\left\{\left(e^{-0.01p}\right)(1) + (p - 40)\left(e^{-0.01p}\right)(-0.01)\right\}$

$= 800\left\{\left(e^{-0.01p}\right)(1.4 - 0.01p)\right\} = 0$ when $0.01p = 1.4,\ p = 140.$

Sell the cameras for $140.

Chapter 6
Additional Applications of the Integral

6.1 Volume: Disks, Washers, and Shells

SURVIVAL HINT

In order to use the method of slices, each cross section must have the same shape, that is, each must use the same formula for its area. Write the *general* formula for the cross sectional area, then substitute the appropriate function values. Finally, sum over the given values.

1.

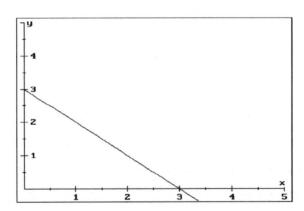

Each square perpendicular to the x-axis will have area of y^2, where $y = 3 - x$.

$$V = \int_0^3 y^2 \, dx = \int_0^3 (3 - x)^2 dx = \int_0^3 (9 - 6x + x^2) \, dx = 9x - 3x^2 + \frac{x^3}{3} \Big|_0^3$$

$$= 27 - 27 + 9 = 9 \text{ cubic units.}$$

This is easily verified with the formula for the volume of a pyramid.

5.

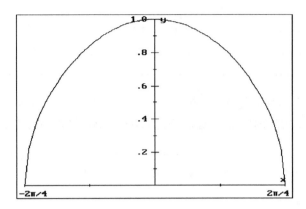

Make use of the symmetry about the y-axis in setting up the integral:

$$V = 2 \int_0^{\pi/2} y^2 \, dx = 2 \int_0^{\pi/2} \cos x \, dx = 2 \sin x \Big|_0^{\pi/2} = 2 \text{ cubic units.}$$

9.

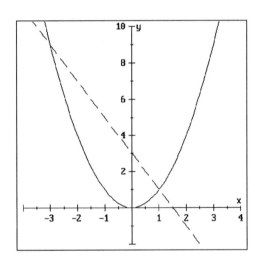

The base of each equilateral triangle is $(3 - 2x) - x^2$, and the area of an equilateral triangle with side s is $\dfrac{s^2\sqrt{3}}{4}$. The curves intersect at $(-3, 9)$ and $(1, 1)$.

$$V = \frac{\sqrt{3}}{4} \int_{-3}^{1} (3 - 2x - x^2)^2 \, dx = \frac{\sqrt{3}}{4} \int_{-3}^{1} (x^4 + 4x^3 - 2x^2 - 12x + 9) \, dx$$

$$= \frac{\sqrt{3}}{4} \left(\frac{x^5}{5} + x^4 - \frac{2x^3}{3} - 6x^2 + 9x \right)\Big|_{-3}^{1}$$

$$= \frac{\sqrt{3}}{4} \left(\frac{1}{5} + 1 - \frac{2}{3} - 6 + 9 + \frac{243}{5} - 81 - 18 + 54 + 27 \right)$$

$$= \frac{\sqrt{3}}{4} \left(\frac{244}{5} - \frac{2}{3} - 14 \right) = \frac{\sqrt{3}}{4} \left(\frac{512}{15} \right)$$

$$= \frac{128\sqrt{3}}{15}$$

SURVIVAL HINT

Success in finding volumes by slices is highly dependent upon good visualization of the three dimensional figure (called spatial perception). Practice drawing a good sketch for each problem. Draw the slice on your sketch and write the appropriate *general* formula; πr^2 for disks, $\pi(R^2 - r^2)$ for washers, and $2\pi rh$ for shells. *Then* substitute the function values for r, R, and h, and sum over the correct limits.

13.   **a.**   The indicated slice will give disks (πr^2). $V = \pi \displaystyle\int_0^4 (4 - x)^2 \, dx$

     b.   The indicated slice will give shells ($2\pi rh$). $V = 2\pi \displaystyle\int_0^4 x(4 - x) \, dx$

     c.   The indicated slice will give washers $\pi(R^2 - r^2)$.

$$V = \pi \int_0^4 [(5 - x)^2 - (1)^2] \, dx = \pi \int_0^4 (x^2 - 10x + 24) \, dx$$

17.   **a.**   A vertical strip will give washers, a horizontal strip will give shells. We will use a vertical strip. The intersection is $(1, 1)$.

$$V = \pi \int_0^1 [\sqrt{x}^2 - (x^2)^2] \, dx = \pi \int_0^1 (x - x^4) \, dx$$

     b.   A vertical strip will give shells, a horizontal strip will give washers. We will use a vertical strip.

$$V = 2\pi \int_0^1 x(\sqrt{x} - x^2) \, dx = 2\pi \int_0^1 (x^{3/2} - x^3) \, dx$$

21.   The base of each square is $2y$, so the area of each square is $4y^2$. Since $y^2 = 9 - x^2$, each square has area $4(9 - x^2)$. These square cross sections are to be summed from -3 to 3, or, if symmetry is used:

$$V = 2 \int_0^3 4(9 - x^2) \, dx = 8 \int_0^3 (9 - x^2) \, dx = 8\left(9x - \frac{x^3}{3}\right)\Big|_0^3 = 8(27 - 9)$$

$$= 144 \text{ cubic units.}$$

25.   The curves intersect when $x + 1 = x^2 - 1$, $x^2 - x - 2 = 0$, $x = -1, 2$.
The base of each square is $(x + 1) - (x^2 - 1)$ or $x + 2 - x^2$. The area of
each square is $(x + 2 - x^2)^2$ or $(x^4 - 2x^3 - 3x^2 + 4x + 4)$. These squares
are to be added from $x = -1$ to $x = 2$.

$$V = \int_{-1}^{2} (x^4 - 2x^3 - 3x^2 + 4x + 4) \, dx = \frac{x^5}{5} - \frac{x^4}{2} - x^3 + 2x^2 + 4x \Big|_{-1}^{2}$$

$$= \frac{32}{5} - 8 - 8 + 8 + 8 - (-\frac{1}{5} - \frac{1}{2} + 1 + 2 - 4)$$

$$= \frac{33}{5} + \frac{1}{2} + 1 = \frac{81}{10}$$

29.   When $y = \sqrt{x}$ is revolved about the x-axis, and vertical slices are used, we get
a set of disks. These disks are added from 0 to 1.

$$V = \pi \int_{0}^{1} (\sqrt{x})^2 \, dx = \pi \left(\frac{x^2}{2}\right)\Big|_{0}^{1} = \frac{\pi}{2}$$

33.   On the interval $[0, \pi]$ the function $y = x^2 + x^3$ is positive, so if it is revolved
about the x-axis, and vertical slices are used, we will have a series of disks.

$$V = \pi \int_{0}^{\pi} (x^2 + x^3)^2 \, dx = \pi\left(\frac{x^5}{5} + \frac{x^6}{3} + \frac{x^7}{7}\right)\Big|_{0}^{\pi} \approx 2555.$$

37.   On the interval $[0, \pi]$ the function $y = \sqrt{\sin x}$ is positive, so if it is revolved
about the x-axis, and vertical slices are used, we will have a series of disks.

$$V = \pi \int_{0}^{\pi} \sqrt{\sin x}\,^2 \, dx = \pi \left(- \cos x\right)\Big|_{0}^{\pi} = \pi \left(1 - (-1)\right) = 2\pi$$

41.   The curves intersect at $(1, 1)$. If they are revolved about the y-axis, and vertical
strips are used, we will have a series of shells to be added from $x = 0$ to $x = 1$.
The radius of each shell is x, and the height of each is $x^2 - x^3$.

$$V = 2\pi \int_{0}^{1} x(x^2 - x^3) \, dx = 2\pi \int_{0}^{1} (x^3 - x^4) \, dx = 2\pi\left(\frac{x^4}{4} - \frac{x^5}{5}\right)\Big|_{0}^{1}$$

$$= 2\pi\left(\frac{1}{20}\right) = \frac{\pi}{10}$$

45. The curves intersect at $\left(\frac{\sqrt{2}}{2}, \frac{1}{2}\right)$. If they are revolved about the y-axis, and vertical strips are used, we will have a series of shells to be added from $x = 0$ to $x = \frac{\sqrt{2}}{2}$. The radius of each shell is x, and the height of each is $(1 - x^2) - x^2$.

$$V = 2\pi \int_0^{\sqrt{2}/2} x(1 - 2x^2)\, dx = 2\pi \int_0^{\sqrt{2}/2} (x - 2x^3)\, dx = 2\pi \left(\frac{x^2}{2} - \frac{x^4}{2}\right)\Big|_0^{\sqrt{2}/2}$$

$$= 2\pi\left(\frac{1}{4} - \frac{1}{8}\right) = \frac{\pi}{4}$$

49.

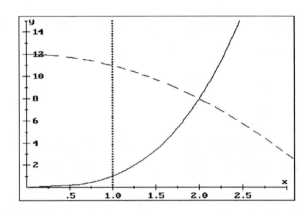

The curves intersect at the point $(2, 8)$.

If the region bounded by the curves is revolved about the the line $x = -1$, and vertical strips are used, we will get a series of shells. The radius of each shell will be $x + 1$, and its height will be $(12 - x^2) - x^3$. These shells will be added from $x = 1$ to $x = 2$.

$$V = 2\pi \int_1^2 (x + 1)(12 - x^2 - x^3)\, dx$$

$$= 2\pi \int_1^2 (12 + 12x - x^2 - 2x^3 - x^4)\, dx$$

$$= 2\pi \left(12x + 6x^2 - \frac{x^3}{3} - \frac{x^4}{2} - \frac{x^5}{5}\right)\Big|_1^2$$

$$= 2\pi\left(24 + 24 - \frac{8}{3} - 8 - \frac{32}{5} - \left[12 + 6 - \frac{1}{3} - \frac{1}{2} - \frac{1}{5}\right]\right)$$

$$= 2\pi\left(22 - \frac{7}{3} - \frac{31}{5} + \frac{1}{2}\right) = \frac{419\pi}{15}$$

53. Let the fixed diameter be the x-axis. The square cross sections will each have a side of $2y$, where $y = \sqrt{1 - x^2}$. The area of each cross section will be $(2y)^2 = 4(1 - x^2)$. If we use the symmetry about the y-axis, we can double the sum of these squares from $x = 0$ to $x = 1$.

$$V = 2\int_0^1 4(1 - x^2)\ dx = 8\left(x - \frac{x^3}{3}\right)\Big|_0^1 = 8\left(1 - \frac{1}{3}\right) = \frac{16}{3}$$

57.

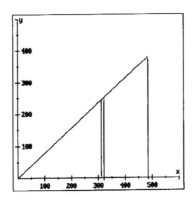

Put the vertex of the pyramid at the origin. Its side will be the the line $y = \frac{375}{480}\, x$. Each rectangular slice perpendicular to the x-axis will have height of y and width $2y$. Twice the sum of all the rectangles from $x = 0$ to $x = 480$ will give the volume.

$$V = 2\int_0^{480} 2y^2\ dx = 4\int_0^{480} \left(\frac{375}{480}\, x\right)^2\ dx = 4\left(\frac{375}{480}\right)^2 \frac{x^3}{3}\Big|_0^{480} = 4(375)^2\frac{480}{3}$$

$$= 90,000,000 \text{ cubic feet.}$$

63. Analysis: As slices are taken perpendicular to the x-axis, each slice is a semi-circle with radius of y. The area of each semi-circle will be $\frac{1}{2}\, \pi y^2 = \frac{\pi}{2}(r^2 - x^2)$. Using the symmetry about the y-axis, twice the sum from $x = 0$ to $x = r$ will give the volume of the hemisphere. Double this result for the volume of the sphere.

65. Analysis: Each slice perpendicular to the x-axis is a circle with radius of y. The area of each of these circles is $\pi y^2 = \pi b^2\left(1 - \frac{x^2}{a^2}\right)$. Twice the sum of these slices from $x = 0$ to $x = a$ will give the volume of the ellipsoid. If the ellipse is revolved about the y-axis, follow the same procedure using slices perpendicular to the y-axis.

67.   Analysis: If slices are taken perpendicular to the y-axis, each slice will be a circle
with radius of x. The area of each circle will be $\pi x^2 = \pi(R^2 - y^2)$. If these
slices are summed from $y = h$ to $y = R$, we will have the volume of the
spherical segment of one base (or "cap"): $V = \frac{\pi}{3}(2R^3 - 3R^2h + h^3)$. If h is
replaced with $-R$, you will get the formula for the volume of a sphere. Compare
your result with the geometry formula for a spherical segment of one base:
$V = \frac{\pi h^2}{3}(3r - h)$, where h is the height of the segment and r is the radius
of the sphere.

69.   Analysis: The "gem" is found by adding the volume of the cone to the volume
of the spherical segment between the planes h_1 and h_2. The volume of the cone
can be found with the elementary geometry formula $V = \frac{\pi}{3} r^2 h$ where $h = h_1$
and $r^2 = R^2 - h_1{}^2$. The volume of the spherical segment can be found with

the integral $V = \pi \displaystyle\int_{h_1}^{h_2} x^2 \, dy$ where $x^2 = R^2 - y^2$.

6.2 Arc Length and Surface Area

SURVIVAL HINT

Formulas do not have to be "memorized" if you understand the concept. The length of
the arc is the sum (integral) of a collection of oblique line segments — each of which can
be considered as the hypotenuse of a right triangle with base 1 and height equal to the
slope of the line, $f'(x)$: $\sqrt{1 + [f'(x)]^2}$.

1. $\sqrt{1 + f'(x)^2} = \sqrt{1 + 3^2} = \sqrt{10}$

$s = \displaystyle\int_{-1}^{2} \sqrt{10} \, dx = 3\sqrt{10}$. This is easily verified as the "arc" is a straight line.

5. $\sqrt{1 + f'(x)^2} = \sqrt{1 + \sqrt{x}^2} = \sqrt{1 + x}$

$s = \displaystyle\int_{0}^{4} \sqrt{1 + x} \, dx = \frac{2}{3}(1 + x)^{3/2} \Big|_0^4 = \frac{2}{3}\left(5\sqrt{5} - 1\right) = \frac{10\sqrt{5}}{3} - \frac{2}{3}$

9. This is one of those algebraic manipulations in which $f'(x)$ is a binomial, which when squared has a middle term of $-\frac{1}{2}$. When the 1 in the formula is added the result is also a trinomial square, where now the middle term is $+\frac{1}{2}$.

$$\sqrt{1 + f'(x)^2} = \sqrt{1 + \left(x^3 - \frac{1}{4x^3}\right)^2} = \sqrt{1 + \left(x^6 - \frac{1}{2} + \frac{1}{16x^6}\right)}$$

$$= \sqrt{\left(x^6 + \frac{1}{2} + \frac{1}{16x^6}\right)} = \sqrt{\left(x^3 + \frac{1}{4x^3}\right)^2} = \left(x^3 + \frac{1}{4x^3}\right)$$

$$s = \int_1^2 \left(x^3 + \frac{1}{4}x^{-3}\right) dx = \frac{x^4}{4} - \frac{1}{8x^2}\Big|_1^2 = 4 - \frac{1}{32} - \frac{1}{4} + \frac{1}{8} = \frac{123}{32}$$

13. $y = \sqrt[3]{\frac{9x^2}{4}}$. Since this does not have a very nice derivative, solve for $x = g(y)$

and use a dy integral. $x = \sqrt{\frac{4y^3}{9}} = \frac{2y^{3/2}}{3}$, $g'(y) = y^{1/2}$.

$$\sqrt{1 + g'(y)^2} = \sqrt{1 + y}. \quad s = \int_0^3 \sqrt{1 + y}\, dy = \frac{2}{3}(1 + y)^{3/2}\Big|_0^3$$

$$= \frac{16}{3} - \frac{2}{3} = \frac{14}{3}$$

SURVIVAL HINT

Understand the *concept* of the surface formula as the sum of small surfaces, each of which is the surface of the frustum of a cone: $2\pi r l$. Where r is the distance from the axis of revolution to the bounding curve, and l is the slant height of the frustum: $\sqrt{1 + [f'(x)]^2}$.

17. $S = 2\pi \int_0^2 (2x + 1)\sqrt{1 + (2)^2}\, dx = 2\pi \sqrt{5}\,(x^2 + x)\Big|_0^2 = 12\pi \sqrt{5}$

This is easily verified using the formula for the lateral surface of a truncated cone.

21. $S = 2\pi \int_0^3 x\sqrt{1 + \left(-\frac{1}{3}\right)^2}\, dx = \frac{2}{3}\sqrt{10}\pi \int_0^3 x\, dx = \frac{2}{3}\sqrt{10}\pi\left(\frac{x^2}{2}\right)\Big|_0^3 = 3\pi \sqrt{10}$

For the lateral surface of a cone $S = \pi r l = \pi(3)(\sqrt{10}) = 3\pi \sqrt{10}$

25. $\quad y' = \dfrac{-(1 - x^{2/3})^{1/2}}{x^{1/3}}, \quad \sqrt{1 + (y')^2} = \sqrt{1 + \dfrac{(1 - x^{2/3})}{x^{2/3}}} = x^{-1/3}$

$$s = 4 \int_0^1 x^{-1/3}\, dx = 6x^{2/3}\Big|_0^1 = 6$$

29. $\quad y = \tan x, \quad y' = \sec^2 x, \quad \sqrt{1 + (y')^2} = \sqrt{1 + \sec^4 x}.$

$$S = 2\pi \int_0^1 \tan x \sqrt{1 + \sec^4 x}\; dx \approx 8.632601 \text{ by Simpson's rule}$$

with $n = 20$.

31. Analysis: Center the cone with the vertex at the origin, and the plane of the base perpendicular to the x-axis. The slant height will be the line $y = \frac{r}{h}\, x$. When it is revolved about the x-axis it will generate the surface of the cone. Integrate from $x = 0$ to $x = h$.

33. Analysis: As indicated in the hint, use the mean value theorem. After verifying the hypotheses, we are assured there exists an $x_k{}^*$ such that
$f'(x_k{}^*) = \dfrac{f(b) - f(a)}{b - a} = \dfrac{\Delta y}{\Delta x}$. Square both sides, add and subtract $(\Delta x)^2$,
to the numerator on the right, and continue the algebra to the desired result.

35. Analysis: Rotate the upper half of the translated circle $(x - R)^2 + y^2 = r^2$ about the y-axis and double the result. This problem becomes trivial if you know and use the Theorem of Pappus for surface area.

37. Analysis: Rotate the first quadrant portion of $y = b^2\left(1 - \dfrac{x^2}{a^2}\right)$ about the y-axis and double the result. Integrate from $x = 0$ to $x = a$.

6.3 Physical Applications: Work, Liquid Force, and Centroids

5. $\quad W = Fd = (850)(15) = 12{,}750 \text{ ft-lbs}$

9. The difference in F is 65 lb, the distance is the same. The additional work for the full bucket: $W = (65)(100) = 6{,}500 \text{ ft-lbs}$

13. $\quad F = \rho h A = 64.0 \displaystyle\int_0^3 (3 - h)\frac{2h}{3}\, dh = \frac{128}{3}\left(\frac{3h^2}{2} - \frac{h^3}{3}\right)\Big|_0^3 = \frac{128}{3}\left(\frac{9}{2}\right) = 192 \text{ lbs}$

17. This works best if the center of the circle is the origin, and we sum from $h = 1$ to $h = 3$.
$$F = \rho h A = 849.0 \int_1^3 h\, 2\sqrt{9 - h^2}\, dh = -849.0 \left(\tfrac{2}{3}\right)(9 - h^2)^{3/2}\Big|_1^3$$

$$= -566\,(0 - 8\sqrt{8}) = 9{,}056\sqrt{2} \approx 12{,}807.1 \text{ lbs}$$

SURVIVAL HINT

Recall that to find $\bar{x}$ you use the moment about the y-axis, and to find $\bar{y}$ you use M_x.

You will not confuse the formulas if you remember that the moment about the y-axis can

be found by placing all the mass at $\bar{x}$: $M_y = m\bar{x}$, and the moment about the x-axis is

found by placing all the mass at $\bar{y}$: $M_x = m\bar{y}$.

21. $M = \displaystyle\int_1^2 x^{-1}\,dx = \ln x \Big|_1^2 = \ln 2.$ $M_y = \displaystyle\int_1^2 (x)x^{-1}\,dx = 2 - 1 = 1.$

$M_x = \frac{1}{2}\displaystyle\int_1^2 (x^{-1})^2\,dx = \frac{1}{2}(-x^{-1})\Big|_1^2 = \frac{1}{2}(-\frac{1}{2} + 1) = \frac{1}{4}$

$\bar{x} = \dfrac{M_y}{M} = \dfrac{1}{\ln 2}$ and $\bar{y} = \dfrac{M_x}{M} = \dfrac{\frac{1}{4}}{\ln 2} = \dfrac{1}{\ln 16}$

25.   The theorem of Pappus says $V = As$, where A is the area of the revolved region,

and s is the distance the centroid travels. In this case $s = 2\pi(\bar{y} + 1)$.

$A = \frac{1}{2}(5)(5) = \frac{25}{2}.$ $\bar{y} = \frac{5}{3}$ since the centroid of a triangle is at the

intersection of the medians, and that point of concurrency divides the median into

segments with ratio $\frac{1}{3}$ to $\frac{2}{3}$. Therefore $V = \frac{25}{2}\,2\pi\left(\frac{5}{3} + 1\right) = \frac{200\pi}{3}$ cu. units.

Using calculus to find $\bar{y}$ is considerably more difficult, as there are two different

curves for the upper bound.

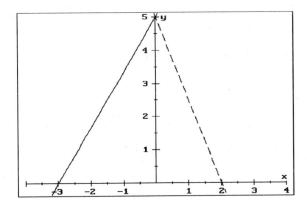

$M = \displaystyle\int_{-3}^0 \frac{5}{3}(x + 3)\,dx + \int_0^2 -\frac{5}{2}(x - 2)\,dx = \frac{5}{3}\left(\frac{x^2}{2} + 3x\right)\Big|_{-3}^0 - \frac{5}{2}\left(\frac{x^2}{2} - 2x\right)\Big|_0^2$

$= \frac{5}{3}\left(\frac{9}{2}\right) - \frac{5}{2}(-2) = \frac{25}{2}$, which we knew since $\rho = 1$ means $M = A.$

25. (con't.)

$$M_x = \frac{1}{2}\int\limits_{-3}^{0} \left(\frac{5}{3}(x+3)\right)^2 dx + \frac{1}{2}\int\limits_{0}^{2}\left(-\frac{5}{2}(x-2)\right)^2 dx$$

$$= \frac{25}{18}\left(\frac{x^3}{3} + 3x^2 + 9x\right)\Big|_{-3}^{0} + \frac{25}{8}\left(\frac{x^3}{3} - 2x^2 + 4x\right)\Big|_{0}^{2}$$

$$= \frac{25}{18}(9) + \frac{25}{8}\left(\frac{8}{3}\right) = \frac{125}{6}$$

$$\bar{y} = \frac{M_x}{M} = \frac{\frac{125}{6}}{\frac{25}{2}} = \frac{5}{3}, \quad \text{(as before)}.$$

29.

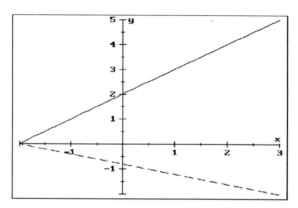

The median from $(-2, 0)$ to $(3, \frac{3}{2})$ has a vertical change of $\frac{3}{2}$.
$\frac{2}{3}$ of this gives $\bar{y} = 1$. The horizontal change is 5. $\frac{2}{3}$ of this is $\frac{10}{3}$. $\frac{10}{3}$ units
from -2 gives $\bar{x} = \frac{4}{3}$. Note that by similar triangles $\frac{2}{3}$ of the horizontal or
vertical changes will also give the values for $\frac{2}{3}$ of the median.
The centroid of the triangle is at $\left(\frac{4}{3}, 1\right)$.

Now to check by using calculus:

$$M = A = \frac{1}{2}(7)(5) = \frac{35}{2}.$$

$$M_x = \frac{1}{2}\int\limits_{-2}^{3}\left((x+2)^2 - \left\{\frac{2}{5}(-x-2)\right\}^2\right)dx = \frac{1}{2}\int\limits_{-2}^{3}\left(\frac{21}{25}(x+2)^2\right)dx$$

$$= \frac{21}{50}\left(\frac{x^3}{3} + 2x^2 + 4x\right)\Big|_{-2}^{3} = \frac{21}{50}\left(39 + \frac{8}{3}\right) = \frac{35}{2}.$$

$$\bar{y} = \frac{M_x}{M} = \frac{\frac{35}{2}}{\frac{35}{2}} = 1$$

29. (con't.)

$$M_y = \int_{-2}^{3} x\left((x + 2) - \tfrac{2}{5}(-x - 2)\right) dx = \int_{-2}^{3} x\left(\tfrac{7}{5}(x + 2)\right) dx$$

$$= \tfrac{7}{5}\left(\tfrac{x^3}{3} + x^2\right)\Big|_{-2}^{3} = \tfrac{7}{5}\left(18 - \tfrac{4}{3}\right) = \tfrac{70}{3}.$$

$$\bar{x} = \frac{M_y}{M} = \frac{\tfrac{70}{3}}{\tfrac{35}{2}} = \tfrac{4}{3}$$

33. No work will be required, as the tank will empty with the force of gravity.

37.

$$W = \rho \int_{a}^{b} h\, dV = 40 \int_{0}^{2} (12 - y)\pi(3)^2 dy = 360\pi \int_{0}^{2} (12 - y)\, dy$$

$$= 360\pi\left(12y - \tfrac{y^2}{2}\right)\Big|_{0}^{2} = 360\pi(22) = 7920\pi \approx 24{,}881 \text{ ft-lbs.}$$

41. Let F_1 be the force on the top, F_2 the force on the larger vertical side, and F_3 be the force on the smaller vertical side, then $F = F_1 + 2F_2 + 2F_3$.

$$F_1 = 62.4\left(\tfrac{2}{12}\right)\left(\tfrac{3}{12}\right)(9.5) = 24.7 \text{ lb}$$

$$F_2 = \int_{9.5}^{10} 62.4\left(\tfrac{3}{12}\right) x\, dx = 62.4\left(\tfrac{1}{4}\right)\tfrac{x^2}{2}\Big|_{9.5}^{10} = 7.8(100 - 90.25) = 76.05 \text{ lb}$$

$$F_3 = \int_{9.5}^{10} 62.1\left(\tfrac{2}{12}\right) x\, dx = 62.4\left(\tfrac{1}{6}\right)\tfrac{x^2}{2}\Big|_{9.5}^{10} = 5.2(100 - 90.25) = 50.7 \text{ lb}$$

$$F = 24.7 + 2(76.05) + 2(50.7) = 278.2 \text{ lb}$$

45.

$$I_x = \int_{0}^{2} y^2(4 - y^2)\, dy = \int_{0}^{2} (4y^2 - y^4)\, dy = \left(\tfrac{4y^3}{3} - \tfrac{y^5}{5}\right)\Big|_{0}^{2} = \tfrac{32}{3} - \tfrac{32}{5} = \tfrac{64}{15}.$$

$$I_y = \int_{0}^{4} x^2(\sqrt{x})\, dx = \frac{x^{7/2}}{\tfrac{7}{2}}\Big|_{0}^{4} = \tfrac{2}{7}(128) = \tfrac{256}{7}.$$

51. Analysis: Taking vertical slices from $x = 0$ to $x = r$, we get disks, each with a radius of y, where $y = x$.

$$V = \int_{0}^{r} \pi x^2\, dx = \tfrac{1}{3}\pi r^3$$

53. Analysis:

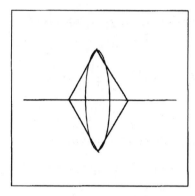

The resulting volume will be congruent cones, base to base. Find the volume of one of the cones by adding the disks from $x = 0$ to $x = \frac{L}{2}$. The radius of each disk is y, where $y = \sqrt{3}\,x$.

$$V = 2\int_0^{L/2} \pi(\sqrt{3}\,x)^2\,dx = \frac{\pi L^3}{4}$$

6.4 Growth, Decay, and First-Order Linear Differential Equations

1. Separating variables: $\frac{dy}{y} = 3\,dx$, $\int \frac{dy}{y} = \int 3\,dx$, $\ln y = 3x + C$,
$y = e^{3x+C}$, $y = Ce^{3x}$

5. Separating variables: $\frac{dy}{y} = \frac{dx}{x}$, $\int \frac{dy}{y} = \int \frac{dx}{x}$, $\ln y = \ln x + C$,
$\ln y = \ln x + \ln e^C = \ln e^C x$, $y = e^C x$, but e^C is just another constant C,
so $y = Cx$.

9. Separating variables: $\frac{dy}{y + 10} = dx$, $\int \frac{dy}{y + 10} = \int dx$,

$\ln(y + 10) = x + C$, $y + 10 = e^{x+C} = Ce^x$, $y = Ce^x - 10$

SURVIVAL HINT

First order linear differential equations must be done carefully. Take the time to write out the values for $P(x)$, $Q(x)$, and $I(x)$, before substituting into the formula.

13. This is a first order linear differential equation with $P(x) = \frac{2}{x}$ and

$$Q(x) = \sqrt{x} + 1, \quad I(x) = e^{\int \frac{2}{x}\,dx} = e^{2\ln x} = e^{\ln x^2} = x^2$$

$$y = \frac{1}{I(x)}\left(\int Q(x)I(x)\,dx + C\right) = \frac{1}{x^2}\left(\int (\sqrt{x} + 1)x^2\,dx + C\right)$$

$$= \frac{1}{x^2}\left(\frac{2}{7}x^{7/2} + \frac{x^3}{3} + C\right) = \frac{2}{7}x^{3/2} + \frac{x}{3} + \frac{C}{x^2}$$

17. This is a first order linear differential equation with $P(x) = 2 + \frac{1}{x}$ and

$$Q(x) = e^{-2x}, \quad I(x) = e^{\int (2 + \frac{1}{x})\,dx} = e^{2x + \ln x} = e^{2x}e^{\ln x} = e^{2x}x$$

$$y = \frac{1}{I(x)}\left(\int Q(x)I(x)\,dx + C\right) = \frac{1}{e^{2x}x}\left(\int e^{-2x}e^{2x}x\,dx + C\right)$$

$$= \frac{1}{xe^{2x}}\left(\frac{x^2}{2} + C\right) = \frac{x}{2e^{2x}} + \frac{C}{xe^{2x}}$$

21. To find the slope of the orthogonal curve we need y'. Using implicit differentiation: $xy' + y = 0$, $y' = -\frac{y}{x}$. The slope of the orthogonal trajectory is the negative reciprocal: $\frac{dy}{dx} = \frac{x}{y}$. This is a separable differential equation: $y\,dy = x\,dx$. Integrating we have: $\frac{y^2}{2} = \frac{x^2}{2} + C$. $y^2 = x^2 + C$

SURVIVAL HINT

Nearly every growth and decay problem, that is, where the rate of change is proportional to the amount present, can be solved with the equation: $Q(t) = Q_0 e^{kt}$. If k is not given it can be found with a data point.

25. $0.28\,Q_0 = Q_0 e^{-kt}$, where for C^{14}, $k = \frac{\ln 2}{5,730}$

$0.28 = e^{-\frac{\ln 2}{5,730}t}$, $\ln 0.28 = -\frac{\ln 2}{5,730}t$, $t = -5,730\frac{\ln 0.28}{\ln 2} \approx 10,523$ years

29. The rate of change proportional to the present quantity gives the relationship:

$Q = Q_0 e^{-kt}$, $1.105\,Q_0 = Q_0 e^{k}$, $k = \ln 1.105$ For 12 years later:

$Q = Q_0 e^{-kt}$, $Q = 2,487,000\,e^{12\ln 1.105}$, $Q = 2,487,000\,(1.105)^{12}$

$Q \approx 8,241,820$ marriages in 1996

33. The solution to the differential equation for current is given on page 496:

$$I(t) = \frac{E}{R}\left(1 - e^{-Rt/L}\right)$$

$$\lim_{t\to\infty} I(t) = \lim_{t\to\infty}\frac{E}{R} - \lim_{t\to\infty}\frac{E}{R}e^{-Rt/L} = \frac{E}{R}$$

If L is doubled but E and R are held constant, there is no change in the long range current in the circuit.

37. Of course he lands safely. If he did not there would be no more spy problems for the rest of the text! To verify this:

For the first 40 seconds of free fall $k = 0.75$, and $m = \dfrac{W}{g} = \dfrac{192}{32} = 6$, $s_0 = 0$, $v_0 = 0$. Putting these values into the formula for motion of a falling body through a resisting medium:

$v(t) = \dfrac{mg}{k} + \left(v_0 - \dfrac{mg}{k}\right) e^{-kt/m}$ we have

$v(40) = \dfrac{192}{0.75} + \left(0 - \dfrac{192}{0.75}\right) e^{-0.75(40)/6} \approx 254.3$ as the velocity when the parachute opens. To find the distance he has fallen during the 40 seconds we need the formula for $s(t)$, which can be found by integrating $v(t)$:

$s(t) = \dfrac{mgt}{k} + \left(\dfrac{mv_0}{k} - \dfrac{gm^2}{k^2}\right)(1 - e^{-kt/m}) + s_0$

$s(40) = \dfrac{192(40)}{0.75} + \left[\dfrac{6(0)}{0.75} - \dfrac{32(6^2)}{(0.75)^2}\right](1 - e^{-0.75(40)/6}) + 0$

$s(40) \approx 8{,}205.8$ ft The parachute now opens and he is $10{,}000 - 8{,}206.8 = 1793.2$ ft from the ground. We will now find how many seconds later the parachute will have slowed his velocity to 20 ft/s. If the distance traveled in this time is less than 1793.2 he will land safely.

$20 = \dfrac{192}{10} + \left(254.3 - \dfrac{192}{10}\right) e^{-10t/6}$

$e^{-10t/6} = 0.00340$

$-\dfrac{5t}{3} = \ln 0.00340$

$t \approx 3.41$ sec Now find the distance traveled in this time using the parachute.

$s(3.41) \approx \dfrac{192(3.41)}{10} + \left[\dfrac{6(254.3)}{10} - \dfrac{32(6^2)}{(10)^2}\right](1 - e^{-10(3.41)/6}) + 0$

≈ 206.1 ft He will land safely, but will be floating down with his parachute much too long and may be spotted. He should have spent longer in free fall.

41. Analysis: **a.** If $u = y^{1-n}$, then $u'(x) = (1 - n)y^{-n}y'(x)$, and

$y'(x) = \dfrac{u'(x)y^n}{(1 - n)}$ Substituting:

$\dfrac{u'(x)y^n}{(1 - n)} + P(x)y = Q(x)y^n$ Now multiply by $\dfrac{1 - n}{y^n}$

b. Let $u = y^{1-n} = y^{1-2} = y^{-1}$

43.   Analysis: **a.** Dividing by m gives a first order linear differential equation.

Solving it you should get:

$$v = \frac{mg}{k} + e^{-kt/m}\left(v_0 - \frac{mg}{k}\right)$$

b. Integrating this function with respect to v gives a function for s.

Use the initial value to find the value for C.

$$s(t) = \frac{mg}{k} t + \frac{m}{k}\left(v_0 - \frac{mg}{k}\right)\left(1 - e^{-kt/m}\right) + s_0$$

Chapter 6 Review

PRACTICE PROBLEMS

13.-18.   The definite integrals could represent the following:

A.   Disks revolved about the y-axis.

B.   Disks revolved about the x-axis.

C.   Slices taken perpendicular to the x-axis.

D.   Slices taken perpendicular to the y-axis.

E.   Mass of a lamina with density π.

F.   Washers taken along the x-axis.

G.   Washers taken along the y-axis.

13.   All but E are formulas for volumes of solids.

14.   A, B, F, G

15.   F, G

16.   C, D

17.   A, F

18.   B, G

19.   **a.**

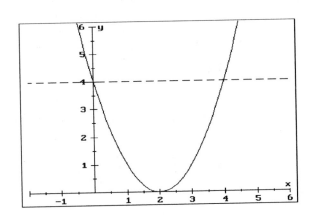

19. (con't.)

Making use of the symmetry about the line $x = 2$:

$$A = 2 \int_0^2 [4 - (x - 2)^2] \, dx = 2 \int_0^2 (4x - x^2) \, dx = 2\left(2x^2 - \frac{x^3}{3}\right)\Big|_0^2$$

$$= 2\left(8 - \frac{8}{3}\right) = \frac{32}{3}$$

b. If slices are taken perpendicular to the x-axis we get washers.

$$V = 2\pi \int_0^2 [R^2 - r^2] \, dx = 2\pi \int_0^2 [4^2 - (x - 2)^4] \, dx$$

$$= 2\pi \int_0^2 (- x^4 + 8x^3 - 24x^2 + 32x) \, dx$$

$$= 2\pi\left(-\frac{x^5}{5} + 2x^4 - 8x^3 + 16x^2\right)\Big|_0^2 = 2\pi\left(-\frac{32}{5} + 32 - 64 + 64\right)$$

$$= 2\pi\left(32 - \frac{32}{5}\right) = \frac{256\pi}{5}$$

c. If slices are taken perpendicular to the x-axis we get cylindrical shells.

$$V = \int_0^4 2\pi rh \, dx = 2\pi \int_0^4 x[4 - (x - 2)^2] \, dx = 2\pi \int_0^4 x(4x - x^2) \, dx$$

$$= 2\pi \int_0^4 (4x^2 - x^3) \, dx = 2\pi\left(\frac{4x^3}{3} - \frac{x^4}{4}\right)\Big|_0^4 = 2\pi\left(\frac{256}{3} - \frac{256}{4}\right)$$

$$= 2\pi\left(\frac{256}{12}\right) = \frac{128\pi}{3}$$

20. Since the base of the solid is a circle with center at the origin, we can use symmetry and compute the volume from $x = 0$ to $x = 2$, where the base $= 2y = 2\sqrt{4 - x^2}$, and the height $= 4\sqrt{4 - x^2}$

$$V = 2 \int_0^2 (\text{base})(\text{height}) \, dx = 2 \int_0^2 2\sqrt{4 - x^2}\, 4\sqrt{4 - x^2} \, dx = 16 \int_0^2 (4 - x^2) \, dx$$

$$= 16\left(4x - \frac{x^3}{3}\right)\Big|_0^2 = 16\left(8 - \frac{8}{3}\right) = \frac{256}{3}$$

21. $s = \int_a^b \sqrt{1 + [f'(x)]^2} \, dx = \int_0^1 \sqrt{1 + \left(-\frac{3\sqrt{x}}{2}\right)^2} \, dx = \int_0^1 \frac{1}{2}\sqrt{4 + 9x}$

$$= \frac{1}{18} \int_0^1 (4 + 9x)^{1/2}(9x \, dx) = \frac{1}{18} \frac{(4 + 9x)^{3/2}}{\frac{3}{2}}\Big|_0^1 = \frac{1}{27}(13\sqrt{13} - 8)$$

22. Each element of revolved surface is a frustum of a cone $= 2\pi r l$, where

$r = y = \sqrt{x}$ and $l = \sqrt{1 + [f'(x)]^2}$. $f'(x) = \frac{1}{2\sqrt{x}}$.

$$s = 2\pi \int_0^1 \sqrt{x} \sqrt{1+\left(\frac{1}{2\sqrt{x}}\right)^2} \, dx = 2\pi \int_0^1 \sqrt{x} \sqrt{\frac{4x+1}{4x}} \, dx = \pi \int_0^1 \sqrt{4x+1} \, dx$$

$$= \frac{\pi}{4} \int_0^1 (4x+1)^{1/2}(4 \, dx) = \frac{\pi}{4} \frac{(4x+1)^{3/2}}{\frac{3}{2}} \Big|_0^1 = \frac{\pi}{6}(5\sqrt{5}-1) \approx 5.33$$

23. $P(x) = \frac{x}{x+1}$, $Q(x) = e^{-x}$. $I(x) = e^{\int \frac{x}{x+1} \, dx} = e^{\int \left(1 - \frac{1}{x+1}\right) dx}$

$$= e^x e^{-\ln|x+1|} = e^x(x+1)^{-1}$$

$$y = \frac{1}{e^x(x+1)^{-1}} \left[\int e^{-x}\left(e^x(x+1)^{-1}\right) dx + C\right]$$

$$= \frac{x+1}{e^x}\left[\int \frac{dx}{x+1} + C\right] = \frac{x+1}{e^x}\left(\ln|x+1| + 1\right)$$

24. $\frac{dS}{dt} = \text{Inflow} - \text{Outflow} = 1.3(5) - \frac{S(t)}{200 + (5-3)t}$ (3)

$\frac{dS}{dt} = 6.5 - \frac{3S(t)}{200 + 2t}$. Write this as: $\frac{dS}{dt} + \frac{3S}{200 + 2t} = 6.5$ and we have

a first order linear differential equation with $P(t) = \frac{3}{200 + 2t}$ and $Q(t) = 6.5$

$I(t) = e^{\int \frac{3}{200 + 2t} \, dt} = e^{\frac{3}{2} \ln(t + 100)} = (t + 100)^{3/2}$

$(t + 100)^{3/2} \frac{dS}{dt} + (t + 100)^{3/2} \frac{3}{2(t + 100)} S = 6.5(t + 100)^{3/2}$

$\frac{d}{dt}\left((t + 100)^{3/2}S\right) = 6.5(t + 100)^{3/2}$,

$\int \frac{d}{dt}\left[(t + 100)^{3/2}S\right] dt = \int 6.5(t + 100)^{3/2} \, dt,$

$(t + 100)^{3/2}S = 6.5(t + 100)^{5/2}\left(\frac{2}{5}\right) + C$

But when $t = 0$, $S = 400$. $C = 400{,}000 - 260{,}000 = 140{,}000$.

$S(t) = 2.6(t + 100) + \frac{140{,}000}{(t + 100)^{3/2}}$.

The concentration of salt in the tank at time t is given by $C(t) = \frac{S(t)}{200 + 2t}$.

$$C(t) = \frac{2.6(t + 100) + \frac{140{,}000}{(t + 100)^{3/2}}}{200 + 2t} = 1.3 + \frac{70{,}000}{(t + 100)^{5/2}}$$

24. (con't.)

Notice that as we continue to dilute the mixture the minimum concentration will approach the incoming concentration of 1.3, and the maximum concentration occurs when $t = 0$. We wish to find the time required for a concentration of two-thirds the original amount of 2 lb per gallon:

$$\frac{4}{3} = 1.3 + \frac{70{,}000}{(t + 100)^{5/2}}$$

$$\frac{1}{30} = \frac{70{,}000}{(t + 100)^{5/2}}$$

$$\frac{1}{30}(t + 100)^{5/2} = 70{,}000$$
$$(t + 100)^{5/2} = 2{,}100{,}000$$
$$t = \left(2{,}100{,}000\right)^{2/5} - 100 \approx 238 \text{ min or 3 hr 58 min}$$

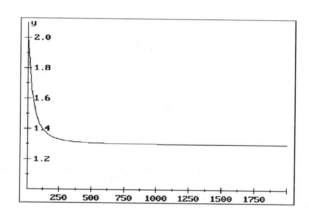

25.

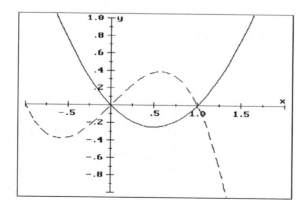

25. (con't.)

Since no density function is given we can assume that the lamina is homogeneous

and $M = \int_0^1 [(x - x^3) - (x^2 - x)]\,dx = \int_0^1 (2x - x^2 - x^3)\,dx$

$= x^2 - \dfrac{x^3}{3} - \dfrac{x^4}{4}\Big|_0^1 = \dfrac{5}{12}$

$M_y = \int_0^1 x(2x - x^2 - x^3)\,dx = \int_0^1 (2x^2 - x^3 - x^4)\,dx = \dfrac{2x^3}{3} - \dfrac{x^4}{4} - \dfrac{x^5}{5}\Big|_0^1$

$= \dfrac{13}{60}$

$M_x = \dfrac{1}{2}\int_0^1 [(x - x^3)^2 - (x^2 - x)^2]\,dx = \dfrac{1}{2}\int_0^1 (x^6 - 3x^4 + 2x^3)\,dx$

$= \dfrac{1}{2}\left(\dfrac{x^7}{7} - \dfrac{3x^5}{5} + \dfrac{x^4}{2}\right)\Big|_0^1 = \dfrac{3}{140}$

$\bar{x} = \dfrac{M_y}{M} = \dfrac{\frac{13}{60}}{\frac{5}{12}} = \dfrac{13}{25}; \quad \bar{y} = \dfrac{M_x}{M} = \dfrac{\frac{3}{140}}{\frac{5}{12}} = \dfrac{9}{175}$

SURVIVAL HINT

The Cumulative Review for Chapters 1-6 can be very valuable to refresh some
of the skills and concepts that you may not have been using often. It is also
a valuable tool to prepare for a final exam. If you do not have the time to actually
do all of the problems, try looking at each one to see if you recall the concept
involved and how to proceed with the solution. If you are confident about your
ability to solve the problem, do not spend the time. If you feel a little uncertain
about the problem, refer back to the appropriate section, review the concepts,
look in your old homework for a similar problem, and then see if you can work it.
Be more concerned about understanding the concept than about getting exactly the
right answer. Do not spend a lot of your time looking for algebra and arithmetic errors.

Chapter 7
Methods of Integration

7.1 Review of Substitution and Integration by Table

SURVIVAL HINT

You will have fewer errors if on each substitution problem you will take the time to write down $u = \ldots$, $du = \ldots$, and carefully make the substitutions. On definite integrals do not forget to change the limits when you make a u substitution.

3. Let $u = \ln x$, $du = \frac{dx}{x}$. $\int u \, du = \frac{u^2}{2} + C = \frac{1}{2} (\ln x)^2 + C$

7. Let $u = t^3$, $du = 3t^2 \, dt$. $\frac{1}{3} \int \frac{du}{3^2 + u^2} = \frac{1}{3} \left(\frac{1}{3} \tan^{-1} \frac{u}{3} \right) + C$
$= \frac{1}{9} \tan^{-1} \frac{t^3}{3} + C$

11. Note that the numerator is the derivative of the denominator, giving $\frac{du}{u}$ form.
$\int \frac{2x + 4}{x^2 + 4x + 3} \, dx = \ln |x^2 + 4x + 3| + C$

15. $\int x \ln x \, dx = \frac{x^2}{2} \left(\ln |x| - \frac{1}{2} \right) + C \quad (\#502)$

19. $\int \frac{x^2 \, dx}{\sqrt{x^2 + 1}} = \frac{x\sqrt{x^2 + 1}}{2} - \frac{1}{2} \ln \left| x + \sqrt{x^2 + 1} \right| + C \quad (\#174)$

23. $\int e^{-4x} \sin 5x \, dx = \int e^{au} \sin bu \, du$ with $a = -4$ and $b = 5$.

$= \frac{e^{-4x}(-4 \sin 5x - 5 \cos 5x)}{41} + C \quad (\#492)$

$= \frac{-e^{-4x}(4 \sin 5x + 5 \cos 5x)}{41} + C$

In problems 27-43 the tables in *Mathematics Handbook for CALCULUS* are used.

27. This could be expanded to give a 4th degree polynomial, but it is easier to

use (#31) with $a = 1$, $b = 1$ and $n = 3$:

$$\int x(1 + x)^3 dx = \frac{(x + 1)^5}{5} - \frac{(x + 1)^4}{4} + C.$$

This can be simplified to: $\dfrac{(x + 1)^4(4x - 1)}{20} + C.$

31. Complete the square in the denominator to get one of the standard forms:

$$\int \frac{dx}{\sqrt{9 - (4 + 4x + x^2)}} = \int \frac{dx}{\sqrt{3^2 - (x + 2)^2}} \text{ which is (#22) with } a = 3$$

and $u = (x + 2)$. $\displaystyle\int \frac{dx}{\sqrt{5 - 4x - x^2}} = \sin^{-1}\left(\frac{x + 2}{3}\right) + C$

35. By formula (#22) with $a = 3$: $\displaystyle\int \frac{dx}{\sqrt{9 - x^2}} = \sin^{-1}\frac{x}{3} + C$

By trigonometric substitution: let $x = 3 \sin \theta$.

Then $\sqrt{9 - x^2} = \sqrt{9 - 9 \sin^2\theta} = 3 \cos \theta$, and $dx = 3 \cos \theta \, d\theta$.

$$\int \frac{dx}{\sqrt{9 - x^2}} = \int \frac{3 \cos \theta}{3 \cos \theta} \, d\theta = \theta + C = \sin^{-1}\left(\frac{x}{3}\right) + C.$$

39. (#179) with $u = 2x$, $du = 2 \, dx$. Write as: $\displaystyle\int \frac{\sqrt{(2x)^2 + 1} \, (2 \, dx)}{2x}$

$$= \sqrt{4x^2 + 1} - \ln\left|\frac{1 + \sqrt{4x^2 + 1}}{2x}\right| + C$$

43. (#245) with $a = 3$, $u = x$.

$$\int (9 - x^2)^{3/2} dx = \frac{x(9 - x^2)^{3/2}}{4} + \frac{27x(9 - x^2)^{1/2}}{8} + \frac{243}{8} \sin^{-1}\left(\frac{x}{3}\right) + C$$

47. Let $u = \sin x$, $du = \cos x \, dx$. We then have form $u^4 \, du$.

$$\int \sin^4 x \cos x \, dx = \frac{u^5}{5} + C = \frac{1}{5} \sin^5 x + C$$

57.   The common denominator for $\frac{1}{2}$ and $\frac{1}{3}$ is 6, so let $x = u^6$, then $dx = 6u^5 du$.

$$\int \frac{4\,dx}{x^{1/3} + 2x^{1/2}} = \int \frac{4(6u^5\,du)}{u^2 + 2u^3} = 24 \int \frac{u^5\,du}{u^2 + 2u^3} = 24 \int \frac{u^3\,du}{2u + 1}$$

Since the degree of the numerator is greater than the degree of the donominator, do the division:

$$= 24 \int \left(\frac{1}{2}u^2 - \frac{1}{4}u + \frac{1}{8} - \frac{\frac{1}{8}}{2u + 1} \right) du$$

$$= 24 \left(\frac{u^3}{6} - \frac{u^2}{8} + \frac{u}{8} - \frac{1}{16} \ln|2u + 1| \right) + C$$

$$= 4x^{1/2} - 3x^{1/3} + 3x^{1/6} - \frac{3}{2} \ln \left| 2x^{1/6} + 1 \right| + C$$

61.

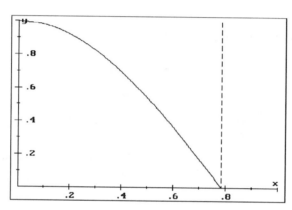

Since $x = \frac{\pi}{4}$ intersects $y = 0$ at the same point as $y = \cos 2x$, we are finding the area under $y = \cos 2x$ from $x = 0$ to $x = \frac{\pi}{4}$.

$$A = \int_0^{\frac{\pi}{4}} \cos 2x \, dx = \frac{1}{2} \int_0^{\frac{\pi}{4}} \cos 2x \, (2 \, dx) = \frac{1}{2} \sin 2x \Big|_0^{\frac{\pi}{4}} = \frac{1}{2} \text{ square units.}$$

65. $\sqrt{1 + (y')^2} = \sqrt{1 + (2x)^2}$. $s = \int_0^1 \sqrt{1 + (2x)^2} \, dx$

This is formula #168 with $a = 1$, $u = 2x$ and $du = 2\,dx$:

$$\frac{1}{2} \int_0^1 \sqrt{1+(2x)^2} \, (2 \, dx) = \frac{1}{2} \left(\frac{2x\sqrt{1+(2x)^2}}{2} + \frac{1}{2} \ln \left| 2x + \sqrt{1+(2x)^2} \right| \right) \Big|_0^1$$

$$= \frac{\sqrt{5}}{2} + \frac{1}{4} \ln(2 + \sqrt{5}) \approx 1.4789$$

69.   Analysis: When the Hint is followed you should have (except for sign) the form $\frac{du}{u}$.

71.   Analysis: Let $u = \pi - x$, $du = -dx$. Substitute and expand, remembering

that $\sin(\pi - u) = \sin u$. Note change to new limits.

$$\int_0^\pi x\, f(\sin x)\ dx = \int_\pi^0 (\pi - u)\, f(\sin u)\ (-du)$$

$$= \int_0^\pi (\pi - u)\, f(\sin u)\, du$$

$$= \int_0^\pi \pi\, f(\sin u)\ du - \int_0^\pi u\, f(\sin u)\ du$$

But the value of the integral is independent of the variable used so

$$= \pi \int_0^\pi f(\sin x)\ dx - \int_0^\pi x\, f(\sin x)\ dx$$

Add the second integral to both sides and divide by 2.

73.   Analysis: If $m = n$ we have $\displaystyle\int_{-\pi}^\pi \cos^2 mx\ dx$. Use an identity to integrate.

If $m \neq n$ we have $\displaystyle\int_{-\pi}^\pi \cos nx \cos mx\ dx = \frac{1}{2}\int_{-\pi}^\pi [\cos(m + n)x + \cos(m - n)x]\, dx$

from the product to sum identities learned in trigonometry. Integrate and

evaluate these two expressions.

7.2 Integration by Parts

SURVIVAL HINT

When selecting u and dv, a necessary criteria is the integrability of dv, a desirable,

but not necessary criteria is that the derivative of u be simpler.

3. Let $u = \ln x$ and $dv = x \, dx$ (note that a differential must always be equal to another differential). Then $du = \frac{1}{x} \, dx$ and $v = \int x \, dx = \frac{x^2}{2}$.

$$\int x \ln x \, dx = uv - \int v \, du$$
$$= \frac{x^2}{2} \ln x - \int \left(\frac{x^2}{2}\right)\left(\frac{1}{x}\right) dx$$
$$= \frac{x^2}{2} \ln x - \frac{x^2}{4} + C$$

7. Let $u = \ln \sqrt{x}$ and $dv = x^{-1/2} \, dx$ (note that a differential must always be equal to another differential). Then $du = \frac{1}{\sqrt{x}}\left(\frac{1}{2\sqrt{x}}\right) dx$ and $v = 2\sqrt{x}$.

$$\int \frac{\ln \sqrt{x}}{\sqrt{x}} \, dx = uv - \int v \, du = 2\sqrt{x} \ln \sqrt{x} - \int 2\sqrt{x} \frac{1}{\sqrt{x}}\left(\frac{1}{2\sqrt{x}}\right) dx$$
$$= 2\sqrt{x} \ln \sqrt{x} - 2\sqrt{x} + C = \sqrt{x} \ln x - 2\sqrt{x} + C$$

11. Let $u = \ln x$ and $dv = x^2 \, dx$ (note that a differential must always be equal to another differential). Then $du = \frac{1}{x} \, dx$ and $v = \int x^2 \, dx = \frac{x^3}{3}$.

$$\int x^2 \ln x \, dx = uv - \int v \, du = \frac{x^3}{3} \ln x - \int \left(\frac{x^3}{3}\right)\left(\frac{1}{x}\right) dx$$
$$= \frac{x^3}{3} \ln x - \frac{x^3}{9} + C = \frac{x^3}{9}(3 \ln x - 1) + C$$

15. Let $u = \ln x$ and $dv = \sqrt{x} \, dx$. Then $du = \frac{1}{x} \, dx$ and $v = \frac{2}{3} x^{3/2}$.

$$\int_1^4 \sqrt{x} \ln x \, dx = \frac{2}{3} x^{3/2} \ln x \Big|_1^4 - \int_1^4 \frac{2}{3} \sqrt{x} \, dx = \frac{2}{3} x^{3/2} \ln x \Big|_1^4 - \frac{4}{9} x^{3/2} \Big|_1^4$$

$$= \frac{16}{3} \ln 4 - \frac{32}{9} + \frac{4}{9} = \frac{16}{3} \ln 4 - \frac{28}{9}$$

19. Either function can be readily integrated or differentiated, so any choice will work. Two differentiations of $\cos 2x$ will return us to a $\cos 2x$, which can then be combined with the original integral on the left of the equation. Let $u = \cos 2x$ and $dv = e^{2x} dx$. Then $du = -2 \sin 2x \, dx$ and $v = \frac{1}{2} e^{2x}$.

$$\int_0^\pi e^{2x} \cos 2x \, dx = \frac{1}{2} e^{2x} \cos 2x \Big|_0^\pi - \int_0^\pi \frac{1}{2} e^{2x}(-2 \sin 2x) \, dx$$

$$= \frac{1}{2} e^{2x} \cos 2x \Big|_0^\pi + \int_0^\pi e^{2x} \sin 2x \, dx. \text{ Again let } u = \sin 2x$$

and $dv = e^{2x} dx$. Then $du = 2 \cos 2x \, dx$ and $v = \frac{1}{2} e^{2x}$. Now we have:

$$\int_0^\pi e^{2x} \cos 2x \, dx = \frac{1}{2} e^{2x} \cos 2x \Big|_0^\pi + \frac{1}{2} e^{2x} \sin 2x \Big|_0^\pi - \int_0^\pi \frac{1}{2} e^{2x}(2 \cos 2x) \, dx$$

19. (con't.)

Now combine this final integral with the original one on the left of the equation:

$$2 \int_0^\pi e^{2x}\cos 2x \, dx \;=\; \tfrac{1}{2} \, e^{2x}\cos 2x \, \Big|_0^\pi \;+\; \tfrac{1}{2} \, e^{2x}\sin 2x \, \Big|_0^\pi. \;\text{ Dividing by 2:}$$

$$\int_0^\pi e^{2x}\cos 2x \, dx \;=\; \tfrac{1}{4} \, e^{2x}\cos 2x \, \Big|_0^\pi \;+\; \tfrac{1}{4} \, e^{2x}\sin 2x \, \Big|_0^\pi$$

$$= \tfrac{1}{4}\Big(e^{2\pi} - 1\Big)$$

23. Make use of the identity $\sin 2x = 2 \sin x \cos x$, and let $u = \cos x$,

$du = -\sin x \, dx$. Our integral now becomes:

$$-2 \int u \ln u \, du \;=\; -2\Big(\tfrac{u^2}{2} \ln u - \tfrac{u^2}{4} + C\Big) \text{ (see problem \#3)}$$

Substituting our x function for the u, and doing some factoring:

$$= \tfrac{1}{2} \cos^2 x - (\cos^2 x) \ln |\cos x| + C$$

27. Let $u = \sin x$, $dv = \sin x \, dx$. Then $du = \cos x \, dx$ and $v = -\cos x$.

$$\int \sin^2 x \, dx \;=\; -\sin x \cos x + \int \cos^2 x \, dx.$$

Now make use of the identity $\cos^2 x = 1 - \sin^2 x$:

$$\int \sin^2 x \, dx \;=\; -\sin x \cos x + \int (1 - \sin^2 x) \, dx$$

$$= -\sin x \cos x + x - \int \sin^2 x \, dx$$

Combining this final integral with the original one on the left of the equation:

$$2 \int \sin^2 x \, dx \;=\; -\sin x \cos x + x + C$$

$$= -\tfrac{1}{2} \sin x \cos x + \tfrac{x}{2} + C \text{ or } -\tfrac{1}{4} \sin 2x + \tfrac{x}{2} + C$$

31. $Q(t) = \int_0^3 100t e^{-0.5t} dt.$ Let $u = 100t$ and $dv = e^{-0.5t} dt.$

Then $du = 100 \, dt$ and $v = -2 \, e^{-0.5t}$

$$\int_0^3 100t e^{-0.5t} \;=\; -200 \, t e^{-0.5t} \, \Big|_0^3 - 400 \int_0^3 e^{-0.5t}\big(-\tfrac{1}{2}\big) dt$$

$$= -200 \, t e^{-0.5t} - 400 \, e^{-0.5t} \, \Big|_0^3$$

$$= -200 \, e^{-0.5t}(t + 2) \, \Big|_0^3$$

$$= -1000 \, e^{-1.5} + 400 \;\approx\; 177 \text{ units}$$

35.   **a.**   When slices are taken perpendicular to the x-axis we get disks.

$$V = \pi \int_1^e (\ln x)^2 dx. \text{ Let } u = (\ln x)^2, \quad dv = dx.$$

Then $du = 2 (\ln x) \left(\frac{dx}{x}\right)$ and $v = x.$

$$\pi \int_1^e (\ln x)^2 dx = \pi x (\ln x)^2 - 2\pi \int_1^e x \ln x \left(\frac{dx}{x}\right)$$

$$= \pi x (\ln x)^2 - 2\pi (x \ln x - x) \Big|_1^e$$

$$= \pi x (\ln x)^2 - 2\pi x \ln x + 2x \Big|_1^e$$

$$= \pi (e - 2e + 2e - 2) = \pi (e - 2)$$

$$V = \pi \int_1^e (\ln x)^2 dx = \pi (e - 2)$$

b.   If vertical slices are taken we get cylindrical shells.

$$V = 2\pi \int_1^e x \ln x \, dx$$

$$= 2\pi \left[\frac{x^2}{4}(2 \ln x - 1)\right] \Big|_1^e \quad (\text{see problem } \#3)$$

$$= 2\pi \left[\frac{e^2}{4} + \frac{1}{4}\right] = \frac{\pi}{2}(e^2 + 1)$$

39.   This is a first order linear differential equation with $P(x) = \dfrac{2x}{1 + x^2}$ and

$Q(x) = \sin x.$ $I(x) = e^{\int \frac{2x}{1 + x^2} dx} = x^2 + 1.$

$$y = \frac{1}{x^2 + 1} \int (x^2 + 1) \sin x \, dx = \frac{1}{x^2 + 1} \left[\int \sin x \, dx + \int x^2 \sin x \, dx\right]$$

The first integral is simple, but the second will require integration by parts twice, or formula #344.

$$y = \frac{1}{x^2 + 1}\left[- \cos x + 2x \sin x + (2 - x^2) \cos x + Cx\right]$$

But $y = 1$ when $x = 0$ gives $C = 0.$

$$y = \frac{-x^2 \cos x + 2x \sin x + \cos x}{x^2 + 1}$$

43. Dividing by m gives a first order linear differential equation with $P(t) = \frac{k}{m}$

and $Q(t) = g$. $I(t) = e^{\int \frac{k}{m} dt} = e^{kt/m}$

$$v = e^{-kt/m} \int e^{kt/m} g \, dt$$

$$= e^{-kt/m} \frac{mg}{k} \int e^{kt/m} \left(\frac{k}{m} dt \right)$$

$$= e^{-kt/m} \left(\frac{mg}{k} e^{kt/m} + C \right)$$

$$= \frac{mg}{k} + e^{-kt/m} + C.$$

But we have initial value of $v = v_0$ when $t = 0$ so $C = v_0 - \frac{mg}{k}$.

$$v = \frac{mg}{k} + \frac{v_0 - \frac{mg}{k}}{e^{kt/m}}$$

47. The average value of a function over an interval is given by: $A = \frac{1}{b-a} \int_a^b f(x) \, dx$

$$W_{avg.} = \frac{1}{9V_1} \int_{V_1}^{10V_1} nRt \ln \frac{V}{V_1} \, dV = \frac{nRt}{9V_1} \int_{V_1}^{10V_1} (\ln V - \ln V_1) \, dV.$$

Use formula #499 on the first term, and note that the second term is a constant:

$$W_{avg.} = \frac{nRt}{9V_1} (V \ln V - V - V \ln V_1) \Big|_{V_1}^{10V_1}$$

$$= \frac{nRt}{9V_1} [10V_1(\ln 10V_1 - \ln V_1) - 9V_1] = \frac{nRt}{9V_1} [10V_1(\ln 10) - 9V_1]$$

$$= \frac{nRt}{9} (10 \ln 10 - 9) \approx 1.558 \, nRt$$

49. Analysis: Integrate by parts letting $u = x^n$ and $dv = e^x dx$.

51. Analysis: Reduction formulas #321 and #352 give a term to be evaluated at upper limit of $\frac{\pi}{2}$ and lower limit of 0, and another integral with the trigonometric function to a degree 2 less. Since the term to be evaluated has the product of $\sin x$ and $\cos x$, it will always be zero at the limits. The reduced integral has a coefficient of $\frac{n-1}{n}$. Its reduction will again have a first term of zero, and the new integral will have a coefficient of $\frac{n-3}{n-2}$. The next reduced integral will now have a coefficient of $\frac{(n-1)(n-3)}{n(n-2)}$. Continued reduction until the exponent of the final function is 0 or 1 will give the desired sequence.

7.3 The Method of Partial Fractions

1. $\dfrac{1}{x(x-3)} = \dfrac{A_1}{x} + \dfrac{A_2}{x-3} = \dfrac{A_1(x-3)+A_2x}{x(x-3)}$. In order for this equality to

be valid, $1 = A_1(x-3) + A_2x$ must be true for all values of x; including those

for which the function is undefined.

If $x = 0$: $1 = -3A_1$, $A_1 = -\frac{1}{3}$

If $x = 3$: $1 = 3A_2$, $A_2 = \frac{1}{3}$

So $\dfrac{1}{x(x-3)} = \dfrac{-\frac{1}{3}}{x} + \dfrac{\frac{1}{3}}{x-3} = \dfrac{-1}{3x} + \dfrac{1}{3(x-3)}$

5. $\dfrac{4}{x(2x+1)} = \dfrac{A_1}{x} + \dfrac{A_2}{2x+1} = \dfrac{A_1(2x+1)+A_2x}{x(2x+1)}$

So $4 = A_1(2x+1) + A_2x$ for all values of x.

If $x = 0$: $4 = A_1$, $A_1 = 4$

If $x = -\frac{1}{2}$: $4 = -\frac{1}{2}A_2$, $A_2 = -8$. $\dfrac{4}{x(2x+1)} = \dfrac{4}{x} + \dfrac{-8}{2x+1}$

9. $\dfrac{x^3 + 3x^2 + 3x - 4}{x^2(x+3)^2} = \dfrac{A_1}{x^2} + \dfrac{A_2}{x} + \dfrac{A_3}{(x+3)^2} + \dfrac{A_4}{x+3}$

$x^3 + 3x^2 + 3x - 4 = A_1(x+3)^2 + A_2x(x+3)^2 + A_3x^2 + A_4x(x+3)$

If $x = 0$: $-4 = 9A_1$, $A_1 = -\frac{4}{9}$

If $x = -3$: $-13 = 9A_3$, $A_3 = -\frac{13}{9}$

Substituting these values and doing some algebra:

$$x^3 + 3x^2 + 3x - 4 = (A_2 + A_4)x^3 + (A_1 + 6A_2 + A_3 + 6A_4)x^2$$
$$+ (6A_1 + 9A_2)x + 9A_1$$

Equating coefficients of like powers:

$A_2 + A_4 = 1$

$A_1 + 6A_2 + A_3 + 6A_4 = 3$

$6A_1 + 9A_2 = 3$

$9A_1 = -4$

Solving this system of equations, using previously found values, gives:

$A_2 = \frac{17}{27}$ and $A_4 = \frac{10}{27}$.   So our partial fractions are:

$$\dfrac{-\frac{4}{9}}{x^2} + \dfrac{\frac{17}{27}}{x} + \dfrac{-\frac{13}{9}}{(x+3)^2} + \dfrac{\frac{10}{27}}{x+3}$$

$$= \dfrac{-4}{9x^2} + \dfrac{17}{27x} - \dfrac{13}{9(x+3)^2} + \dfrac{10}{27(x+3)}$$

13. $\int \frac{dx}{x(x-3)} = -\frac{1}{3}\int \frac{dx}{x} + \frac{1}{3}\int \frac{dx}{x-3}$ (see #1 for partial fraction solution.)

$$= -\frac{1}{3}\ln|x| + \frac{1}{3}\ln|x-3| + C = \frac{1}{3}\ln\left|\frac{x-3}{x}\right| + C$$

17. $\int \frac{4\,dx}{2x^2+x} = 4\int \frac{dx}{x} - 8\int \frac{dx}{2x+1}$ (see #5 for partial fraction solution.)

$$= 4\ln|x| - 4\ln|2x+1| + C$$

$$= 4\ln\left|\frac{x}{2x+1}\right| + C$$

21. $\int \frac{x^2+1}{x^2+x-2}\,dx$ has the degree of the numerator equal to the degree of the

denominator, so we begin by dividing: $\int 1\,dx + \int \frac{-x+3}{x^2+x-2}\,dx.$

Using partial fractions on the second integral:

$$\frac{-x+3}{x^2+x-2} = \frac{A_1}{x+2} + \frac{A_2}{x-1} = \frac{A_1(x-1) + A_2(x+2)}{(x+2)(x-1)}.$$

So $-x+3 = A_1(x-1) + A_2(x+2)$.

If $x = 1$: $2 = 3A_2$, $A_2 = \frac{2}{3}$

If $x = -2$: $5 = -3A_1$, $A_1 = -\frac{5}{3}$

So $\int \frac{x^2+1}{x^2+x-2} = \int 1\,dx + -\frac{5}{3}\int \frac{dx}{x+2} + \frac{2}{3}\int \frac{dx}{x-1}$

$$= x - \frac{5}{3}\ln|x+2| + \frac{2}{3}\ln|x-1| + C$$

25. Setting up a $\frac{du}{u}$ function:

$$\int \frac{x\,dx}{(x+1)^2} = \frac{1}{2}\int \frac{2x+2-2}{(x+1)^2} = \frac{1}{2}\int \frac{2x+2}{(x+1)^2}\,dx - \int \frac{1}{(x+1)^2}\,dx$$

The first integral is of the form $\frac{du}{u}$ and the second is of the form $u^{-2}\,du$.

$$= \frac{1}{2}\ln(x+1)^2 + \frac{1}{x+1} + C = \ln|x+1| + \frac{1}{x+1} + C$$

29. Using partial fractions:

$$\frac{x}{(x+1)(x+2)^2} = \frac{A_1}{x+1} + \frac{A_2}{(x+2)^2} + \frac{A_3}{x+2}$$

$$= \frac{A_1(x+2)^2 + A_2(x+1) + A_3(x+1)(x+2)}{(x+1)(x+2)^2}$$

So $x = A_1(x+2)^2 + A_2(x+1) + A_3(x+1)(x+2)$

If $x = -1$: $-1 = A_1$

If $x = -2$: $-2 = -A_2$, $A_2 = 2$. Equating the coefficients of the x^2 term

to find the remaining value: $0 = A_1 + A_3$, $0 = -1 + A_3$, $A_3 = 1$.

29. (con't.)

$$\int \frac{x \, dx}{(x + 1)(x + 2)^2} = -\int \frac{dx}{x + 1} + 2\int \frac{dx}{(x + 2)^2} + \int \frac{dx}{x + 2}$$

$$= -\ln|x + 1| - \frac{2}{x + 2} + \ln|x + 2| + C$$

$$= \ln\left|\frac{x + 2}{x + 1}\right| - \frac{2}{x + 2} + C$$

33. This has form $\frac{du}{u}$, so $\int \frac{3x^2 - 2x + 4}{x^3 - x^2 + 4x - 4} \, dx = \ln|x^3 - x^2 + 4x - 4| + C$

39. $\int \frac{\sin x \, dx}{(1 + \cos x)^2} = -\int (1 + \cos x)^{-2}(-\sin x \, dx) = \frac{1}{1 + \cos x} + C$

SURVIVAL HINT

In general, to eliminate fractional exponents let $x = u^n$, where n is the least common denominator for all the fractional exponents.

43. To eliminate the fractional exponents let $x = u^4$, $dx = 4u^3 \, du$.

$$\int \frac{4u^3 \, du}{u - u^5} = \int \frac{-4u^2 \, du}{u^4 - 1} = \int \frac{-4u^2 \, du}{(u^2 + 1)(u + 1)(u - 1)}$$

Using partial fractions: $\frac{-4u^2}{u^4 - 1} = \frac{A_1 u + A_2}{u^2 + 1} + \frac{A_3}{u + 1} + \frac{A_4}{u - 1}$

So $-4u^2 = (A_1 u + A_2)(u^2 - 1) + A_3(u^2 + 1)(u - 1)$
$\qquad\qquad + A_4(u^2 + 1)(u + 1)$ ·

If $u = 1$: $-4 = 4A_4$, $A_4 = -1$

If $u = -1$: $-4 = -4A_3$, $A_3 = 1$

Equating coefficients: $0 = A_1 + A_3 + A_4$, $0 = A_1 + 4 - 4$, $A_1 = 0$
$\qquad\qquad\qquad -4 = A_2 - A_3 + A_4$, $-4 = A_2 - 1 - 1$, $A_2 = -2$

$$\int \frac{4u^3 \, du}{u - u^5} = \int \frac{-2 \, du}{u^2 + 1} + \int \frac{du}{u + 1} + \int \frac{-du}{u - 1}$$

$$= -2 \tan^{-1} u + \ln|u + 1| - \ln|u - 1| + C$$

Returning to the original independent variable of x:

$$\int \frac{dx}{x^{1/4} - x^{5/4}} = -2 \tan^{-1} x^{1/4} + \ln|x^{1/4} + 1| - \ln|x^{1/4} - 1| + C$$

47. If you are observant you will notice that the numerator is the opposite of the derivative of the denominator. So $-\int \dfrac{(\cos x - \sin x)\, dx}{\sin x + \cos x}$ is of the form $\dfrac{du}{u}$ and will equal $-\ln|\sin x + \cos x| + C.$

If you do not notice this, the expression can be simplified multiplying the numerator and denominator by $(\sin x + \cos x)$:

$$\int \frac{\sin^2 x - \cos^2 x}{\sin^2 x + 2\sin x \cos x + \cos^2 x}\, dx = \int \frac{-\cos 2x}{1 + \sin 2x}dx = -\frac{1}{2}\int \frac{2\cos 2x}{1 + \sin 2x}dx$$

This is now form $\dfrac{du}{u}$ so $\displaystyle\int \frac{\sin x - \cos x}{\sin x + \cos x}\, dx = -\frac{1}{2}\ln|1 + \sin 2x| + C$

If you do not remember your trigonometric identities you can use the Weierstrauss substitutions:

$$\int \frac{\sin x - \cos x}{\sin x + \cos x}\, dx = \int \frac{\left(\dfrac{2u}{1 + u^2} - \dfrac{1 - u^2}{1 + u^2}\right)\left(\dfrac{2}{1 + u^2}\right) du}{\dfrac{2u}{1 + u^2} + \dfrac{1 - u^2}{1 + u^2}}$$

Multiply numerator and denominator by $(1 + u^2)^2$:

$$= \int \frac{2(2u - 1 + u^2)\, du}{(2u + 1 - u^2)(1 + u^2)}$$

We now use partial fractions:

$$\frac{2u^2 + 4u - 2}{(2u + 1 - u^2)(1 + u^2)} = \frac{A_1 u + A_2}{1 + 2u - u^2} + \frac{A_3 u + A_4}{1 + u^2}$$

$$2u^2 + 4u - 2 = (A_1 u + A_2)(1 + u^2) + (A_3 u + A_4)(1 + 2u - u^2)$$

Expanding and equating coefficients we have the following system:

$$0 = A_1 - A_3$$
$$2 = A_2 + 2A_3 - A_4$$
$$4 = A_1 + A_3 + 2A_4$$
$$-2 = A_2 + A_4$$

The solution of this system gives: $A_1 = 2,\ A_2 = -2,\ A_3 = 2,\ A_4 = 0.$

$$\int \frac{2(2u - 1 + u^2)\, du}{(2u + 1 - u^2)(1 + u^2)} = \int \frac{2u - 2}{1 + 2u - u^2}\, du + \int \frac{2u\, du}{1 + u^2}$$

which are form $\dfrac{du}{u}$

$$= -\ln|1 + 2u - u^2| + \ln(1 + u^2) + C$$

$$= \ln\left|\frac{u^2 + 1}{1 + 2u - u^2}\right| + C.$$

Returning to our original function of x:

$$= \ln\left|\frac{\tan^2 \frac{x}{2} + 1}{1 + 2\tan \frac{x}{2} - \tan^2 \frac{x}{2}}\right| + C$$

It is a good trigonometry exercise to show that these answers are equivalent.

SURVIVAL HINT

Unless your instructor give open book exams you will find it necessary to memorize the Weierstrauss substitutions.

51. Using the Weierstrauss substitutions: $\displaystyle\int \frac{\dfrac{2}{1+u^2}\,du}{\dfrac{8u}{1+u^2} - \dfrac{3(1-u^2)}{1+u^2} - 5}$

$$= \int \frac{2\,du}{8u - 3 - 3u^2 - 5(1+u^2)} = \int \frac{2\,du}{-2u^2 + 8u - 8}$$

$$= -\int \frac{du}{u^2 - 4u + 4} = -\int \frac{du}{(u-2)^2} = \frac{1}{u-2} + C$$

$$= \frac{1}{\tan\frac{x}{2} - 2} + C$$

55. $\displaystyle A = \int_{4/3}^{7/4} \frac{dx}{x^2 - 5x + 6} = \int_{4/3}^{7/4} \frac{dx}{(x-3)(x-2)}.$

Using integration by partial fractions:

$1 = A_1(x-2) + A_2(x-3)$. If $x = 2$: $1 = -A_2$, $A_2 = -1$

If $x = 3$: $1 = A_1$

$$\int_{4/3}^{7/4} \frac{dx}{x^2 - 5x + 6} = \int_{4/3}^{7/4}\left(\frac{1}{x-3} - \frac{1}{x-2}\right)dx$$

$$= \ln\left|\frac{x-3}{x-2}\right|\Big|_{4/3}^{7/4} = \ln 5 - \ln\frac{5}{2} = \ln 2 \approx 0.6931$$

59. $\displaystyle V = \pi\int_0^4 y^2\,dx = \pi\int_0^4 \frac{x^2(4-x)}{4+x}\,dx.$ Since the degree of the numerator is greater than the degree of the denominator we will divide:

$$= \pi\int_0^4\left(-x^2 + 8x - 32 + \frac{128}{x+4}\right)dx$$

$$= \pi\left[-\frac{x^3}{3} + 4x^2 - 32x + 128\ln(x+4)\right]\Big|_0^4$$

$$= \pi\left(-\frac{64}{3} + 64 - 128 + 128\ln 8 - 128\ln 4\right) = \pi\left(-\frac{256}{3} + 128\ln 2\right)$$

$$= 128\pi\left(\ln 2 - \frac{2}{3}\right) \approx 10.648 \text{ cu units}$$

63.   Analysis: The partial fractions give the desired result without any difficulty.

Note that the a in the numerator term of the $ax + b$ fraction is necessary for

the du.

65.   Analysis: The Weierstrass substitution will give a $\frac{du}{u}$ form, but then there are

some tricky trigonometric substitutions: $\ln \left| \tan \frac{x}{2} \right| = -\ln \left| \cot \frac{x}{2} \right| = -\ln \left| \dfrac{\cos \frac{x}{2}}{\sin \frac{x}{2}} \right|$

$= -\ln \left| \dfrac{2 \cos^2 \frac{x}{2}}{2 \sin \frac{x}{2} \cos \frac{x}{2}} \right| = -\ln \left| \dfrac{\cos x + 1}{\sin x} \right| = -\ln \left| \cot x + \csc x \right|$

67.   Analysis: **a.** This says that the change is the difference between the original

concentration and the concentration at time t.

b. If we let z be the amount at time t, and let the original amounts of

α and β be A and B, then $\dfrac{dz}{dt} = k(A - z)(B - z)$. Separate the

variables and use partial fractions.

7.4 Summary of Integration Techniques

SURVIVAL HINT

It would take an unreasonable amount of time to do all of the problems in set **A**.

However it would be an excellent use of time for you, or a small study group, to look

at each problem and decide *how* to proceed. Identify one or more techniques that look

promising.

1.   Let $u = (x - x^2)$, $du = (1 - 2x)\ dx$. Introducing a negative sign gives

$u^n\ du$ form. $-\displaystyle\int \frac{(1 - 2x)\,dx}{(x - x^2)^3} = \int \frac{-dv}{u^3} = \frac{1}{2(x - x^2)^2} + C$

5.   Let $u = e^x$, $du = e^x dx$. We then have $\cot u\ du$ form.

$\displaystyle\int e^x \cot e^x\ dx = \ln \left| \sin e^x \right| + C$

9.   Do the division:

$\displaystyle\int 3 \sec t\ dt + \int 2 \tan t\ dt = 3 \ln \left| \sec t + \tan t \right| + 2 \ln \left| \sec t \right| + C$

13. The degree of the numerator is the same as the degree of the denominator, so begin by dividing:

$$\int \left(1 + \frac{x - 8}{x^2 + 9}\right) dx = \int dx + \int \frac{x \, dx}{x^2 + 9} - 8\int \frac{dx}{x^2 + 9}$$

$$= x + \tfrac{1}{2} \ln (x^2 + 9) - \tfrac{8}{3} \tan^{-1} \tfrac{x}{3} + C$$

17. Let $u = t^3$, $du = 3t^2 \, dt$. Set up $\dfrac{du}{\sqrt{1 - u^2}}$ form.

$$\tfrac{2}{3}\int \frac{3t^2 \, dt}{\sqrt{1 - (t^3)^2}} = \tfrac{2}{3} \sin^{-1} t^3 + C$$

21. Complete the square: $\displaystyle\int \frac{dx}{(x + 1)^2 + 1} = \tan^{-1}(x + 1) + C$

25. Formula #457. $\displaystyle\int \tan^{-1}x \, dx = x \tan^{-1}x - \tfrac{1}{2} \ln (x^2 + 1) + C$

29. Introduce a negative sign to have the form of formula (#445):

$$\int \cos^{-1}(-x) \, dx = -\left[(-x) \cos^{-1}(-x) - \sqrt{1 - x^2}\right] + C$$
$$= x \cos^{-1}(-x) + \sqrt{1 - x^2} + C$$

33. Let $u = \cos x$, $du = -\sin x$, and convert the remaining $\sin^2 x$ to $1 - \cos^2 x$:

$$\int \sin^3 x \cos^2 x \, dx = -\int (1 - \cos^2 x)\cos^2 x(-\sin x \, dx)$$

$$= \int (\cos^4 x - \cos^2 x)(-\sin x \, dx)$$

$$= \frac{\cos^5 x}{5} - \frac{\cos^3 x}{3} + C$$

37. Let $u = \cos x$, $du = -\sin x$, and convert the remaining $\sin^4 x$ to $(1 - \cos^2 x)^2$:

$$\int \sin^5 x \cos^4 x \, dx = -\int (1 - \cos^2 x)^2 \cos^4 x(-\sin x \, dx)$$

$$= -\int (\cos^8 x - 2 \cos^6 x + \cos^4 x)(-\sin x \, dx)$$

$$= -\frac{\cos^9 x}{9} + \frac{2 \cos^7 x}{7} - \frac{\cos^5 x}{5} + C$$

41. Use formula #235:

$$\int \frac{\sqrt{1 - x^2}}{x} \, dx = \sqrt{1 - x^2} - \ln \left|\frac{1 + \sqrt{1 - x^2}}{x}\right| + C$$

45.   Use formula (#176):

$$\int \frac{dx}{x\sqrt{x^2 + 1}} = -\ln\left|\frac{1 + \sqrt{x^2 + 1}}{x}\right| + C$$

$$= \ln|x| - \ln\left|1 + \sqrt{x^2 + 1}\right| + C$$

49.   Let $u = \sin x$, $du = \cos x \, dx$ and we will have form $\dfrac{du}{\sqrt{1 + u^2}}$.

By formula #172: $\int \dfrac{\cos x \, dx}{\sqrt{1 + \sin^2 x}} = \ln\left|\sin x + \sqrt{1 + \sin^2 x}\right| + C$

53.   This matches formula #182 with $a = \sqrt{2}$.

$$\int_0^1 \frac{dx}{(x^2 + 2)^{3/2}} = \frac{x}{2\sqrt{x^2 + 2}}\bigg|_0^1 = \frac{1}{2\sqrt{3}} = \frac{\sqrt{3}}{6}$$

57.   This matches formula #171 with $a = 2$:

$$\int_{-2}^{2\sqrt{3}} = \frac{(x^2 + 4)^{5/2}}{5} - \frac{4(x^2 + 4)^{3/2}}{3}\bigg|_{-2}^{2\sqrt{3}} = \frac{1024}{5} - \frac{256}{3} - \frac{128\sqrt{2}}{5} + \frac{64\sqrt{2}}{3}$$

$$= -\frac{64\sqrt{2}}{15} + \frac{1{,}792}{15} = \frac{1{,}792 - 64\sqrt{2}}{15} \approx 113.43$$

61.   Let $u = \cos x$, $du = -\sin x \, dx$:

$$\int_0^{\pi/4} \sin^5 x \, dx = -\int_0^{\pi/4} (\sin^4 x)(-\sin x \, dx)$$

$$= -\int_0^{\pi/4} (1 - \cos^2 x)^2(-\sin x \, dx)$$

$$= -\int_0^{\pi/4} (\cos^4 x - 2\cos^2 x + 1)(-\sin x \, dx)$$

$$= \left(-\frac{\cos^5}{5} + \frac{2\cos^3 x}{3} - \cos x\right)\bigg|_0^{\frac{\pi}{4}}$$

$$= -\frac{\frac{\sqrt{2}}{8}}{5} + \frac{(2)\frac{\sqrt{2}}{4}}{3} - \frac{\sqrt{2}}{2} - \left(-\frac{1}{5} + \frac{2}{3} - 1\right)$$

$$= -\frac{43\sqrt{2}}{120} + \frac{8}{15} \approx 0.0266$$

65. Let $u = e^x$, $du = e^x dx$. We then have form $\dfrac{du}{\sqrt{1 + u^2}}$:

$$\int \frac{e^x dx}{\sqrt{1 + e^{2x}}} = \ln\left(\sqrt{1 + e^{2x}} + e^x\right) + C$$

69. Using integration by partial fractions:
$$\frac{5x^2 + 18x + 34}{(x - 7)(x + 2)^2} = \frac{A_1}{x - 7} + \frac{A_2}{(x + 2)^2} + \frac{A_3}{x + 2}$$

$$= \frac{A_1(x + 2)^2 + A_2(x - 7) + A_3(x - 7)(x + 2)}{(x - 7)(x + 2)^2}$$

So $5x^2 + 18x + 34 = A_1(x + 2)^2 + A_2(x - 7) + A_3(x - 7)(x + 2)$

If $x = 7$: $405 = 81A_1$, $A_1 = 5$

If $x = -2$: $18 = -9A_2$, $A_2 = -2$

Equating the coefficients of the constant term: $34 = 4A_1 - 7A_2 - 14A_3$,

$34 = 20 + 14 - 14A_3$, $A_3 = 0$.

$$\int \frac{5x^2 + 18x + 34}{(x - 7)(x + 2)^2}\, dx = \int \frac{5\, dx}{x - 7} - \int \frac{2\, dx}{(x + 2)^2}$$

$$= 5 \ln |x - 7| + \frac{2}{x + 2} + C$$

73. Using integration by partial fractions:
$$\frac{5x^2 + 3x - 2}{x^3 + 2x^2} = \frac{Ax + B}{x^2} + \frac{C}{x + 2}$$

$$= \frac{(A + C)x^2 + (2A + B)x + 2B}{x^2(x + 2)}$$

$A + C = 5$, $2A + B = 3$, $2B = -2$, $B = -1$, $A = 2$, $C = 3$

$$\int \frac{5x^2 + 3x - 2}{x^3 + 2x^2}\, dx = \int \frac{2x - 1}{x^2}\, dx + \int \frac{3}{x + 2}\, dx$$

$$= 2 \ln|x| + \frac{1}{x} + 3 \ln|x + 2| + C$$

75. Using integration by parts:
$$\frac{x}{(x + 1)(x + 2)(x + 3)} = \frac{A_1}{x + 1} + \frac{A_2}{x + 2} + \frac{A_3}{x + 3}$$

$$= \frac{A_1(x + 2)(x + 3) + A_2(x + 1)(x + 3) + A_3(x + 1)(x + 2)}{(x + 1)(x + 2)(x + 3)}$$

If $x = -1$: $-1 = 2A_1$, $A_1 = -\frac{1}{2}$

If $x = -2$: $-2 = -A_2$, $A_2 = 2$

If $x = -3$: $-3 = 2A_3$, $A_3 = -\frac{3}{2}$

75. (con't.)

$$\int \frac{x\,dx}{(x\,+\,1)(x\,+\,2)(x\,+\,3)} \;=\; \int \frac{-\frac{1}{2}\,dx}{x\,+\,1} \;+\; \int \frac{2\,dx}{x\,+\,2} \;-\; \int \frac{\frac{3}{2}\,dx}{x\,+\,3}$$

$$= \;-\tfrac{1}{2}\ln|x\,+\,1| \;+\; 2\ln|x\,+\,2| \;-\; \tfrac{3}{2}\ln|x\,+\,3| \;+\; C$$

$$= \;\ln\sqrt{\frac{(x\,+\,2)^4}{(x\,+\,1)(x\,+\,3)^3}} \;+\; C$$

79.
$$f(x)_{AVG} \;=\; \frac{1}{1\,-\,0}\int_0^1 x\sin^3 x^2\,dx$$

$$= \;\tfrac{1}{2}\int_0^1 (1\,-\,\cos^2 x^2)\sin x^2\,(2x\,dx)$$

$$= \;\tfrac{1}{2}\left[-\cos x^2 \;+\; \frac{\cos^3 x^2}{3}\right]\Big|_0^1$$

$$= \;\tfrac{1}{2}\left[-\cos 1 \;+\; \frac{\cos^3 1}{3} \;-\; \left(-\cos 0 \;+\; \frac{\cos^3 0}{3}\right)\right]$$

$$= \;-\tfrac{1}{2}\cos 1 \;+\; \tfrac{1}{6}\cos^3 1 \;+\; \tfrac{1}{3} \;\approx\; 0.089\,470\,3$$

83.
$$f(x)_{AVG} \;=\; \frac{1}{\frac{\pi}{4}\,-\,0}\int_0^{\frac{\pi}{4}} \sec x\,dx$$

$$= \;\tfrac{4}{\pi}\ln|\sec x \;+\; \tan x|\,\Big|_0^{\frac{\pi}{4}}$$

$$= \;\tfrac{4}{\pi}\left(\ln|\sqrt{2}\,+\,1| \;-\; \ln|1\,+\,0|\right)$$

$$= \;\tfrac{4}{\pi}\ln(\sqrt{2}\,+\,1) \;\approx\; 1.122\,199\,7$$

87. The function has symmetry about the origin, so the volume generated from
-3 to 0 will be the same as that from 0 to 3. Double the integral from 0 to 3.
Slices perpendicular to the x-axis will be disks.

$$V \;=\; 2\pi\int_0^3 \left(x\sqrt{9\,-\,x^2}\,\right)^2\,dx$$

$$= \;2\pi\int_0^3 (9x^2\,-\,x^4)\,dx$$

$$= \;2\pi\left[3x^3 \;-\; \frac{x^5}{5}\right]\Big|_0^3$$

$$= \;2\pi[81\,-\,\tfrac{3}{5}(81)]$$

$$= \;\frac{324\,\pi}{5} \;\approx\; 203.575\,204$$

91.   Analysis: Let $u = e^{ax}$, $dv = \cos bx\, dx$. Two integrations by parts will return to the original integral, combine with the one on the left and divide by the coefficient.

93.   Analysis: Using integration by parts; let $u = (\ln x)^n$, $dv = x^m\, dx$, then

$$du = \frac{n\,(\ln x)^{n-1}}{x}\, dx \quad \text{and} \quad v = \frac{x^{m+1}}{m+1}\,. \quad \text{This will yield the desired formula.}$$

To evaluate $\displaystyle\int x^2\,(\ln x)^3\, dx$ let $m = 2$ and $n = 3$ in the just derived formula.

7.5 Improper Integrals

3.   We will evaluate the integral for some finite upper limit, t, and then see if a limit exists as t approaches infinity.

$$\int_1^t \frac{dx}{x^3} = -\frac{1}{2x^2}\bigg|_1^t = -\frac{1}{2t^2} + \frac{1}{2}$$

$$\lim_{t\to\infty}\left(-\frac{1}{2t^2} + \frac{1}{2}\right) = \frac{1}{2}$$

7.   We will evaluate the integral for some finite upper limit, t, and then see if a limit exists as t approaches infinity.

$$\int_1^t \frac{dx}{x^{1.1}} = -\frac{1}{0.1x^{0.1}}\bigg|_1^t = -\frac{10}{t^{0.1}} + 10$$

$$\lim_{t\to\infty}\left(-\frac{10}{t^{0.1}} + 10\right) = 10$$

11.   We will evaluate the integral for some finite upper limit, t, and then see if a limit exists as t approaches infinity.

$$\int_3^t \frac{dx}{(2x-1)^2} = \frac{1}{2}\int_3^t \frac{2\,dx}{(2x-1)^2}$$

$$= -\frac{1}{2(2x-1)}\bigg|_3^t = -\frac{1}{4t-2} + \frac{1}{10}\,,$$

$$\lim_{t\to\infty}\left(-\frac{1}{4t-2} + \frac{1}{10}\right) = \frac{1}{10}$$

15. $\frac{1}{3}\int_{1}^{t} \frac{3x^2\, dx}{(x^3 + 2)^2} = \frac{1}{3}\left(-\frac{1}{x^3 + 2}\right)\Big|_{1}^{t} = -\frac{1}{3t^3 + 6} + \frac{1}{9}.$

$\lim_{t\to\infty}\left(-\frac{1}{3t^3 + 6} + \frac{1}{9}\right) = \frac{1}{9}$

19. Let $u = -\sqrt{x}$, then $du = -\frac{1}{2\sqrt{x}}$. Set up $e^u\, du$ form:

$-2\int_{1}^{t} -\frac{e^{-\sqrt{x}}dx}{2\sqrt{x}} = -2e^{-\sqrt{x}}\Big|_{1}^{t} = -2\left(\frac{1}{e^{\sqrt{t}}} - \frac{1}{e}\right),$

$\lim_{t\to\infty} -2\left(\frac{1}{e^{\sqrt{t}}} - \frac{1}{e}\right) = \frac{2}{e}$

23. Let $u = \ln x,\ du = \frac{dx}{x}$. We now have $u^{-1}\, du$ form.

$\int_{2}^{t} (\ln x)^{-1} \frac{dx}{x} = \ln(\ln x)\Big|_{2}^{t} = \ln(\ln t) - \ln(\ln 2),$

$\lim_{t\to\infty}[\ln(\ln t) - \ln(\ln 2)] = \infty$ The integral diverges.

27. We will evaluate the integral for some finite lower limit, t, and then see if a limit exists as t approaches negative infinity.

$\int_{t}^{0} \frac{2x\, dx}{x^2 + 1} = \ln(x^2 + 1)\Big|_{t}^{0} = \ln 1 - \ln(t^2 + 1) = -\ln(t^2 + 1)$

$\lim_{t\to-\infty} -\ln(t^2 + 1) = -\infty$ The integral diverges.

31. This is an improper integral at both the upper and lower limits. So pick some value in between the undefined points (0 is usually a good number) and consider two improper integrals. In order to have a convergent value both must converge. If either is divergent then the original integral is divergent.

$\int_{-\infty}^{\infty} = \int_{-\infty}^{0} + \int_{0}^{\infty}$ We will evaluate each separately. Remember that $|x| = -x$ for $x < 0$.

$\int_{t}^{0} xe^x dx = e^x(x - 1)\Big|_{t}^{0}$ (Integration by parts or formula #484)

$= [-1 - e^t(t - 1)],\quad \lim_{t\to-\infty}[-1 - e^t(t - 1)] = -1$

31. (con't.)

$$\int_0^\infty xe^{-x}dx = \int_0^\infty (-x)e^{-x}(-dx) = e^{-x}(-x-1)\Big|_0^t \ (\ \#484 \text{ with } u = -x.)$$

$$= [e^{-t}(-t-1) - (-1)]$$

$$\lim_{t\to\infty} [e^{-t}(-t-1) - (-1)] = 1$$

$$\int_{-\infty}^\infty xe^{-|x|}\,dx = -1 + 1 = 0$$

We might have suspected this, as the function is symmetric about the origin.

$f(x) = -f(-x).$

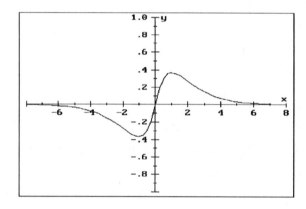

35.   This function is undefined at the upper limit of 1. We will find the integral

from 0 to t, and then find the limit (if it exists) as t approaches 1.

$$-\int_0^t \frac{-dx}{(1-x)^{1/2}} \text{ has form } u^{-1/2}du = -2\sqrt{1-x}\ \Big|_0^t = -2(\sqrt{1-t}-1)$$

$$\lim_{t\to 1^-} -2(\sqrt{1-t}-1) = 2$$

39. $\displaystyle\int_e^t \frac{dx}{x(\ln x)^2} \text{ has form } u^{-2}du = -\frac{1}{\ln x}\ \Big|_e^t = -\frac{1}{\ln t} + 1$

$$\lim_{t\to\infty} \left(-\frac{1}{\ln t} + 1\right) = 1$$

43. $\displaystyle A = \int_6^t \frac{2\,dx}{(x-4)^3} = -\frac{1}{(x-4)^2}\ \Big|_6^t = -\frac{1}{(t-4)^2} + \frac{1}{4}$

$$\lim_{t\to\infty} \left(-\frac{1}{(t-4)^2} + \frac{1}{4}\right) = \frac{1}{4}$$

47.

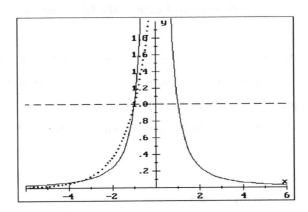

We will divide the region into three parts and evaluate each separately:

$$\int_{-\infty}^{\infty} = \int_{-\infty}^{-1} + 2 + \int_{1}^{\infty}$$

$$\int_{t}^{-1} \frac{dx}{x^2} = -\frac{1}{x} \Big|_{t}^{-1} = 1 + \frac{1}{t}. \quad \lim_{t \to -\infty}\left(1 + \frac{1}{t}\right) = 1$$

$$\int_{1}^{t} \frac{dx}{x^2} = -\frac{1}{x} \Big|_{1}^{t} = -\frac{1}{t} + 1. \quad \lim_{t \to \infty}\left(-\frac{1}{t} + 1\right) = 1$$

So $\displaystyle\int_{-\infty}^{\infty} f(x) \, dx = 1 + 2 + 1 = 4$

51. The integration is correct, but has overlooked the fact that the function is undefined (has a vertical asymptote) at $x = 0$. It should be written as:

$$\lim_{t \to 0^-} \int_{-1}^{t} \frac{dx}{x^2} + \lim_{t \to 0^+} \int_{t}^{1} \frac{dx}{x^2} .$$ Both of these integrals diverge (see Example 3).

55. Analysis: For each of the Laplace transforms we need to find $\displaystyle\lim_{n \to \infty} \int_{0}^{n}$ for the specified function.

    a. Adding the exponents of e we have $\displaystyle\lim_{n \to \infty} \int_{0}^{n} e^{(a-s)t} dt$

    b. $\displaystyle\lim_{n \to \infty} -\frac{a}{s} \int_{0}^{n} e^{-st}(-s \, dt)$

    c. Integrate by parts or use formula #492.

    d. Integrate by parts or use formula #493.

57.   Analysis: Find the antiderivative of each of the functions, and then use ex. 55.

    a.   Using ex. 55b, $\ell^{-1}\{\frac{5}{s}\} = 5$

    b.   Write as two fractions: $\dfrac{s}{s^2 + 4} + \dfrac{2}{s^2 + 4}$ and use ex. 55d and ex. 55c.

    c.   Decompose into partial fractions, finish by using 55a and 55d.

    d.   Also requires partial fractions.

59.   Analysis: In each of these make the appropriate substitution for $f(t)$ and a, and apply the result of ex. 58.

    a. $f(t) = t^3, \quad a = -2$

    b. $f(t) = \cos 2t, \quad a = -3$

    c.   Substitute $(s - 1)$ for s in 55c. $a = 1$, so we get $e^t 5t$

    d.   Complete the square, giving $\dfrac{4s}{(s + 2)^2 + 1}$. $a = -2$. Use 55d.

7.6 The Hyperbolic and Inverse Hyperbolic Functions

SURVIVAL HINT

The hyperbolic functions are good news and bad news. The good news is that the identities, derivatives, and integrals are *almost* the same as the trigonometric functions. The bad news is the *almost*. Try to learn these function definitions by concentrating on which ones are different by a sign.

3. $\tanh(-1) = \dfrac{e^{-1} - e^1}{e^{-1} + e^1} = \dfrac{1 - e^2}{1 + e^2} \approx -0.7616$

7. $\cosh^{-1} 1.5 = \ln\left(1.5 + \sqrt{(1.5)^2 - 1}\right) \approx 0.9624$

11. $\text{sech } 1 = \dfrac{2}{e^x + e^{-x}} \approx 0.6481$

15. $y' = \sinh u \; du = (4x + 3)\sinh(2x^2 + 3x)$

19. $y' = \dfrac{1}{\sqrt{1 + u^2}} \; du = \dfrac{3x^2}{\sqrt{1 + x^6}}$

23.   If $|\sin x| < 1$, $y' = \dfrac{1}{1 - u^2} \; du = \dfrac{\cos x}{1 - \sin^2 x} = \dfrac{\cos x}{\cos^2 x} = \sec x$

27.   We need the product rule on the first term:

    $y' = x\left(\dfrac{1}{\sqrt{x^2 - 1}}\right) + \cosh^{-1}x - \dfrac{2x}{2\sqrt{x^2 - 1}} = \cosh^{-1}x$

SURVIVAL HINT

Although the inverse hyperbolic functions are a quick and easy way to integrate the forms on p. 546, most can also be done by trigonometric substitution.

31.   Let $u = \frac{1}{x}$, the $du = -\frac{1}{x^2}$. Set up $\sinh u \, du$ form:

$$-\int \sinh \frac{1}{x}\left(-\frac{1}{x^2}\, dx\right) = -\cosh \frac{1}{x} + C$$

35.   To set up $\dfrac{du}{\sqrt{u^2 - 1}}$ form, factor 16 out of the radical and let $u = \frac{3}{4} t$.

$$\int \frac{dt}{4\sqrt{\frac{9}{16}t^2 - 1}} = \frac{1}{3}\int \frac{\frac{3}{4}\, dt}{\sqrt{(\frac{3}{4}t)^2 - 1}} = \frac{1}{3}\cosh^{-1}(\tfrac{3}{4}t) + C$$

39.   Let $u = x^3$, $du = 3x^2\, dx$. Set up $\dfrac{du}{1 - u^2}$ form:

$$\frac{1}{3}\int \frac{3x^2\, dx}{1 - (x^3)^2} = \frac{1}{3}\tanh^{-1} x^3 + C$$

43.   Let $u = e^x$, $du = e^x dx$. We now have $\dfrac{du}{\sqrt{u^2 - 1}}$ form.

$$\int_1^2 \frac{e^x dx}{\sqrt{(e^x)^2 - 1}} = \cosh^{-1} e^x \Big|_1^2 = \cosh^{-1} e^2 - \cosh^{-1} e$$

$$\text{or } \ln\!\left(e^2 + \sqrt{e^4 - 1}\right) - \ln\!\left(e + \sqrt{e^2 - 1}\right) \approx 1.0311$$

47.   Use the definitions and do the algebra;

a. $\sinh 2x = \dfrac{e^{2x} - e^{-2x}}{2}$

$2 \sinh x \cosh x = 2\left(\dfrac{e^x - e^{-x}}{2}\right)\left(\dfrac{e^x + e^{-x}}{2}\right) = \dfrac{e^{2x} - e^{-2x}}{2}$

b. $\cosh 2x = \dfrac{e^{2x} + e^{-2x}}{2}$

$\cosh^2 x + \sinh^2 x = \left(\dfrac{e^x + e^{-x}}{2}\right)^2 + \left(\dfrac{e^x - e^{-x}}{2}\right)^2$

$$= \frac{e^{2x} + 2 + e^{-2x}}{4} + \frac{e^{2x} - 2 + e^{-2x}}{4} = \frac{2e^{2x} + 2e^{-2x}}{4}$$

$$= \frac{e^{2x} + e^{-2x}}{2}$$

51. $\cosh x + \sinh x = \left(\dfrac{e^x + e^{-x}}{2}\right) + \left(\dfrac{e^x - e^{-x}}{2}\right) = \dfrac{2e^x}{2} = e^x$

So $(\sinh x + \cosh x)^n = (e^x)^n = e^{nx}$. Now by definition

$$\cosh nx + \sinh nx = \frac{e^{nx} + e^{-nx}}{2} + \frac{e^{nx} - e^{-nx}}{2} = \frac{2e^{nx}}{2} = e^{nx}$$

55. $s = \int\limits_{-a}^{a} \sqrt{1 + (y')^2}\, dx$. Using symmetry and $y' = \sinh \frac{x}{a}$, we have:

$$s = 2\int\limits_{0}^{a} \sqrt{1 + \sinh^2 \frac{x}{a}}\, dx = 2a\int\limits_{0}^{a} \cosh \frac{x}{a}\left(\frac{1}{a}\, dx\right) = 2a \sinh \frac{x}{a}\, \Big|_{0}^{a}$$

$$= 2a(\sinh 1 - \sinh 0) = 2a \sinh 1 = (e - e^{-1})a \approx 2.3504a$$

59.   Analysis: For $\cosh u$ use the definition. For $\tanh u$ use $\dfrac{\sinh u}{\cosh u}$ and the quotient rule, since these have now been derived. For $\operatorname{sech} u$ use $(\cosh u)^{-1}$ and the power rule.

61.   Analysis: Follow the pattern in the proof for $\sinh^{-1} u$ given in Theorem 7.4. If $y = \cosh^{-1} x$, then $x = \cosh y = \dfrac{e^y + e^{-y}}{2}$. Solve this for e^y as a function of x, then take ln of both sides to get y as a function of x.

Chapter 7 Review

PRACTICE PROBLEMS

14.   **a.** $\tanh^{-1}(0.5) = \frac{1}{2}\ln \dfrac{1 + 0.5}{1 - 0.5} \approx 0.5493$

   b. $\sinh(\ln 3) = \dfrac{e^{\ln 3} - e^{-\ln 3}}{2} = \dfrac{3 - \frac{1}{3}}{2} = \dfrac{4}{3}$

   c. $\coth^{-1} 2 = \frac{1}{2}\ln \dfrac{2 + 1}{2 - 1} = \frac{1}{2}\ln 3 = \ln \sqrt{3} \approx 0.5493$

15.   Write as two separate integrals: $\displaystyle\int \dfrac{2x\, dx}{\sqrt{x^2 + 1}} + \int \dfrac{3\, dx}{\sqrt{x^2 + 1}}$

   The first is $\frac{du}{u}$ form and the second is a trigonometric substitution or #172.
   $\displaystyle\int \dfrac{2x + 3}{\sqrt{x^2 + 1}}\, dx = 2\sqrt{x^2 + 1} + 3\sinh^{-1} x + C$

16.   Two different type of factors in a function often suggest integration by parts.
   Let $u = x$, $dv = \sin 2x\, dx$. Then $du = dx$, $v = -\frac{1}{2}\cos 2x$.
   $\displaystyle\int x \sin 2x\, dx = -\frac{x}{2}\cos 2x - \int -\frac{1}{2}\cos 2x\, dx$

   $$= -\frac{x}{2}\cos 2x + \frac{1}{4}\sin 2x + C$$

17.   Set up $\sinh u\, du$ form: $-\frac{1}{2}\displaystyle\int \sinh(1 - 2x)(-2\, dx) = -\frac{1}{2}\cosh(1 - 2x) + C$

18. This is the form for $\sin^{-1}x$, with $a = 2$. (#224)

$$\int \frac{dx}{\sqrt{4 - x^2}} = \sin^{-1}\frac{x}{2} + C$$

19. Break into partial fractions: $\dfrac{x^2}{(x^2 + 1)(x - 1)} = \dfrac{A_1x + A_2}{x^2 + 1} + \dfrac{A_3}{x - 1}$.

So $x^2 = (A_1x + A_2)(x - 1) + A_3(x^2 + 1)$. Equating coefficients:

$1 = A_1 + A_3$, $0 = -A_1 + A_2$, $0 = -A_2 + A_3$.

If $x = 1$: $1 = 2A_3$, $A_3 = \frac{1}{2}$. Substituting, $A_1 = \frac{1}{2}$, $A_2 = \frac{1}{2}$.

$$\int \frac{x^2\,dx}{(x^2 + 1)(x - 1)} = \frac{1}{2}\int \frac{x + 1}{x^2 + 1}\,dx + \frac{1}{2}\int \frac{dx}{x - 1}$$

$$= \frac{1}{4}\int \frac{2x\,dx}{x^2 + 1} + \frac{1}{2}\int \frac{dx}{x^2 + 1} + \frac{1}{2}\int \frac{dx}{x - 1}$$

$$= \frac{1}{4}\ln(x^2 + 1) + \frac{1}{2}\tan^{-1}x + \frac{1}{2}\ln|x - 1| + C$$

20. The degree of the numerator is greater than the degree of the denominator, so

begin by dividing: $\displaystyle\int \frac{x^3\,dx}{x^2 - 1} = \int\left(x + \frac{x}{x^2 - 1}\right)dx$

$$= \int x\,dx + \frac{1}{2}\int \frac{2x\,dx}{x^2 - 1}$$

$$= \frac{x^2}{2} + \frac{1}{2}\ln|x^2 - 1| + C$$

21. Recall that $\ln x^3 = 3\ln x$, and use integration by parts: $3\displaystyle\int_1^2 x\ln x\,dx$.

Let $u = \ln x$, $dv = x\,dx$. Then $du = \frac{1}{x}\,dx$, $v = \frac{x^2}{2}$.

$$3\int_1^2 x\ln x\,dx = 3\left(\frac{x^2}{2}\ln x - \int_1^2\frac{x^2}{2}\frac{1}{x}\,dx\right) = 3\left(\frac{x^2}{2}\ln x - \frac{x^2}{4}\right)\Big|_1^2$$

$$= \frac{3x^2}{2}(\ln x - \frac{1}{2})\Big|_1^2 = 6\ln 2 - 3 + \frac{3}{4}$$

$$= 6\ln 2 - \frac{9}{4} \approx 1.9089$$

22. Break into partial fractions:

$$\frac{1}{(x - 1)^2(x + 2)} = \frac{A_1}{(x - 1)^2} + \frac{A_2}{x - 1} + \frac{A_3}{x + 2}$$

$1 = A_1(x + 2) + A_2(x - 1)(x + 2) + A_3(x - 1)^2$.

If $x = 1$: $1 = 3A_1$, $A_1 = \frac{1}{3}$

If $x = -2$: $1 = 9A_3$, $A_3 = \frac{1}{9}$ Equating the coefficients of the constant term:

$1 = 2A_1 - 2A_2 + A_3$, $1 = \frac{2}{3} - 2A_2 + \frac{1}{9}$, $A_2 = -\frac{1}{9}$

22. (con't.)

$$\int_2^3 \frac{dx}{(x-1)^2(x+2)} = \frac{1}{9} \int_2^3 \left(\frac{3}{(x-1)^2} - \frac{1}{x-1} + \frac{1}{x+2} \right) dx$$

$$= \frac{1}{9} \left(-\frac{3}{x-1} - \ln(x-1) + \ln(x+2) \right) \Big|_2^3$$

$$= \frac{1}{9} \left(-\frac{3}{2} - \ln 2 + \ln 5 + 3 - \ln 4 \right)$$

$$= \frac{1}{9} \left(\ln \frac{5}{8} + \frac{3}{2} \right) \approx 0.1144$$

23.   Complete the square in the denominator to give $\dfrac{du}{1-u^2}$ form:

$$\int_3^4 \frac{dx}{1-1+2x-x^2} = \int_3^4 \frac{dx}{1-(x^2-2x+1)} = \int_3^4 \frac{dx}{1-(x-1)^2}$$

$$= \coth^{-1}(x-1) \Big|_3^4$$

$$= \coth^{-1}3 - \coth^{-1}2 \approx -0.2027$$

or by the definition: $= \frac{1}{2} \ln \dfrac{(x-1)+1}{(x-1)-1} \Big|_3^4 = \frac{1}{2}\ln 2 - \frac{1}{2}\ln 3 \approx -0.2027$

The problem can also be done by partial fractions:

$$\int_3^4 \frac{dx}{2x-x^2} = \int_3^4 \left(\frac{\frac{1}{2}}{x} + \frac{\frac{1}{2}}{2-x} \right) dx = \frac{1}{2}\ln x - \frac{1}{2}\ln|2-x| \Big|_3^4 = \frac{1}{2}\ln\frac{2}{3}$$

24.   Let $u = \sec x$, $du = \sec x \tan x\, dx$. Set up $u^2 du$ form:

$$\int_0^{\frac{\pi}{4}} (\sec^2 x)(\sec x \tan x\, dx) = \frac{\sec^3 x}{3} \Big|_0^{\frac{\pi}{4}} = \frac{2\sqrt{2}}{3} - \frac{1}{3} = \frac{2\sqrt{2}-1}{3} \approx 0.6095$$

25. $\displaystyle\int_0^t x\, e^{-2x}\, dx$ has different type factors in the function, so try integration by parts.

Let $u = x$, $dv = e^{-2x}dx$, then $du = dx$, $v = -\frac{1}{2}e^{-2x}$.

$$\int_0^t x\, e^{-2x}\, dx = -\frac{x}{2}e^{-2x} - \int -\frac{1}{2}e^{-2x}dx$$

$$= -\frac{x}{2}e^{-2x} - \frac{1}{4}e^{-2x} \Big|_0^t$$

$$= -\frac{1}{4}e^{-2x}(2x+1) \Big|_0^t$$

$$= -\frac{1}{4}e^{-2t}(2t+1) + \frac{1}{4}$$

$$\lim_{t\to\infty} \left(-\frac{1}{4}e^{-2t}(2t+1) + \frac{1}{4} \right) = \frac{1}{4}$$

26.   The function is undefined at $x = 0$, replace by t, then take limit.

"Degree" of numerator and denominator are the same so begin by dividing:

$$\int_t^\pi \left(-1 + \frac{1}{1 - \cos x}\right)dx = -x + \int_t^\pi \frac{1(1 + \cos x)}{(1 - \cos x)(1 + \cos x)}dx$$

$$= -x + \int_t^\pi \frac{dx}{\sin^2 x} + \int_t^\pi \frac{\cos x\, dx}{\sin^2 x}$$

$$= -x - \cot x - \frac{1}{\sin x}\bigg|_t^\pi$$

$$= -\pi - \infty \quad \text{(we can stop here).}$$

The integral diverges.

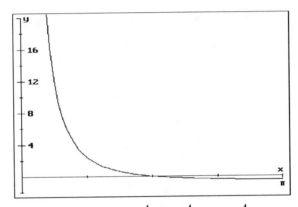

27.   By partial fractions: $\dfrac{2x + 3}{x^2(x - 2)} = \dfrac{A_1}{x^2} + \dfrac{A_2}{x} + \dfrac{A_3}{x - 2}$

$2x + 3 = A_1(x - 2) + A_2 x(x - 2) + A_3 x^2$

If $x = 0$: $3 = -2A_1$, $A_1 = -\dfrac{3}{2}$

If $x = 2$: $7 = 4A_3$, $A_3 = \dfrac{7}{4}$

Equating the coefficients for the x^2 term: $0 = A_2 + A_3$, $A_2 = -\dfrac{7}{4}$

Now the function is undefined at the lower limit, so integrate from $x = t$ to 1, then see if a limit exists as t approaches 0.

$$\int_t^1 \frac{2x + 3}{x^2(x - 2)}dx = \int_t^1 \left(\frac{-\frac{3}{2}}{x^2} + \frac{-\frac{7}{4}}{x} + \frac{\frac{7}{4}}{x - 2}\right)dx$$

$$= \frac{3}{2x} + \frac{7}{4}\ln\left|\frac{x - 2}{x}\right|\,\bigg|_t^1$$

$$= \frac{3}{2} + \frac{7}{4}\ln 1 - \frac{3}{2t} - \ln\left|\frac{t - 2}{t}\right|$$

$$\lim_{t\to 0^+}\left(\frac{3}{2} + \frac{7}{4}\ln 1 - \frac{3}{2t} - \ln\left|\frac{t - 2}{t}\right|\right) = \infty. \quad \text{The integral diverges.}$$

27. (con't.)

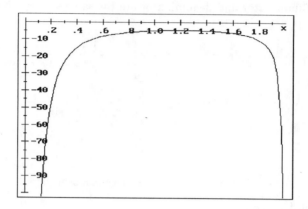

28. Integration by parts twice will return us to the original integral, which can be combined with the one on the left, or formula #492, with $a = -1$, gives us:

$$\int_0^t e^{-x}(\sin x)(-dx) = \left.\frac{e^{-x}(-\sin x - \cos x)}{2}\right|_0^t$$

$$= \frac{e^{-t}(-\sin t - \cos t)}{2} + \frac{1}{2}.$$

$$\lim_{t \to \infty}\left(\frac{e^{-t}(-\sin t - \cos t)}{2} + \frac{1}{2}\right) = \frac{1}{2}$$

29. Use the chain rule several times, as this is a function of a function of a function.

$$y' = \frac{1}{2\sqrt{\tanh^{-1}2x}}\left(\frac{2}{1 - 4x^2}\right) = \frac{1}{(1 - 4x^2)\sqrt{\tanh^{-1}2x}}$$

30. Vertical slices gives the method of cylindrical shells:

$$2\pi\int_0^2 x\left(\frac{1}{\sqrt{9 - x^2}}\right)dx = -\pi\int_0^2\frac{-2x\,dx}{\sqrt{9 - x^2}} \quad \text{which is form } u^{-\frac{1}{2}}du$$

$$= -2\pi\left(\sqrt{9 - x^2}\right)\Big|_0^2 = -2\pi\left(\sqrt{5} - 3\right)$$

$$= 2\pi(3 - \sqrt{5}) \approx 4.7999$$

Chapter 8
Infinite Series

8.1 Sequences and Their Limits

3. $1 + (-1)^1, 1 + (-1)^2, \ldots, 1 + (-1)^5 = 0, 2, 0, 2, 0$

7. $\dfrac{3(1) + 1}{1 + 2}, \dfrac{3(2) + 1}{2 + 2}, \ldots, \dfrac{3(5) + 1}{5 + 2} = \dfrac{4}{3}, \dfrac{7}{4}, \dfrac{10}{5}, \dfrac{13}{6}, \dfrac{16}{7}$

11. $1, 1^2 + 1 + 1, 3^2 + 3 + 1, 13^2 + 13 + 1, 183^2 + 183 + 1$

$= 1, 3, 13, 183, 33673$

15. Dividing by n: $\displaystyle\lim_{n \to \infty}\left(\dfrac{\frac{4}{n} - 7}{\frac{8}{n} + 1}\right) = -7$

19. Dividing by n^3: $\displaystyle\lim_{n \to \infty}\left(\dfrac{1 - \frac{6}{n} + \frac{85}{n^3}}{2 - \frac{5}{n^2} + 1702n^3}\right) = \dfrac{1}{2}$

SURVIVAL HINT

Never use l'Hôpital's rule without first verifying that the limit has form of $\frac{0}{0}$ or $\frac{\infty}{\infty}$.

23. Use l'Hôpital's rule twice: $\displaystyle\lim_{n \to \infty} \dfrac{\ln n}{n^2} = \lim_{n \to \infty} \dfrac{\frac{1}{n}}{2n} = \lim_{x \to \infty} \dfrac{1}{4n} = 0$

27. Taking the limit as n goes to infinity gives form ∞^0.

Let $y = (n + 4)^{1/n}$ then $\ln y = \ln (n + 4)^{1/n} = \frac{1}{n} \ln (n + 4)$.

$\displaystyle\lim_{n \to \infty} \ln y = \lim_{n \to \infty} \frac{1}{n} \ln (n + 4)$ Using l'Hôpital's rule:

$\displaystyle\lim_{n \to \infty} \dfrac{\ln (n + 4)}{n} = \lim_{n \to \infty} \dfrac{\frac{1}{n + 4}}{1} = 0.$ $\ln \displaystyle\lim_{n \to \infty} y = 0,$ $\displaystyle\lim_{n \to \infty} y = e^0 = 1$

31. Since this is $\infty - \infty$ form, we will rationalize by multiplying by the conjugate:

$$\left(\sqrt{n^2 + n} - n\right)\frac{\sqrt{n^2 + n} + n}{\sqrt{n^2 + n} + n} = \frac{n^2 + n - n^2}{\sqrt{n^2 + n} + n}. \text{ Now divide numerator}$$

and denominator by n: $\dfrac{1}{\sqrt{1 + \frac{1}{n}} + 1}$.

So now $\displaystyle\lim_{n\to\infty}\left(\sqrt{n^2 + n} - n\right) = \lim_{n\to\infty} \dfrac{1}{\sqrt{1 + \frac{1}{n}} + 1} = \dfrac{1}{2}$

35. $\ln n - \ln(n + 1) = \ln \dfrac{n}{n + 1}$. Dividing by $n = \ln \dfrac{1}{1 + \frac{1}{n}}$.

$\displaystyle\lim_{n\to\infty} \ln \dfrac{1}{1 + \frac{1}{n}} = \ln 1 = 0$

39. $a_n = 4 + \frac{5}{n}$ is a decreasing sequence. $\displaystyle\lim_{n\to\infty} \dfrac{4n + 5}{n} = 4$.

This is a decreasing sequence bounded below by 4, and therefore converges.

43. It alternates between -1 and 1, and therefore is divergent.

47. $a_1 = \frac{1}{2}$, $a_2 = \frac{1}{4}$, $a_3 = \frac{1}{8}$, $a_4 = \frac{1}{16}$, ..., $a_n = \left(\frac{1}{2}\right)^n$

$a_4 = 6.25\%$, $a_n = 100\left(\frac{1}{2}\right)^n\%$

51. Analysis: Solving $\dfrac{n^2 + 1}{n^3} = 0.001$ we can see that the denominator needs to be

1000 times larger than the numerator, so try $n = 1000$: $\dfrac{1000^2 + 1}{1000^3} > 0.001$,

but $\dfrac{1001^2 + 1}{1001^3} < 0.001$. So let $N \geq 1001$.

53. Analysis:

a. Recall the Binomial Theorem: $(a + b)^n = \displaystyle\sum_{k=0}^{n} \binom{n}{k} a^{n-k} b^k$. So

$(1 + \frac{1}{n})^n = \frac{n!}{n!} + \frac{n!}{(n-1)!}\frac{1}{n} + \frac{n!}{(n-2)!2!}\frac{1}{n^2} + \frac{n!}{(n-3)!3!}\frac{1}{n^3} + \cdots$

Write out the factorials, and distribute the powers of n to each factor

to get the required result. Note that each term is positive, so the

sequence is monotonic increasing.

b. Beginning with the third term, each numerator is less than one and each

denominator is greater than or equal to 2^n. The sequence from the third

term onward is, therefore, less than the infinite geometric series $\left(\frac{1}{2}\right)^n$.

If you recall your precalculus, this infinite series is equal to 1.

c. Parts **a** and **b** give the required hypotheses for *BMCT* in order to show that

the series converges.

8.2 Introduction to Infinite Series: Geometric Series

SURVIVAL HINT

The last section dealt with sequences (range values of a function whose domain is the natural numbers), and this section deals with series (the sum of the terms of a sequence). Read test or quiz questions carefully so as not to confuse the two. A sequence may converge and its series diverge. The classic example is the harmonic series. See p. 512.

3. $r = \frac{4}{5}$ so $S = \frac{a}{1-r} = \frac{1}{1-\frac{4}{5}} = 5$

7. This is a geometric series with $r = \frac{3}{2} > 1$, so it diverges.

11. $\sum_{1}^{\infty}\left(e^{-0.2}\right)^k$ is a geometric series with $r = e^{-0.2} < 1$, so $S = \frac{e^{-0.2}}{1 - e^{-0.2}}$

$= \frac{1}{e^{0.2} - 1} \approx 4.5167$

15. This is a geometric series with $r = -\frac{1}{2}$. $S = \frac{\frac{1}{2}}{1 - \left(-\frac{1}{2}\right)} = \frac{1}{3}$

19. This is a geometric series with $r = \frac{1}{\sqrt{2}}$.

$S = \frac{2}{1 - \frac{1}{\sqrt{2}}} = \frac{2\sqrt{2}}{\sqrt{2} - 1} = 4 + 2\sqrt{2}$

23. $\sum_{1}^{\infty}\left(\frac{1}{k^{0.1}} - \frac{1}{(k+1)^{0.1}}\right)$

$= \frac{1}{1^{0.1}} - \frac{1}{(2)^{0.1}} + \frac{1}{2^{0.1}} - \frac{1}{(3)^{0.1}} + \frac{1}{3^{0.1}} - \frac{1}{(4)^{0.1}} + \dots + \frac{1}{k^{0.1}} - \frac{1}{(k+1)^{0.1}}$

$= \frac{1}{1^{0.1}} - \frac{1}{(n+1)^{0.1}} \cdot \lim_{n\to\infty}\left(\frac{1}{1^{0.1}} - \frac{1}{(n+1)^{0.1}}\right) = 1$

27. $\ln\left(1 + \frac{1}{k}\right) = \ln\left(\frac{k+1}{k}\right) = \ln(k+1) - \ln k$

$\sum_{1}^{\infty}[\ln(k+1) - \ln k]$

$= \ln 2 - \ln 1 + \ln 3 - \ln 2 + \ln 4 - \ln 3 + \dots + \ln(n+1) - \ln n$

$\lim_{n\to\infty} S_n = \lim_{n\to\infty}[\ln(n+1) - \ln 1] = \infty$. The series diverges.

31. $0.01010101\dots = \frac{1}{100} + \frac{1}{10000} + \frac{1}{1000000} + \dots$, is a geometric series with

$r = \frac{1}{100}$. $S = \frac{\frac{1}{100}}{1 - \frac{1}{100}} = \frac{\frac{1}{100}}{\frac{99}{100}} = \frac{1}{99}$

35. **a.** Adding on the right: $\dfrac{2Ak - (Bk + 1)}{2^{k+1}}$. So $k - 1 = 2Ak - Bk - 1$,

$1 = 2A - B$, which has numerous solutions. Choose $A = B = 1$

b. $S_n = \sum_1^\infty \dfrac{k}{2^k} - \sum_1^\infty \dfrac{k+1}{2^{k+1}}$

$$= \lim_{n\to\infty}\left(\dfrac{1}{2} - \dfrac{2}{2^2} + \dfrac{2}{2^2} - \dfrac{3}{2^3} + \dfrac{3}{2^3} - \dfrac{4}{2^4} + \dots + \dfrac{n}{2^n} - \dfrac{n+1}{2^{n+1}}\right)$$

$$= \lim_{n\to\infty}\left(\dfrac{1}{2}\right) - \lim_{n\to\infty}\left(\dfrac{n+1}{2^{n+1}}\right). \text{ Using l'Hôpital's rule:}$$

$$= \dfrac{1}{2} - \lim_{n\to\infty}\dfrac{1}{2^{n+1}\ln 2} = \dfrac{1}{2}$$

39. $2\sum_0^\infty a_k + \sum_0^\infty \dfrac{1}{2^k} = 2(3.57) + 2 = 9.14$ (Note lower limit of 0.)

43. By partial fractions: $\dfrac{1}{(a + k)(a + k + 1)} = \dfrac{A}{a + k} + \dfrac{B}{a + k + 1}$

$1 = A(a + k + 1) + B(a + k)$. Equating coefficients:

$0 = A + B, \ 1 = A$. Substituting, $B = -1$.

$S_n = \sum_0^\infty\left(\dfrac{1}{a + k} - \dfrac{1}{a + k + 1}\right)$. Which telescopes to: $\dfrac{1}{a} - \dfrac{1}{a + n + 1}$

$\lim_{n\to\infty}\left(\dfrac{1}{a} - \dfrac{1}{a + n + 1}\right) = \dfrac{1}{a}$.

47. $a_1 = 500, \ r = \dfrac{2}{3}, \ S = \dfrac{500}{1 - \frac{2}{3}} = 1,500$

51. $a_1 = 10,000(0.20), \ a_2 = 10,000(0.80)(0.20), \ a_3 = 10,000(0.80)^2(0.20)$

$r = 0.80 \ \ S = \dfrac{10,000(0.20)}{1 - 0.80} = \dfrac{2,000}{0.20} = \$10,000.$

55. Let a_n be the number of trustees on 12-31 of the nth year.

$a_1 = 6$

$a_2 = 6e^{-0.2} + 6,$

$a_3 = a_2 e^{-0.2} + 6 = 6(1 + e^{-0.2})e^{-0.2} + 6,$

$a_4 = a_3 e^{-0.2} + 6 = 6\left((1 + e^{-0.2})e^{-0.2} + 1\right)e^{-0.2} + 6$

$a_n = 6 + 6e^{-0.2} + 6e^{-0.2(2)} + 6e^{-0.2(3)} + \dots.$

$S_\infty = \dfrac{6}{1 - e^{-0.2}} \approx 33$

59. Analysis:

a. The difference between the nth sum and the $(n - 1)$ sum is just the last term, a_n.

b. Subtract S_{n-1} from S_n.

61. Analysis: Use linearity and ∞ + a constant $= \infty$.

63.   Analysis: If $r > 1$ then $\lim\limits_{n \to \infty} a_n = \infty$ and S_n must diverge.

      If $r = 1$, $S_n = an$ and $\lim\limits_{n \to \infty} an = \infty$.

65.   Analysis: The original square has side, s, of 1. The first inscribed square has

      $s_1 = \frac{1}{\sqrt{2}}$. The nth inscribed square has $s_n = \left(\frac{1}{\sqrt{2}}\right)^n$. The first shaded area is

      $\frac{1}{4}$ of $1 - s_1{}^2$. The second shaded area is $\frac{1}{4}(1 - s_1{}^2) + \frac{1}{4}(s_1{}^2 - s_2{}^2)$, etc.

      Take the limit of this telescoping series.

8.3 The Integral Test: *p*-series

SURVIVAL HINT

The convergence of the *sequence* a_n to 0 is a necessary, but not sufficient, condition for the convergence of a *series*. If the *sequence* does not converge to 0, then the *series* must diverge. However if the *sequence* does converge to 0 the *series* may either converge or diverge. Developing tests to decide the question of convergence or divergence is one of the main objectives of this chapter.

3. $p = 3$. Converges.

7. $\lim\limits_{k \to \infty} \dfrac{\ln k}{k^2}$ by l'Hôpital's rule $= \lim\limits_{k \to \infty} \dfrac{\frac{1}{k}}{2k} = 0$, so the necessary condition for

      convergence is met. The hypotheses for the integral test are met so evaluate:

      $\lim\limits_{n \to \infty} \displaystyle\int_1^n \dfrac{\ln x}{x^2}\, dx$. Integration by parts or formula #505 gives

      $\lim\limits_{n \to \infty} \left(-\frac{1}{x}(\ln x + 1)\right)\Big|_1^n = \lim\limits_{n \to \infty} \left(-\frac{\ln n}{n} - \frac{1}{n} + 1\right) = 1$,

      so the series converges.

11.   This is a *p*-series with $p = 4$, so the series converges.

15.   Using the integral test:

      $\lim\limits_{n \to \infty} \displaystyle\int_1^n \dfrac{x\, dx}{x^2 + 1}$ can be done by $\frac{du}{u}$ form: $\lim\limits_{n \to \infty} \frac{1}{2} \ln (x^2 + 1)\Big|_1^n$

      $= \lim\limits_{n \to \infty} \frac{1}{2}[\ln (n^2 + 1) - \ln 2] = \infty$. The series diverges.

19. $\dfrac{1}{(\frac{1}{4})^k} = 4^k$. $\lim\limits_{k \to \infty} 4^k = \infty$. The necessary condition is not met. The series diverges.

23. $\lim\limits_{k\to\infty} \dfrac{k}{e^k}$ by l'Hôpital's rule $= \lim\limits_{k\to\infty} \dfrac{1}{e^k} = 0$, so the necessary condition is met.

By the integral test: $\lim\limits_{n\to\infty} \displaystyle\int_1^n \dfrac{x}{e^x}\, dx$ suggests integration by parts.

Let $u = x$, $dv = e^{-x}dx$. Then $du = dx$, $v = -e^{-x}$.

$$\int_1^n \dfrac{x}{e^x}\, dx = -\dfrac{x}{e^x} + \int_1^n e^{-x}dx = -\dfrac{1}{e^x}(x+1)\Big|_1^n = -\dfrac{n}{e^n} - \dfrac{1}{e^n} + \dfrac{2}{e}$$

$\lim\limits_{n\to\infty}\left(-\dfrac{n}{e^n} - \dfrac{1}{e^n} + \dfrac{2}{e}\right) = \dfrac{2}{e}$. The series converges.

27. Recall that $\cosh x = \dfrac{e^x + e^{-x}}{2}$, and use the integral test.

$$\lim\limits_{n\to\infty} \int_1^n \dfrac{\frac{1}{2}}{\cosh x}\, dx = \lim\limits_{n\to\infty} \dfrac{1}{2}\int_1^n \operatorname{sech} x\, dx = \lim\limits_{n\to\infty} \dfrac{1}{2}\sin^{-1}(\tanh x)\Big|_1^n$$

$= \dfrac{1}{2}\lim\limits_{n\to\infty}[\sin^{-1}(\tanh n) - \sin^{-1}(\tanh 1)] = \dfrac{1}{2}\left(\dfrac{\pi}{2} - \sin^{-1}\left(\dfrac{e^2-1}{e^2+1}\right)\right)$.
This is a finite value, so the series converges.

31. Doing the division: $\dfrac{k^2+1}{k^3} = \dfrac{1}{k} + \dfrac{1}{k^3}$. The second term is a convergent p-series, but the first term is the harmonic series which diverges.

The series diverges.

35. Doing the division: $\dfrac{n^{\sqrt{3}}+1}{n^{2.7321}} = \dfrac{1}{n^{\sqrt{3}-2.7321}} + \dfrac{1}{n^{2.7321}}$. Both of these are p-series with $|p| > 1$, so the series converges.

39. Since $\lim\limits_{k\to\infty}\left(\dfrac{1}{2^k} + \dfrac{2k+3}{3k+4}\right) = \dfrac{2}{3}$ the series fails the necessary condition and is therefore divergent.

43 The function is positive, continuous and decreasing (since k increases faster than $\ln k$. The hypotheses are met.

$\lim\limits_{n\to\infty} \displaystyle\int_2^n \dfrac{\ln x\, dx}{x}$. Let $u = \ln x$, $du = \dfrac{1}{x}dx$, and this is $u\, du$ form.

$= \lim\limits_{n\to\infty} \dfrac{1}{2}(\ln x)^2\Big|_2^n = \dfrac{1}{2}\lim\limits_{n\to\infty}[(\ln n)^2 - (\ln 2)^2] = \infty$. The series diverges.

47. The hypotheses for the integral test are met, so the series will converge if the integral test gives a finite value.

$$\lim\limits_{n\to\infty} \dfrac{1}{2}\int_2^n (x^2-1)^{-p}(2x\, dx) = \lim\limits_{n\to\infty} \dfrac{1}{2}\dfrac{(x^2-1)^{-p+1}}{-p+1}\Big|_2^n$$

$= \dfrac{1}{2(1-p)}\lim\limits_{n\to\infty}\left((n^2-1)^{1-p} - 3^{1-p}\right)$ which converges if $p > 1$.

51. The hypotheses for the integral test are met, so the series will converge if the integral test gives a finite value.

$$\lim_{n\to\infty} \int_3^n [\ln (\ln x)]^{-p} \frac{dx}{x \ln x} \text{ has form } u^n du \ = \ \lim_{n\to\infty} \frac{[\ln (\ln x)]^{-p+1}}{1 - p} \Big|_3^n$$

$$= \ \frac{1}{1 - p} \lim_{n\to\infty} \Big([\ln (\ln n)]^{1-p} \ - \ [\ln (\ln 3)]^{1-p} \Big) \text{ which converges if } p > 1.$$

55. Analysis: This is the same as the proof of the integral test as shown on p. 602.

8.4 Comparison Tests

3. Geometric series with $r = \cos \frac{\pi}{6}$. It converges since $|\cos \frac{\pi}{6}| < 1$.

7. p-series with $p = 1$. It diverges.

11. Geometric with $r = 1$, or p-series with $p = 0$. It diverges.

15. p-series with $p = \frac{1}{2}$. It diverges.

SURVIVAL HINT

In order to use the comparison tests we need to know some convergent and divergent series with which to compare. It is a good idea to keep a list of series for that purpose. In addition to the p, geometric, and harmonic add other general types as you determine their convergence or divergence.

19. $\dfrac{1}{\sqrt{k^3 + 2}} < \dfrac{1}{k^{3/2}}$ which is a convergent p-series. So by direct comparison our series converges.

23. Divide numerator and denominator by k^2: $\dfrac{1 + \frac{5}{k} + \frac{6}{k^2}}{k^{3/2}} < \dfrac{12}{k^{3/2}}$ which is a convergent p-series. So by direct comparison our series converges.

27. Divide by k: $\dfrac{1}{(1 + \frac{2}{k})2^k} < \dfrac{1}{2^k}$ which is a convergent geometric series.

So by direct comparison our series converges.

31. $\dfrac{1}{\sqrt{k}\, 2^k} < \dfrac{1}{2^k}$ which is a convergent geometric series. So by direct comparison our series converges.

35. $\lim\limits_{k\to\infty} a_k = 2$. It does not meet the necessary condition for convergence so it must diverge.

39. Divide numerator and denominator by $k^2 + 1$: $\dfrac{1}{\left(1 + \dfrac{1}{k^2 + 1}\right)k^2} < \dfrac{1}{k^2}$

which is a convergent p-series. Our series converges.

43. The denominator is similar to $k^{3/4}k^{1/8} = k^{7/8}$, which would be a divergent

p-series. So we will establish a "greater than" inequality:

$$\frac{k^{1/6}}{(k^3 + 2)^{1/4}k^{1/8}} > \frac{1}{(k^3 + 2)^{1/4}k^{1/8}} > \frac{1}{(k + 2)^{3/4}k^{1/8}}$$

$$> \frac{1}{(k + 2)^{3/4}(k + 2)^{1/8}} = \frac{1}{(k + 2)^{7/8}}.$$ So by not so direct comparison our

series is divergent.

47. Since $q > 1$, this is a convergent log-power quotient series.

51. Checking the necessary condition for convergence that $\lim\limits_{k \to \infty} a_k = 0$:

$$\lim_{k \to \infty} k^{1/(k-1)} = \lim_{k \to \infty} \frac{1}{k^{1-1/k}} = 0.$$ So it could converge.

Now notice the similarity to a p-series: $\dfrac{1}{k^{1-1/k}} > \dfrac{1}{k^1}$ which is the divergent

harmonic series. So our series diverges. Note that k to a power less than one

is smaller than k, and a smaller denominator means a larger fraction.

55. $\ln k > e^2$ if $e^{\ln k} > e^{e^2}$, $k > e^{e^2}$, $k > 1619$. Replacing the values in

the denominator with smaller values gives a larger fraction, so:

$$\frac{1}{(\ln k)^{\ln k}} < \frac{1}{(e^2)^{\ln k}} = \frac{1}{e^{2 \ln k}} = \frac{1}{e^{\ln k^2}} = \frac{1}{k^2}$$ which is a convergent p-series.

So our series converges.

57. Analysis:

a. If $\lim\limits_{k \to \infty} \dfrac{a_k}{k^p a_k} = $ some positive finite value for certain values of p,

then since the denominator is given as convergent, the limit-comparison test

guarantees the numerator is also convergent. What p will cause

the series to converge?

b. $k^2 e^{-k^2} = \dfrac{k^2}{e^{k^2}}$. Use l'Hôpital's rule twice.

59. Analysis: If $b_k > k^2$ then $\dfrac{a_k}{b_k} < \dfrac{a_k}{k^2} < \dfrac{A}{k^2}$.

61. Analysis: The square of a number less than 1 is smaller than the number.

$a_k{}^2 < a_k$.

8.5 The Ratio Test and the Root Test

SURVIVAL HINT

Page 527 gives an excellent summary of the methods discussed to this point. It should be studied carefully and thoroughly understood. Note that these methods apply to series with non-negative terms.

3. $\lim\limits_{k\to\infty} \dfrac{\frac{1}{(k\,+\,1)!}}{\frac{1}{k!}} = \lim\limits_{k\to\infty} \dfrac{k!}{(k\,+\,1)!} = \lim\limits_{k\to\infty} \dfrac{1}{k\,+\,1} = 0 < 1,$

so the series converges by the ratio test.

7. $\lim\limits_{k\to\infty} \dfrac{\frac{k\,+\,1}{2^{k+1}}}{\frac{k}{2^k}} = \lim\limits_{k\to\infty} \dfrac{2^k(k\,+\,1)}{2^{k+1}\,k} = \lim\limits_{k\to\infty} \tfrac{1}{2}\Big(1\,+\,\tfrac{1}{k}\Big) = \tfrac{1}{2} < 1,$

so the series converges by the ratio test.

11. $\lim\limits_{k\to\infty} \sqrt[k]{k(\tfrac{4}{3})^k} = \lim\limits_{k\to\infty} \tfrac{4}{3}\,k^{1/k}$ which is form ∞^0.

Let $y = k^{1/k}$, $\ln y = \ln k^{1/k} = \tfrac{1}{k}\ln k.$ $\lim\limits_{k\to\infty} \ln y = \lim\limits_{k\to\infty} \dfrac{\ln k}{k}$

which by l'Hôpital's rule $= \lim\limits_{k\to\infty} \dfrac{1}{k} = 0.$

$\lim\limits_{k\to\infty} \ln y = \ln \lim\limits_{k\to\infty} y = 0, \quad \lim\limits_{k\to\infty} y = e^0 = 1.$

So $\lim\limits_{k\to\infty} \tfrac{4}{3}\,k^{1/k} = \tfrac{4}{3} > 1.$ By the ratio test our series diverges.

15. $\lim\limits_{k\to\infty} \dfrac{\frac{(k\,+\,1)^5}{10^{k+1}}}{\frac{k^5}{10^k}} = \lim\limits_{k\to\infty} \dfrac{(k\,+\,1)^5}{10k^5} = \dfrac{1}{10}.$ The series converges by the ratio test.

19. $\lim\limits_{k\to\infty} \dfrac{\frac{(k\,+\,1)!}{(k\,+\,3)^4}}{\frac{k!}{(k\,+\,2)^4}} = \lim\limits_{k\to\infty} \dfrac{(k\,+\,1)(k\,+\,2)^4}{(k\,+\,3)^4} = \infty.$ The series diverges.

23. $\lim\limits_{k\to\infty} \dfrac{\frac{[(k\,+\,1)!]^2}{[(2k\,+\,2)!]^2}}{\frac{(k!)^2}{[(2k)!]^2}} = \lim\limits_{k\to\infty} \dfrac{(k\,+\,1)^2[(2k)!]^2}{[(2k\,+\,2)!]^2} = \lim\limits_{k\to\infty} \dfrac{(k\,+\,1)^2}{(2k\,+\,2)^2(2k\,+\,1)^2}$

$= 0.$ The series converges by the ratio test.

27. 1000 times the divergent harmonic will be divergent. Direct comparison test.

31.
$$\lim_{k \to \infty} \frac{\dfrac{\sqrt{(k+1)!}}{2^{k+1}}}{\dfrac{\sqrt{k!}}{2^k}} = \lim_{k \to \infty} \frac{\sqrt{(k+1)!}}{2\sqrt{k!}} = \lim_{k \to \infty} \frac{1}{2}\sqrt{\frac{(k+1)!}{k!}}$$

$$= \lim_{k \to \infty} \frac{1}{2}\sqrt{k+1} = \infty. \text{ The series diverges by the ratio test.}$$

35.
$$\lim_{k \to \infty} \frac{\dfrac{\sqrt{k+2}}{k^{k+1.5}}}{\dfrac{\sqrt{k+1}}{k^{k+0.5}}} = \lim_{k \to \infty} \left(\frac{1}{k}\right)\sqrt{\frac{k+2}{k+1}} = 0. \text{ Convergent by the ratio test.}$$

39. Write the function as: $\left(\dfrac{1}{1+\frac{1}{k}}\right)^{k^2} = \left\{\left(\dfrac{1}{1+\frac{1}{k}}\right)^k\right\}^k = \left(\dfrac{1}{\left(1+\frac{1}{k}\right)^k}\right)^k$

Recall that a^{b^c} means $(a)^{\left(b^c\right)}$, and not $\left(a^b\right)^c = a^{bc}$.

Now two applications of the root test will give a limit of 1, which is inconclusive.

So make use of the definition of e:

$$\lim_{k \to \infty} \left(\frac{1}{\left(1+\frac{1}{k}\right)^k}\right)^k = \lim_{k \to \infty} \frac{1}{\left(1+\frac{1}{k}\right)^k} \text{ (by the root test)} = \left(\frac{1}{e}\right) < 1.$$

So the series is convergent.

43. By the root test: $\lim_{k \to \infty} \left(\dfrac{\ln k}{k}\right)^k = \lim_{k \to \infty} \left(\dfrac{\ln k}{k}\right).$ Use l'Hôpital's rule:

$$= \lim_{k \to \infty} \frac{1}{k} = 0. \text{ So the series converges.}$$

47. By the ratio test we need $\lim_{k \to \infty} \dfrac{\dfrac{(x+0.5)^{k+1}}{(k+1)^{3/2}}}{\dfrac{(x+0.5)^k}{k^{3/2}}}$

$$= \lim_{k \to \infty} \frac{(x+0.5)k^{3/2}}{(k+1)^{3/2}} = x + 0.5$$

to be less than 1 for convergence. $x + 0.5 < 1$, $x < 0.5$.

We have convergence if $0 < x < 0.5$

51. By the root test we need $\lim_{k \to \infty} ax < 1.$ $x < \frac{1}{a}.$ Also, if $a > 0$ then x must be ≥ 0. We have convergence if $0 < x < \frac{1}{a}.$

53. Analysis: Take the k th root and then investigate $\lim_{k \to \infty} \dfrac{k^{p/k}}{e}.$ The numerator is ∞^0 form, so use logarithms. The integral and the series both converge or diverge.

55.   **a.**   This is a necessary condition for a convergent series.

   b.   Let $a_k = \dfrac{x^k}{k!}$ and establish the convergence of $\left\{\dfrac{x^k}{k!}\right\}$ by the ratio test, then apply the result of **a.**

8.6 Alternating Series; Absolute and Conditional Convergence

SURVIVAL HINT

When using the alternating series test it does not matter whether the first term is positive or negative. However when writing a particular series in sigma notation take care to use $(-1)^k$ or $(-1)^{k+1}$, whichever gives the first term the correct sign.

5. This series is alternating, decreasing, and $\lim\limits_{k\to\infty} a_k = 0$, so it converges.

To test for absolute convergence use the integral test:

$$\lim_{n\to\infty} \frac{1}{3} \int_1^n \frac{3x^2\, dx}{x^3 + 1} = \lim_{n\to\infty} \frac{1}{3} \ln(x^3 + 1) = \infty.$$ The series of absolute values

does not converge. The given series is conditionally convergent.

9. Applying the ratio test to the series of absolute values:

$$\lim_{k\to\infty} \frac{\dfrac{k+1}{2^{k+1}}}{\dfrac{k}{2^k}} = \lim_{k\to\infty} \frac{1}{2} \frac{k+1}{k} = \frac{1}{2} < 1.$$ The series is absolutely convergent.

13. This series is alternating, decreasing, and $\lim\limits_{k\to\infty} a_k = 0$, so it converges.

Applying the ratio test to the series of absolute values:

$$\lim_{k\to\infty} \frac{\dfrac{(k+1)!}{(k+1)^{k+1}}}{\dfrac{k!}{k^k}} = \lim_{k\to\infty} \frac{(k+1)k^k}{(k+1)^{k+1}} = \lim_{k\to\infty} \frac{k^k}{(k+1)^k}$$

$$= \lim_{k\to\infty} \frac{1}{\dfrac{(k+1)^k}{k^k}} = \lim_{k\to\infty} \frac{1}{\left(1 + \frac{1}{k}\right)^k} = \frac{1}{e} < 1.$$

The series is absolutely convergent. (This is similar to example 2 page 615.)

17. Using the ratio test on the series of absolute values:

$$\lim_{k\to\infty} \frac{\dfrac{2^{k+1}}{(k+1)!}}{\dfrac{2^k}{k!}} = \lim_{k\to\infty} \frac{2}{k+1} = 0.$$ The series is absolutely convergent.

21. This series is alternating, decreasing, and $\lim\limits_{k \to \infty} a_k = 0$, so it converges. Considering the series of absolute values:

$$\frac{1}{(\ln k)^4} > \frac{1}{(k \ln k)^4} > \frac{1}{(k \ln k)^k}.$$ Using the root test:

$$\lim_{k \to \infty} \sqrt[k]{\frac{1}{(k \ln k)^k}} = \lim_{k \to \infty} \frac{1}{(k \ln k)}, \text{ which diverges by the integral test.}$$

Our series converges, but does not converge absolutely. It is conditionally convergent.

25. Testing for the necessary conditions: $\lim\limits_{k \to \infty} \dfrac{k}{\ln k}$ by l'Hôpital's rule $=$

$\lim\limits_{k \to \infty} \dfrac{1}{\frac{1}{k}} = \infty.$ This series diverges.

29. This series is alternating, decreasing, and $\lim\limits_{k \to \infty} a_k = 0$, so it converges. We now consider the series of absolute values:

$\dfrac{\ln (\ln k)}{k \ln k} > \dfrac{1}{k \ln k}$ which is known to diverge. The series diverges.

Or by the integral test: Let $u = \ln (\ln x)$, then $du = \dfrac{dx}{x \ln x}$, so

$$\lim_{n \to \infty} \int_2^n \frac{\ln (\ln x) \, dx}{x \ln x} \text{ is of form } u \, du \text{ and } = \lim_{n \to \infty} \tfrac{1}{2}[\ln (\ln x)]^2 \Big|_2^n = \infty.$$

The series converges, but not absolutely, so it is conditionally convergent.

33. **a.** $S_4 = 1 - \dfrac{1}{2} + \dfrac{1}{6} - \dfrac{1}{24} = \dfrac{5}{8}.$ $|E_4| < a_5 = \dfrac{1}{120}$

   b.   We need $E_n < 0.0005$, $\dfrac{1}{n!} < 0.0005$, $n! > 2{,}000$, $n \geq 7$

   $S_7 = 0.63214\ldots$, so accurate to three decimal places $S = 0.632$

37. **a.** $S_4 = -\dfrac{1}{5} + \dfrac{1}{25} - \dfrac{1}{125} + \dfrac{1}{625} = -\dfrac{104}{625} = -0.1664$

   $|E_n| < a_5 = \dfrac{1}{3125} = 0.000\ 32$

   b.   We need $E_n < 0.0005$, $\dfrac{1}{5^n} < 0.000\ 5$, $5^n > 2\ 000$, $n \geq 5$

   $S_5 = -\dfrac{521}{3125} = -0.166\ 72$, so accurate to 3 decimal places $S = -0.167$

SURVIVAL HINT

If a series is not strictly alternating or does not have all non-negative terms
(like example 6), then it can be tested for absolute convergence. If it fails to be
absolutely convergent then it is either conditionally convergent or divergent.
The only test given to determine which, is the generalized ratio test. This exercise
set does not have any series with non-alternating negatives which are conditionally
convergent, but your instructor might find one for an exam! Use the generalized
ratio test.

41. $\lim\limits_{k \to \infty} \dfrac{\dfrac{(k + 3)x^{k+1}}{(k + 1)^2(k + 4)}}{\dfrac{(k + 2)x^k}{k^2(k + 3)}} = \lim\limits_{k \to \infty} \dfrac{k^2(k + 3)^2}{(k + 1)^2(k + 2)(k + 4)} x = x$

So we need $|x| < 1$, $-1 < x < 1$ will guarantee convergence. We now need
to check the endpoints. If $x = -1$, the series is alternating, decreasing,
and $\lim\limits_{k \to \infty} a_k = 0$, so it converges. If $x = 1$, the series is less than $\dfrac{1}{k^2}$,
a convergent p-series, so it converges. Our series converges at both endpoints
so the interval of convergence is $[-1, 1]$.

49. This is a series of positive values, in which $\sin 2^{1/k}$ is always less than 1.
So the series is dominated by the convergent p-series $\dfrac{1}{k^2}$ and is therefore
convergent.

51. Analysis: $\lim\limits_{n \to \infty}\left(\sum\limits_1^n \dfrac{1}{k} - \ln n\right)$ is form $\infty - \infty$. Show convergence by
rewriting it in a form that is alternating and decreasing, and therefore convergent.
Replace $\ln n$ with $\int\limits_1^n \dfrac{1}{x}\, dx$ written in the form: $\int\limits_1^2 \dfrac{1}{x}\, dx + \int\limits_2^3 \dfrac{1}{x}\, dx + \ldots + \int\limits_{n-1}^n \dfrac{1}{x}\, dx$
Alternate these terms with the harmonic and the result is alternating
and decreasing.

53. Analysis: The new series uses enough of the positive terms for a sum
greater than 1, then a negative term to make the sum less than 1, then
enough positive terms so that the sum will again be greater than 1, then
negative terms so that the sum is less than 1, etc. This series will converge to 1.
But the alternating harmonic has a sum of ln 2. Explain.

55.   Analysis:

     a.   Test the general term for the necessary condition of $\lim\limits_{k \to \infty} a_k = 0$.

     b.   Verify that this series meets the conditions for convergence of alternating

          series, then try the integral test to see if it converges absolutely.

59.   Analysis: If the necessary condition of $\lim\limits_{k \to \infty} a_k = 0$ is met, then at some k,

     $a_k < 1$. Now use the comparison test for $a_k{}^2$.

61.   Try the ratio test on the harmonic and on a p-series.

8.7 Power series

SURVIVAL HINT

Power series and their associated intervals of convergence are the basis for the
development of Taylor series in the next section. Taylor series are a powerful tool
for the computation of values for various transcendental functions. To gain a
better perspective about our interest in power series and their convergence it might
be a good idea to read the next section at this point. It is always a good idea to
take a quick look at the material of the next section before the professor lectures on it.
It increases your understanding and allows you to ask perceptive questions.

SURVIVAL HINT

The interval of convergence may be open, closed, or half-open. Always remember to test
both endpoints.

3. $\lim\limits_{k \to \infty} \left| \dfrac{a_{k+1}}{a_k} \right| = |x|$. $R = 1$. Interval of absolute convergence: $-1 < x < 1$.

     $S(1)$ diverges, $S(-1)$ lacks the necessary conditions for convergence and also

     diverges. The interval of convergence is: $(-1, 1)$.

7. $\lim\limits_{k \to \infty} \left| \dfrac{a_{k+1}}{a_k} \right| = \frac{3}{4} |x + 3|$. $R = \frac{4}{3}$. Interval of absolute convergence:

     $|x + 3| < \frac{4}{3}$, $-\frac{13}{3} < x < -\frac{5}{3}$. Testing the endpoints:

     $S(-\frac{13}{3}) = \sum\limits_{1}^{\infty} \dfrac{3^k (\frac{4}{3})^k (-1)^k}{4^k} = -1 + 1 - 1 + 1...$ which is divergent.

     $S(\frac{5}{3}) = 1 + 1 + 1 + ...$ which is also divergent.

     The interval of convergence is $(-\frac{13}{3}, -\frac{5}{3})$.

11. $\lim\limits_{k\to\infty}\left|\dfrac{a_{k+1}}{a_k}\right| = \tfrac{1}{2}|x-1|$. $R = 2$. Interval of absolute convergence:

$|x-1| < 2$, $-1 < x < 3$. Testing the endpoints:

$S(-1) = \sum\limits_{1}^{\infty}(-1)^k k^2$, which diverges. $S(3) = \sum\limits_{1}^{\infty} k^2$, which diverges.

The interval of convergence is: $(-1, 3)$.

15. $\lim\limits_{k\to\infty}\left|\dfrac{a_{k+1}}{a_k}\right| = \dfrac{|x|}{7}$. $R = 7$. Interval of absolute convergence: $(-7, 7)$.

Testing the endpoints: $S(-7) = \sum\limits_{1}^{\infty} k(-1)^k$, which diverges.

$S(7) = \sum\limits_{1}^{\infty} k$, which diverges. The interval of convergence is $(-7, 7)$.

19. $\lim\limits_{k\to\infty}\left|\dfrac{a_{k+1}}{a_k}\right| = |x|$. $R = 1$. Interval of absolute convergence: $(-1, 1)$.

$S(-1) = \sum\limits_{1}^{\infty}\dfrac{1}{k(\ln k)^2}$. Using the integral test: $\lim\limits_{n\to\infty}\int\limits_{1}^{n}(\ln x)^{-2}\,\tfrac{1}{x}\,dx$

is $u^n du$ form, $= \lim\limits_{n\to\infty}[-(\ln x)^{-1}]\Big|_{1}^{n} = \lim\limits_{n\to\infty}\left(-\dfrac{1}{\ln n} + \dfrac{1}{\ln 1}\right) = 0 + \infty$,

and so it diverges. $S(1) = \sum\limits_{1}^{\infty}\dfrac{(-1)^k}{k(\ln k)^2}$ which is alternating, decreasing, and

has $\lim\limits_{k\to\infty} a_k = 0$, so it converges. The interval of convergence is $(-1, 1]$.

23. $\lim\limits_{k\to\infty}\left|\dfrac{a_{k+1}}{a_k}\right| = \lim\limits_{k\to\infty}\left|\dfrac{(k+1)(3x)^3}{2}\right| = \infty$, $R = 0$. The series converges

only in the trivial case when $x = 0$.

27. $\lim\limits_{k\to\infty}\left|\dfrac{a_{k+1}}{a_k}\right| = |3(5x)^4| < 1$, $|(5x)^4| < \tfrac{1}{3}$, $|x| < \sqrt[4]{\dfrac{1}{3(5^4)}} = \dfrac{\sqrt[4]{27}}{15}$.

$R = \dfrac{\sqrt[4]{27}}{15}$. It is divergent at both endpoints, so the interval of convergence

is: $\left(-\dfrac{\sqrt[4]{27}}{15}, \dfrac{\sqrt[4]{27}}{15}\right)$.

31. $\lim\limits_{k\to\infty}\left|\dfrac{a_{k+1}}{a_k}\right| = \left|2^{(\sqrt{k+1}-\sqrt{k})}(x-1)\right| = |x-1|$. (Rationalize the radicals

to find their limit of 0.) $R = 1$. The interval of absolute convergence:

$0 < x < 2$. It is easy to verify that both endpoints diverge, so the interval

of convergence is $(0, 2)$.

35. $f(x) = 1 + \dfrac{x}{2} + \dfrac{x^2}{4} + \dfrac{x^3}{8} + \dfrac{x^4}{16} + \cdots$

$f'(x) = (1)\dfrac{1}{2} + (2)\dfrac{x}{4} + (3)\dfrac{x^2}{8} + (4)\dfrac{x^3}{16} + \cdots = \sum\limits_{1}^{\infty}\dfrac{kx^{k-1}}{2^k}$

39. $f(x) = 1 + \frac{x}{2} + \frac{x^2}{4} + \frac{x^3}{8} + \frac{x^4}{16} + \cdots$

$$\int_0^\infty f(x)\,dx = x + \frac{x^2}{2(2)} + \frac{x^3}{3(2)^2} + \frac{x^4}{4(2)^3} + \cdots = \sum_1^\infty \frac{x^k}{k(2)^{k-1}}$$

45. Analysis: Apply the generalized ratio test to this series, and substitute $\frac{1}{R}$ for $\left|\frac{a_{k+1}}{a_k}\right|$.

47. Analysis: Let $u = \frac{1}{x}$. The resulting series converges for $|u| < 1$, $\left|\frac{1}{x}\right| < 1$, $|x| > 1$. Do not forget to check endpoints.

49. Analysis: Apply the generalized ratio test and carefully reduce the resulting factorials.

51. Analysis: Write out several term for this series, then differentiate twice. Compare the result to the original function.

8.8 Taylor and Maclaurin Series

SURVIVAL HINT

The power and the utility of the Taylor series is well illustrated by the list of functions on page 561, which we can find values for, to any desired accuracy. From thils exercise set you might add other functions to the list; such as $\sqrt{x}$, cosh x, and sinh x.

3. Since $e^x = 1 + x + \frac{x^2}{2!} + \frac{x^3}{3!} + \frac{x^4}{4!} + \cdots = \sum_0^\infty \frac{x^k}{k!}$,

$e^{2x} = \sum_0^\infty \frac{(2x)^k}{k!}$

7. Since $\sin x = x - \frac{x^3}{3!} + \frac{x^5}{5!} - \cdots = \sum_0^\infty \frac{(-1)^k x^{2k+1}}{(2k+1)!}$

$\sin x^2 = \sum_0^\infty \frac{(-1)^k x^{4k+2}}{(2k+1)!}$

11. Since $\cos x = 1 - \frac{x^2}{2!} + \frac{x^4}{4!} - \frac{x^6}{6!} + \cdots = \sum_0^\infty \frac{(-1)^k x^{2k}}{(2k)!}$

$\cos 2x^2 = \sum_0^\infty \frac{(-1)^k (2x^2)^{2k}}{(2k)!} = \sum_0^\infty \frac{(-1)^k (4)^k (x)^{4k}}{(2k)!}$

15. $(1 + x)^2 = 1 + 2x + \frac{2(2-1)}{2!}x^2 = 1 + 2x + x^2$

19. $e^x + \sin x = \sum_0^\infty \frac{x^k}{k!} + \sum_0^\infty \frac{(-1)^k x^{2k+1}}{(2k+1)!} = \sum_0^\infty \left(\frac{x^k}{k!} + \frac{(-1)^k x^{2k+1}}{(2k+1)!} \right)$

23. Since $\dfrac{1}{1-x} = 1 + x + x^2 + x^3 + \ldots = \displaystyle\sum_0^\infty x^k$, factor out the a and substitute $\left(-\dfrac{x}{a}\right)$ for x:

$$\frac{1}{a\left(1 + \frac{x}{a}\right)} = \frac{1}{a\left(1 - \left\{-\frac{x}{a}\right\}\right)} = \frac{1}{a}\sum_0^\infty \left(-\frac{x}{a}\right)^k$$

27. Since $\tan^{-1}x = x - \dfrac{x^3}{3} + \dfrac{x^5}{5} - \dfrac{x^7}{7} + \ldots = \displaystyle\sum_0^\infty \frac{(-1)^k x^{2k+1}}{2k+1}$,

$$\tan^{-1}(2x) = \sum_0^\infty \frac{(-1)^k (2x)^{2k+1}}{2k+1}$$

31. e^x at $c = 1$ is given by $\displaystyle\sum_0^\infty \frac{(x-1)^k}{k!}$, with $R_n = \dfrac{f^{(n+1)}(z)(x-1)^{n+1}}{(n+1)!}$

$$= \frac{e^z(x-1)^{n+1}}{(n+1)!}. \quad \lim_{n\to\infty} \left| \frac{\dfrac{e^z(x-1)^{n+2}}{(n+2)!}}{\dfrac{e^z(x-1)^{n+1}}{(n+1)!}} \right| = \lim_{n\to\infty}\left|\frac{x-1}{n+2}\right| = 0.$$

$R = \infty$, and it converges for all x.

For $f(x) = e^x$ at 1 note that $f^{(n)}(1) = e^1 = e$, so

e^x at $c = 1$ is given by: $e\left\{ 1 + (x-1) + \dfrac{(x-1)^2}{2!} + \dfrac{(x-1)^3}{3!} + \ldots\right\}$

35. $\tan x = \dfrac{\sin x}{\cos x} = \dfrac{\displaystyle\sum_0^\infty \dfrac{(-1)^k x^{2k+1}}{(2k+1)!}}{\displaystyle\sum_0^\infty \dfrac{(-1)^k x^{2k}}{(2k)!}} = x + \dfrac{x^3}{3} + \dfrac{2x^5}{15} + \dfrac{17x^7}{315} + \ldots$

(The division can not be done term by term. Use the long division algorithm.)

The series is convergent since it is the quotient of two convergent series.

39. $f(x) = (2-x)^{-1}$ $f(5) = -\dfrac{1}{3}$

$f'(x) = (2-x)^{-2}$ $f'(5) = \dfrac{1}{9}$

$f''(x) = 2(2-x)^{-3}$ $f''(5) = -\dfrac{2}{27}$

$f'''(x) = 6(2-x)^{-4}$ $f'''(5) = \dfrac{6}{81}$

$-\dfrac{1}{3} + \dfrac{1}{9}(x-5) - \dfrac{1}{27}(x-5)^2 + \dfrac{1}{81}(x-5)^3 - \ldots$

$= -\dfrac{1}{3}\left[1 + \left(\dfrac{x-5}{-3}\right)^1 + \left(\dfrac{x-5}{-3}\right)^2 + \left(\dfrac{x-5}{-3}\right)^3 + \ldots \right]$

$= \displaystyle\sum_0^\infty \frac{(-1)^{k+1}(x-5)^k}{3^{k+1}}$

$R_n = f^{(n+1)}(z)\left(\dfrac{x-5}{-3}\right)^{n+1}. \quad \lim_{n\to\infty} R_n = 0$ for $\left|\dfrac{x-5}{-3}\right| < 1,\ 2 < x < 8.$

43. $f(x) = (1 + x)^{\frac{1}{2}} = 1 + \frac{1}{2}x + \dfrac{\frac{1}{2}(\frac{1}{2} - 1)x^2}{2!} + \dfrac{\frac{1}{2}(\frac{1}{2} - 1)(\frac{1}{2} - 2)x^3}{3!} + \cdots,$

$\qquad = 1 + \frac{1}{2}x - \frac{1}{8}x^2 + \frac{1}{16}x^3 - \frac{5}{128}x^4 + \cdots$

If p is greater than 0 and not an integer the interval of absolute convergence
is $(-1, 1)$.

47. $f(x) = x(1 - x^2)^{-\frac{1}{2}} = x\left(1 - (-\frac{1}{2})x^2 + \dfrac{(-\frac{1}{2})(-\frac{1}{2} - 1)x^4}{2!}\right.$

$\qquad \left. - \dfrac{(-\frac{1}{2})(-\frac{1}{2} - 1)(-\frac{1}{2} - 2)x^6}{3!} + \cdots \right)$

$\qquad = x - (-\frac{1}{2})x^3 + \dfrac{(-\frac{1}{2})(-\frac{1}{2} - 1)x^5}{2!} - \dfrac{(-\frac{1}{2})(-\frac{1}{2} - 1)(-\frac{1}{2} - 2)x^7}{3!} + \cdots$

$\qquad = x + \frac{1}{2}x^3 + \frac{3}{8}x^5 + \frac{5}{16}x^7 + \cdots$

Since p is in the interval $(-1, 0)$, the interval of absolute convergence for the
series will be $(-1, 1)$.

51. $\sinh x = \dfrac{e^x - e^{-x}}{2} = \dfrac{\displaystyle\sum_0^\infty \frac{x^k}{k!} - \sum_0^\infty \frac{(-x)^k}{k!}}{2}$

$\qquad = \frac{1}{2}\left(1 + x + \frac{x^2}{2!} + \frac{x^3}{3!} + \frac{x^4}{4!} + \cdots - \left(1 - x + \frac{x^2}{2!} - \frac{x^3}{3!} + \frac{x^4}{4!} - \cdots\right)\right)$

$\qquad = \frac{1}{2}\left(2x + \frac{2x^3}{3!} + \frac{2x^5}{5!} + \cdots\right) = \sum_0^\infty \dfrac{x^{2k+1}}{(2k + 1)!}$

55. $f(x) = \dfrac{6 - x}{(2 - x)(2 + x)} = \dfrac{1}{2(1 - \frac{x}{2})} + \dfrac{2}{2(1 + \frac{x}{2})}$ by partial fractions

$\qquad = \frac{1}{2}\sum_0^\infty \left(\frac{x}{2}\right)^k + \sum_0^\infty \left(-\frac{x}{2}\right)^k = \sum_0^\infty \left(\frac{1}{2} + (-1)^k\right)\left(\frac{x}{2}\right)^k$

59. $f(x) = \dfrac{-x^2}{(2 + x)(1 - x^2)} = \dfrac{\frac{4}{3}}{2(1 + \frac{x}{2})} + \dfrac{-\frac{1}{6}}{1 - x} + \dfrac{-\frac{1}{2}}{1 + x}$

$\qquad = \frac{2}{3}\sum_0^\infty \left(-\frac{x}{2}\right)^k - \frac{1}{6}\sum_0^\infty x^k - \frac{1}{2}\sum_0^\infty (-x)^k = \sum_0^\infty \left(\frac{2}{3}(-\frac{1}{2})^k - \frac{1}{6} - \frac{1}{2}(-1)^k\right)x^k$

63.   In the identity let $x = 2x$, $y = x$:

$\qquad f(x) = [\cos \frac{1}{2}(2x + x)][\cos \frac{1}{2}(2x - x)] = \frac{1}{2}\cos 2x + \frac{1}{2}\cos x$

$\qquad = \frac{1}{2}\sum_0^\infty \dfrac{(-1)^k(2x)^{2k}}{(2k)!} + \frac{1}{2}\sum_0^\infty \dfrac{(-1)^k(x)^{2k}}{(2k)!} = \frac{1}{2}\sum_0^\infty \dfrac{(-1)^k}{(2k)!}\left(4^k + 1\right)x^{2k}.$

67. $g(x) = \dfrac{e^x - 1}{x} = \dfrac{\displaystyle\sum_0^\infty \dfrac{x^k}{k!} - 1}{x} = 1 + \dfrac{x}{2!} + \dfrac{x^2}{3!} + \dfrac{x^3}{4!} + \ldots = \displaystyle\sum_0^\infty \dfrac{x^k}{(k+1)!}$

$\lim\limits_{x \to 0} g(x) = 1$. This can be easily checked by using l'Hôpital's rule on $g(x)$.

73. Analysis: The second derivative of $(1-x)^{-1} = 2(1-x)^{-3}$. So

$$f(x) = \frac{1}{2} \frac{d^2}{dx^2} \sum_0^\infty x^k.$$

75. Analysis: Write the series for e^x, multiply by $x^{0.2}$, then integrate the resulting infinite series term by term.

77. Analysis: From **74b** find J_0'' and substitute into the given expression to verify the equality.

Chapter 8 Review

PRACTICE PROBLEMS

31. The sequence has an upper bound of 4, a lower bound of 0, and after $n = 4$ is monotonic decreasing:

$$\lim_{n \to \infty} \frac{\dfrac{e^{n+1}}{(n+1)!}}{\dfrac{e^n}{n!}} = \lim_{n \to \infty} \frac{e}{n+1} = 0.$$ The sequence converges.

Note that from a_k to a_{k+1} the numerator increases by a factor of e, while the denominator increases by a factor of n. The denominator is growing more rapidly than the numerator, so the limit will be 0.

32. Divide the numerator and denominator by n^2. The limit $= -\dfrac{3}{2}$.

33. This is a definition of e given in Chapter 5.

34. Different type of expressions suggest the ratio test:

$$\lim_{k \to \infty} \frac{\dfrac{e^{k+1}}{(k+1)!}}{\dfrac{e^k}{k!}} = \lim_{k \to \infty} \frac{e}{k+1} = 0 < 1.$$ The series converges.

The convergence can also be verified by finding the value to which it converges:

$$\sum_0^\infty \frac{x^k}{k!} = e^x, \quad \text{so} \quad \sum_0^\infty \frac{e^k}{k!} = e^e, \quad \text{and} \quad \sum_1^\infty \frac{e^k}{k!} = e^e - 1$$

35. This series does not meet the necessary convergence condition of $\lim\limits_{k \to \infty} a_k = 0$. (see #32) The series diverges.

36. Use the integral test: $\displaystyle\lim_{n\to\infty}\int_2^n (\ln x)^{-1}\dfrac{dx}{x} = \lim_{n\to\infty}\ln(\ln x)\Big|_2^n = \infty.$
The series diverges.

37. The series is alternating, decreasing, and has a limit of 0. It is convergent.
Testing the series of absolute values: $\displaystyle\lim_{k\to\infty}\sum_1^\infty \dfrac{1}{k^2}$ is a convergent p-series.
The series converges absolutely.

38. $\displaystyle S = \sum_0^\infty (-1)^k k x^k.$ $\displaystyle\lim_{k\to\infty}\left|\dfrac{(k+1)\,x^{k+1}}{kx^k}\right| = |x| < 1.$ The series is

absolutely convergent on $(-1, 1)$. Testing the endpoints: $\displaystyle S(-1) = \sum_0^\infty k,$

which is divergent. $\displaystyle S(1) = \sum_0^\infty (-1)^k k = 1 - 1 + 1 - 1 + ...,$

which is also divergent. The interval of convergence is $(-1, 1)$.

39. $\displaystyle \sin x = \sum_0^\infty \dfrac{(-1)^k x^{2k+1}}{(2k+1)!}. \quad \sin 2x = \sum_0^\infty \dfrac{(-1)^k (2x)^{2k+1}}{(2k+1)!}.$

40. $f(x) = \dfrac{1}{x-3}$ at $c = \dfrac{1}{2}$

$$= \dfrac{1}{x - \dfrac{1}{2} - \dfrac{5}{2}}$$

$$= \dfrac{-1}{\dfrac{5}{2} - \left(x - \dfrac{1}{2}\right)}$$

$$= \dfrac{-\dfrac{2}{5}}{1 - \dfrac{2}{5}\left(x - \dfrac{1}{2}\right)}$$

$$= -\dfrac{2}{5}\left\{1 + \dfrac{2}{5}\left(x - \dfrac{1}{2}\right) + ... \right\}$$

$$= -\dfrac{2}{5}\sum_0^\infty \left(\dfrac{2}{5}\right)^k\left(x - \dfrac{1}{2}\right)^k$$

Or, using the $\dfrac{1}{1-x}$ series first, and then translating for $c = \dfrac{1}{2}$

$$f(x) = -\dfrac{1}{3-x} = -\dfrac{1}{3}\dfrac{1}{1-\dfrac{x}{3}} = -\dfrac{1}{3}\sum_0^\infty \left(\dfrac{x}{3}\right)^k.$$

$f(x)$ expanded about $c = \dfrac{1}{2}:$ $-\dfrac{1}{3}\displaystyle\sum_0^\infty \left(\dfrac{x - \dfrac{1}{2}}{3}\right)^k = -\dfrac{1}{3}\sum_0^\infty \left(\dfrac{2x-1}{6}\right)^k.$

Chapter 9
Polar Coordinates and the Conic Sections

9.1 The Polar Coordinate System

SURVIVAL HINT

When a problem asks for exact values, rather than calculator approximations, it will most likely involve an isosceles right triangle or a 30-60-90° triangle. The sides are in the ratio of 1, 1, $\sqrt{2}$; or 1, 2, $\sqrt{3}$. Proper placement of the triangle on a coordinate system usually solves the problem.

3. $(4, \frac{\pi}{4})$ and $(-4, \frac{5\pi}{4})$. $x = r\cos\theta = 4\frac{\sqrt{2}}{2} = 2\sqrt{2}$, $y = r\sin\theta = 2\sqrt{2}$.

Rectangular representation: $(2\sqrt{2}, 2\sqrt{2})$

7. $(\frac{3}{2}, \frac{7}{6}\pi)$ and $(-\frac{3}{2}, \frac{\pi}{6})$. $x = \frac{3}{2}\left(-\frac{\sqrt{3}}{2}\right) = -\frac{3\sqrt{3}}{4}$, $y = \frac{3}{2}\left(-\frac{1}{2}\right) = -\frac{3}{4}$

Rectangular representation: $\left(-\frac{3\sqrt{3}}{4}, -\frac{3}{4}\right)$

11. $(0, \pi - 3)$ and $(0, 2\pi - 3)$. $x = 0$, $y = 0$. Rectangular representation: $(0, 0)$.

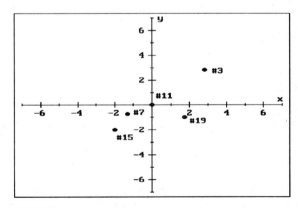

15. $r = \sqrt{(-2)^2 + (-2)^2} = 2\sqrt{2}$, $\theta = \tan^{-1}\frac{-2}{-2} = \frac{5\pi}{4}$ (Quadrant III)

Polar representation: $(-2\sqrt{2}, \frac{\pi}{4})$ and $(2\sqrt{2}, \frac{5\pi}{4})$

19. $r = \sqrt{(\sqrt{3})^2 + (-1)^2} = 2, \ \theta = \tan^{-1} \dfrac{-1}{\sqrt{3}} = \dfrac{11\pi}{6}$ (Quadrant IV)

Polar representation: $(-2, \frac{5\pi}{6})$ and $(2, \frac{11\pi}{6})$

23. $r = 4 \sin \theta, \ r^2 = 4 \, r \sin \theta, \ x^2 + y^2 = 4y, \ x^2 + y^2 - 4y + 4 = 4,$

$x^2 + (y - 2)^2 = 4$

27. $r = \sec \theta, \ r \cos \theta = 1, \ x = 1$

31. Since θ is unspecified and can be all values, this is a circle with radius of $\frac{3}{2}$

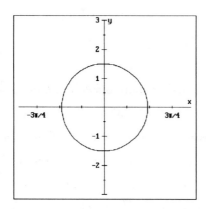

35. Since r is unspecified and can be all values, this is a line in the direction $\theta = 1$.

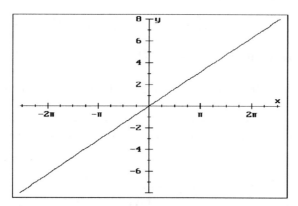

39. $10 = \dfrac{5}{1 - \sin \frac{\pi}{6}}, \ \ 10 = \dfrac{5}{1 - \frac{1}{2}}, \ \ 10 = 10.$ The point is on the graph.

43. $20 + 10\sqrt{3} = \dfrac{5}{1 - \sin \frac{\pi}{3}}, \ \ 10(2 + \sqrt{3}) = \dfrac{5}{1 - \frac{\sqrt{3}}{2}}, \ \ 10(2 + \sqrt{3}) = \dfrac{10}{2 - \sqrt{3}},$

$10(2 + \sqrt{3})(2 - \sqrt{3}) = 10, \ \ 10 = 10.$ The point is on the graph.

47. $-1 = 2(1 - \cos\frac{\pi}{3}) = 2(1 - \frac{1}{2}) = 1$, does not satisfy equation. We need to try the other primary representation $(1, \frac{4\pi}{3})$: $1 = 2(1 - \cos\frac{4\pi}{3})$ $= 2(1 + \frac{1}{2}) = 3$. Also does not satisfy the equation.

The point is not on the graph.

51. $0 = 2(1 - \cos\frac{\pi}{4}) = 2(1 - \frac{\sqrt{2}}{2}) \neq 0$. Try $(0, \frac{5\pi}{4})$; also fails.

The point is not on the curve.

55. If $\theta = \frac{\pi}{2}$: $(\frac{3\pi}{2}, \frac{\pi}{2})$ and $(-\frac{3\pi}{2}, \frac{3\pi}{2})$. If $\theta = \frac{\pi}{4}$: $(\frac{3\pi}{4}, \frac{\pi}{4})$ and $(-\frac{3\pi}{4}, \frac{5\pi}{4})$.

If $\theta = \pi$: $(3\pi, \pi)$ and $(-3\pi, 0)$.

59. If $\theta = 0$: $(2, 0)$ and $(-2, \pi)$. If $\theta = \frac{\pi}{6}$: $(1, \frac{\pi}{6})$ and $(-1, \frac{7\pi}{6})$.

If $\theta = \frac{7\pi}{6}$: $(\frac{4}{3}, \frac{7\pi}{6})$ and $(-\frac{4}{3}, \frac{\pi}{6})$.

61. Analysis:

   a. Consider the distance formula for the points $(3, \pi)$ and $(3, \frac{\pi}{2})$.

      Is the answer $3\sqrt{2}$? Why not?

   b. For $P_1 = (r_1, \theta_1)$, and $P_2 = (r_2, \theta_2)$, $x_1 = r_1\cos\theta_1$, $y_1 = r_1\sin\theta_1$, $x_2 = r_2\cos\theta_2$, $y_2 = r_2\sin\theta_2$. Now use the distance formula.

   c. The Law of Cosines is derived by equating two distance formulas. Check a trigonometry book for this equality.

63. Analysis: Multiply through by r and make the appropriate rectangular substitutions.

9.2 Graphing in Polar Coordinates

SURVIVAL HINT

Study page 583 until you can identify a circle, limacon, cardioid, rose and lemniscate from its equation. Once the general shape is identified, a decent sketch can usually be drawn by plotting the four intercepts.

5.   **a.** A four leafed rose with length of petal $= 2$.

     b. A lemniscate with length $\sqrt{2}$.

     c. A circle, since r is equal to a constant.

     d. A 16 leafed rose with length of petal $= 5$.

     e. None. An inward spiral, $r = \frac{3}{\theta}$. r decreases as θ increases.

     f. A lemniscate, length 3, rotated $\frac{\pi}{8}$.

     g. A 3 leafed rose, rotated $-\frac{\pi}{6}$.

     h. A cardioid symmetric about the x-axis.

9. The y-axis from 0 to -3.

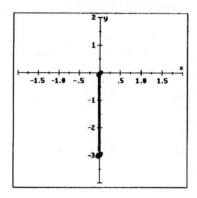

13. A spiral.

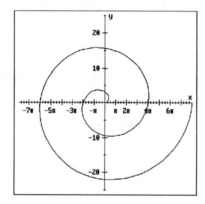

17. A 3 leafed rose, leaf length $=5$. First leaf tip when $\sin 3\theta = 1$, $\theta = \frac{\pi}{6}$.

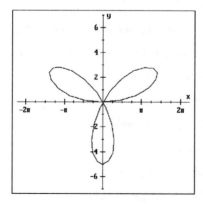

21. A 3 leafed rose, leaf length $= 3$. First leaf tip when $\cos 3(\theta - \frac{\pi}{4}) = 1$, $\theta = \frac{\pi}{4}$.

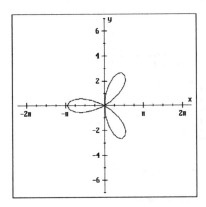

25. A 4 leafed rose, leaf length $= 1$. First leaf tip when $\sin 2(\theta + \frac{\pi}{6}) = 1$, $\theta = \frac{\pi}{12}$

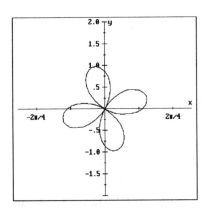

29. Limaçon, symmetric about the x-axis.

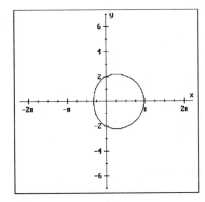

33.   Vertical line. $x = 2$

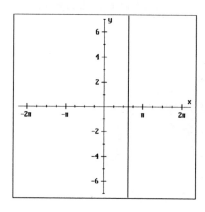

37.   Circle, symmetric about the negative y-axis.

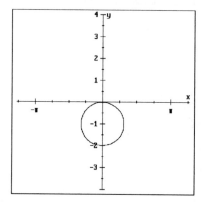

41.   On the θ interval $\left(0, \frac{\pi}{2}\right)$ we get a semi-circle. On $\left(\frac{\pi}{2}, \pi\right)$ we get a semi-circle in quadrant II. On $\left(\pi, \frac{3\pi}{2}\right)$ we get a semi-circle in quadrant III. On $\left(\frac{3\pi}{2}, 2\pi\right)$ we get another semi-circle in quadrant IV.

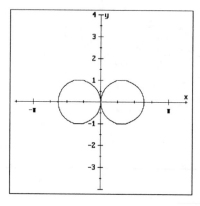

45. The interior of the circle with radius 1. Boundary included

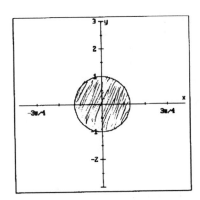

49. The exterior of the circle with radius 1. Boundary not included.

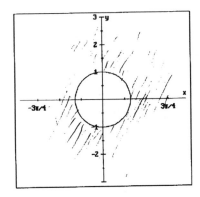

53. A section of a ring. Boundaries included.

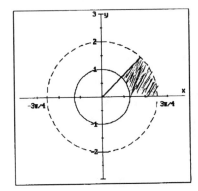

57.   The strophoid with $a = 2$: $r = 2 \cos 2\theta \sec \theta$

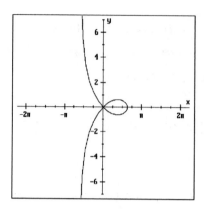

61. $y = \cos x$ is the cosine wave and $r = \cos \theta$ is a circle.

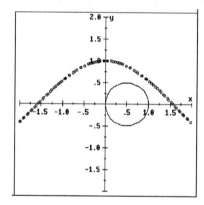

65.   The trigonometric function $y = \csc x$ and $r = \csc \theta$, $r \sin \theta = 1$, $y = 1$.

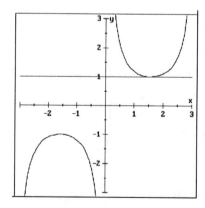

69.   Analysis: Make the substitutions to show the definition of x-symmetry:
            $r = f(\theta) = f(-\theta)$. Draw a graph.

71.   Analysis: Show that the graphs have the same amount of rotation, one clock-wise and one counter clock-wise from $\theta = \frac{\pi}{4}$.

73.   Analysis:

   a.   Use a sketch to show $r = f(\theta)$ and $r = f(\theta - \alpha)$.

   b.   Note that $r = 2 \sec \theta$ is $r \cos \theta = 2$ or $x = 2$. It's rotation will be an oblique rather than a vertical line.

9.3 Area and Tangent Lines in Polar Coordinates

SURVIVAL HINT

Graphing in polar coordinates is not a strict one-to-one correspondence, as is a Cartesian graph. Any value of θ for which $r = 0$ designates the pole. Remember to treat the pole as singleton point. If r can be zero for *any* value of θ, the pole is a point of the graph.

5. $8 \cos \theta = 8 \sin \theta$, $\tan \theta = 1$, $\theta = \frac{\pi}{4}$, $r = 8 \cos \frac{\pi}{4} = 4\sqrt{2}$.

The pole also satisfies both equations. $(0, 0)$, $(4\sqrt{2}, \frac{\pi}{4})$

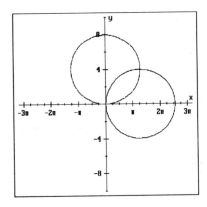

9. $4 = 4 \sin^2 2\theta$, $\sin^2 2\theta = 1$, $\sin 2\theta = \pm 1$, $r^2 \neq -1$, so $2\theta = \frac{\pi}{2}, \frac{5\pi}{2}$, $\theta = \frac{\pi}{4}, \frac{5\pi}{4}$. $(2, \frac{\pi}{4})$, $(2, \frac{5\pi}{4})$

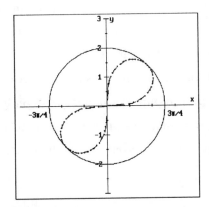

13. $8 \cos^2 \theta = 4 \sin 2\theta$, $2 \cos^2 \theta = 2 \sin \theta \cos \theta$, $\cos \theta (\cos \theta - \sin \theta) = 0$,
$\cos \theta = 0$, $\tan \theta = 1$, $\theta = \frac{\pi}{2}$, $r = 0$. $\theta = \frac{\pi}{4}$, $r = 2$. $(0, 0)$, $(2, \frac{\pi}{4})$

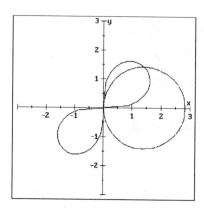

17. $2 \sin^2 \theta = \cos 2\theta$, $2 \sin^2 \theta = 1 - 2 \sin^2 \theta$, $4 \sin^2 \theta = 1$, $\sin^2 \theta = \frac{1}{4}$,
$\sin \theta = \pm \frac{1}{2}$, $\theta = \frac{\pi}{6}, \frac{5\pi}{6}$, $r = \frac{\sqrt{2}}{2}$. The pole also satisfies both equations.
$(0, 0)$, $(\frac{\sqrt{2}}{2}, \frac{\pi}{6})$, $(\frac{\sqrt{2}}{2}, \frac{5\pi}{6})$

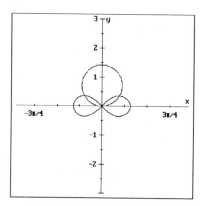

21. $\sin \theta = 2 \cos \theta + 1$, Squaring: $\sin^2 \theta = 4 \cos^2 \theta + 4 \cos \theta + 1$,
$1 - \cos^2 \theta = 4 \cos^2 \theta + 4 \cos \theta + 1$, $5 \cos^2 \theta + 4 \cos \theta = 0$,
$\cos \theta (5 \cos \theta + 4) = 0$, $\cos \theta = 0$, $\cos \theta = -\frac{4}{5}$, $\theta = \frac{\pi}{2}$, $\cos^{-1}(-\frac{4}{5})$.
$r = 1, \pm \frac{3}{5}$. The pole also satisfies both equations. The point corresponding to
$\cos^{-1}(-\frac{4}{5})$ has θ in quadrants II or III. We want the quadrant III value,
$(-\frac{3}{5}, \approx 3.785)$, as the cardioid has traced past π before arriving at this
point (although it appears in the first quadrant of the graph). However
the instructions call for a positive r, so write it as $(\frac{3}{5}, \pi - \cos^{-1}\frac{4}{5})$.
Note that although this is a point of intersection on the graph it is not a
simultaneous solution to the two equations. It is passed through with
different values of θ. The "points of intersection" are $(0, 0)$, $(1, \frac{\pi}{2})$,
and $(\frac{3}{5}, \approx .6435)$.

21. (con't.)

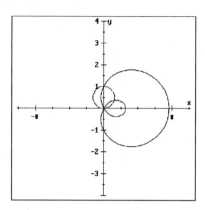

25. $2 \cos \theta = \dfrac{1}{1 - \cos \theta}$, $2 \cos \theta - 2 \cos^2\theta = 1$, $2 \cos^2\theta - 2 \cos \theta + 1 = 0$ has no real roots. The pole is not a solution. There are no intersections.

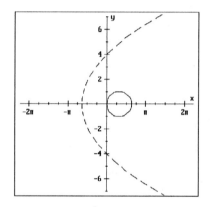

29. $4 \sin^2 \theta = 1$, $\sin \theta = \pm\dfrac{1}{2}$, $\theta = \dfrac{\pi}{6}$, $\dfrac{5\pi}{6}$, $r = 2$. The pole is not a solution.
$\left\{(2, \dfrac{\pi}{6}), (2, \dfrac{5\pi}{6})\right\}$

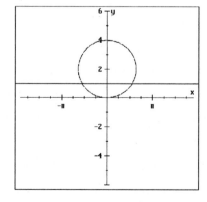

SURVIVAL HINT

Be careful to include the coefficient of $\frac{1}{2}$ in the area formula. A full circle is 2π radians and its area is πr^2. Notice the $\frac{1}{2}$ ratio. The area of a sector is $\frac{1}{2}\theta\,r^2$.

33. $A = \dfrac{1}{2}\displaystyle\int_{\frac{\pi}{6}}^{\frac{\pi}{2}} \sin\theta\,d\theta = -\dfrac{1}{2}\cos\theta\,\Big|_{\frac{\pi}{6}}^{\frac{\pi}{2}} = 0 + \dfrac{1}{2}\left(\dfrac{\sqrt{3}}{2}\right) = \dfrac{\sqrt{3}}{4}$

37. $A = \dfrac{1}{2}\displaystyle\int_{0}^{2\pi}\left(\dfrac{\theta^2}{\pi}\right)^2 d\theta = \dfrac{1}{2\pi^2}\dfrac{\theta^5}{5}\,\Big|_{0}^{2\pi} = \dfrac{(2\pi)^5}{10\pi^2} = \dfrac{16\pi^3}{5}$

41. $f(\theta) = r = 2,\ x = 2\cos\theta,\ \dfrac{dx}{d\theta} = -2\sin\theta,\ y = 2\sin\theta,\ \dfrac{dy}{d\theta} = 2\cos\theta$

$m = \dfrac{\frac{dy}{d\theta}}{\frac{dx}{d\theta}} = \dfrac{2\cos\theta}{-2\sin\theta} = -\cot\theta.\ m\left(\dfrac{\pi}{3}\right) = -\cot\dfrac{\pi}{3} = -\dfrac{\sqrt{3}}{3}$

45. $2\,r\cos\theta + 3\,r\sin\theta = 3,\ 2x + 3y = 3,\ m = -\dfrac{2}{3}$ for all values of θ,

as this is a straight line.

49.   For the tangent to be 0 we need $m = 0$. This requires $\dfrac{dy}{d\theta} = 0$, and $\dfrac{dx}{d\theta} \neq 0$.

$y = r\sin\theta = \sin\theta[a(1 + \cos\theta)] = a\sin\theta + a\sin\theta\cos\theta$.

$\dfrac{dy}{d\theta} = a\cos\theta - a\sin^2\theta + a\cos^2\theta$. (product rule)

$= a(2\cos^2\theta + \cos\theta - 1) = a(2\cos\theta - 1)(\cos\theta + 1) = 0$.

$\cos\theta = \dfrac{1}{2}, -1.\ \theta = \dfrac{\pi}{3}, \dfrac{5\pi}{3}, \pi.\ (0, \pi)$ is not a solution because $\dfrac{dx}{d\theta} = 0$ there

and the slope is undefined. $\left(\dfrac{3a}{2}, \dfrac{\pi}{3}\right),\ \left(\dfrac{3a}{2}, \dfrac{5\pi}{3}\right)$

53.

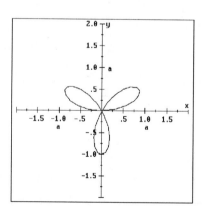

$r = a(\sin 3\theta)$ has a tip of a leaf when $\sin 3\theta = 1,\ 3\theta = \dfrac{\pi}{2},\ \theta = \dfrac{\pi}{6}$.

53. (con't.)

We will find the area from $\theta = 0$ to $\theta = \frac{\pi}{6}$ and multiply by 6.

$$A = (6)\frac{1}{2}\int_0^{\frac{\pi}{6}} a^2 \sin^2 3\theta \, d\theta = 3a^2\int_0^{\frac{\pi}{6}} \left(\frac{1 - \cos 6\theta}{2}\right) d\theta = \frac{3a^2}{2}\int_0^{\frac{\pi}{6}} (1 - \cos 6\theta)\, d\theta$$

$$= \frac{3a^2}{2}\left(\theta - \frac{1}{6}\sin 6\theta\right)\Big|_0^{\frac{\pi}{6}} = \frac{3a^2}{2}\left(\frac{\pi}{6}\right) = \frac{\pi a^2}{4}$$

57.

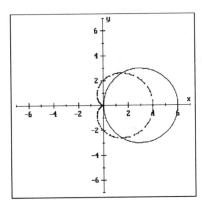

Solving simultaneously: $6\cos\theta = 2 + 2\cos\theta$, $4\cos\theta = 2$, $\cos\theta = \frac{1}{2}$, $\theta = \frac{\pi}{3}, \frac{5\pi}{3}$. We will find the area from $\theta = 0$ to $\theta = \frac{\pi}{3}$ and multiply by 2.

$$A = (2)\frac{1}{2}\int_0^{\frac{\pi}{3}} [36\cos^2\theta - 4(1 + \cos\theta)^2]\, d\theta$$

$$= 4\int_0^{\frac{\pi}{3}} [9\cos^2\theta - 1 - 2\cos\theta - \cos^2\theta]\, d\theta$$

$$= 4\int_0^{\frac{\pi}{3}} [8\cos^2\theta - 2\cos\theta - 1]\, d\theta = 4\int_0^{\frac{\pi}{3}} [4(1 + \cos 2\theta) - 2\cos\theta - 1]\, d\theta$$

$$= 4\int_0^{\frac{\pi}{3}} [3 + 4\cos 2\theta - 2\cos\theta]\, d\theta = 4\left(3\theta + 2\sin 2\theta - 2\sin\theta\right)\Big|_0^{\frac{\pi}{3}}$$

$$= 4\left(\pi + \sqrt{3} - \sqrt{3}\right) = 4\pi$$

61. Analysis: Find x, y, $\frac{dy}{d\theta}$, $\frac{dx}{d\theta}$, and m for each of the curves. Then show that $m_1 m_2 = -1$.

63. Analysis:

a, b, c. Simply apply the formula. Very easy to use.

9.4. Parametric Representation of Curves

SURVIVAL HINT

In these problems the domain of t is given, but this is often not the case. When you eliminate the parameter to obtain an explicit equation it may not have the same domain as the original parametric equations. Always use the domain of the original equations. For instance problem 19 gives an explicit equation of a line, but the parametric equations give only positive values. Use of the entire line would be incorrect.

3. $t = x - 1, t = y + 1$; $x - 1 = y + 1$; $x - y = 2$, $1 \le x \le 3$

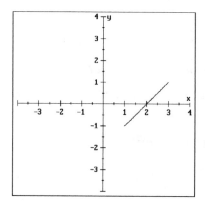

7. $y = 2 + \frac{2}{3}(x - 1)$; $y = \frac{2}{3}x + \frac{4}{3}$, $2 \le x \le 5$

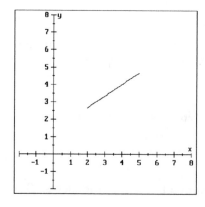

11. $t = x^{1/3}, t = y^{1/2}; x^{1/3} = y^{1/2}; y = x^{2/3}, x \geq 0$

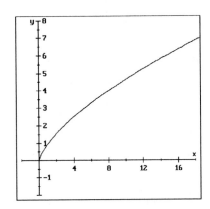

15. $\sin t = x - 1, \cos t = y + 2; \sin^2 t + \cos^2 t = 1;$
$(x - 1)^2 + (y + 2)^2 = 1$

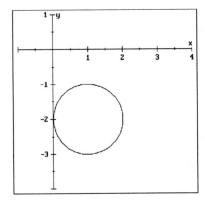

19. $3(3^t) = y, 3^t = x; y = 3x, x \geq 1$

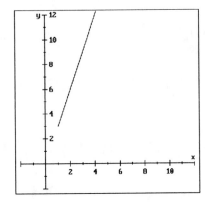

23. $\ln x = \ln t^3 = 3 \ln t; \quad y = \ln x$

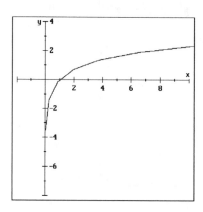

27. $\dfrac{dx}{dt} = 4\,e^{4t}, \ \dfrac{dy}{dt} = 2\cos 2t; \quad \dfrac{dy}{dx} = \dfrac{\dfrac{dy}{dt}}{\dfrac{dx}{dt}} = \dfrac{\cos 2t}{2\,e^{4t}}$

$\dfrac{dy'}{dt} = \dfrac{d}{dt}\!\Big(\dfrac{dy}{dx}\Big) = \dfrac{2e^{4t}(-2\sin 2t) - (\cos 2t)8\,e^{4t}}{4e^{8t}} = \dfrac{-\sin 2t - 2\cos 2t}{e^{4t}}$

$\dfrac{d^2 y}{dt^2} = \dfrac{\dfrac{dy'}{dt}}{\dfrac{dx}{dt}} = \dfrac{\dfrac{-\sin 2t - 2\cos 2t}{e^{4t}}}{4e^{4t}} = \dfrac{-\sin 2t - 2\cos 2t}{4e^{8t}}$

31. $\dfrac{dx}{dt} = 2t, \ \dfrac{dy}{dt} = 4t^3; \quad \dfrac{dy}{dx} = \dfrac{\dfrac{dy}{dt}}{\dfrac{dx}{dt}} = \dfrac{4t^3}{2t} = 2t^2 = 2(x - 1)$

35. $A = \displaystyle\int_a^b \Big(y(t)\dfrac{dx}{dt}\Big)dt = \int_0^1 (t^2)(4t^3)\,dt = \dfrac{2t^6}{3}\Big|_0^1 = \dfrac{2}{3}$

39. $A = \displaystyle\int_0^1 \Big(y(t)\dfrac{dx}{dt}\Big)dt = \int_0^1 (u^3)\Big(\dfrac{1}{1 + u^2}\Big)du = \int_0^1\Big(u - \dfrac{u}{u^2 + 1}\Big)du \ \text{(Division)}$

$= \Big(\dfrac{u^2}{2} - \dfrac{1}{2}\ln(u^2 + 1)\Big)\Big|_0^1 = \dfrac{1}{2} - \dfrac{1}{2}\ln 2 = \dfrac{1}{2}(1 - \ln 2)$

43. $L = \displaystyle\int_a^b \sqrt{\Big(\dfrac{dx}{dt}\Big)^2 + \Big(\dfrac{dy}{dt}\Big)^2}\ dt = \int_3^7 \sqrt{\Big(\dfrac{t}{t^2 - 1}\Big)^2 + \Big(\dfrac{t}{\sqrt{t^2 - 1}}\Big)^2}\ dt$

$= \displaystyle\int_3^7 \sqrt{\dfrac{t^4}{(t^2 - 1)^2}}\ dt = \int_3^7 \dfrac{t^2}{t^2 - 1}dt = \int_3^7\Big(1 + \dfrac{1}{t^2 - 1}\Big)\,dt$

$= \Big(t + \dfrac{1}{2}\ln\dfrac{t - 1}{t + 1}\Big)\Big|_3^7 = 7 + \dfrac{1}{2}\ln\dfrac{3}{4} - 3 - \dfrac{1}{2}\ln\dfrac{1}{2} = 4 + \ln\sqrt{\dfrac{3}{2}}$

47. $\sin t = \frac{x}{4a}$, $y = b \cos^2 t = b(1 - \sin^2 t) = b\left(1 - \frac{x^2}{16a^2}\right)$,

$y = \frac{b}{16a^2}(16a^2 - x^2)$

51. $r = f(\theta) = 2 \cos \theta$, $f'(\theta) = -2 \sin \theta$

$$L = \int_0^{\frac{\pi}{3}} \sqrt{(2 \cos \theta)^2 + (-2 \sin \theta)^2} \, d\theta = \int_0^{\frac{\pi}{3}} 2 \, d\theta = 2\theta \Big|_0^{\frac{\pi}{3}} = \frac{2\pi}{3}$$

61. Analysis: $x = r \cos \theta$, $y = r \sin \theta$. Substitute.

63. Analysis: x and y each have two components; the distance to the center of the rolling circle, and position of the point on that circle. It is easy to see, with a carefully drawn figure (make it large), that the first of these is $x_1 = (a - R) \cos t$, $y_1 = (a - R) \sin t$. The second component, the placement of P on the smaller circle, depends on the rotation, θ, of the smaller circle. Show that its rotation is proportional to the ratio of the radii of the two circles.

65. Analysis: Our previous formula for area uses $\int_a^b y(x) \, dx$. By the chain rule,

$y \, dx = y \frac{dx}{dt} \, dt = y(t) \frac{dx}{dt} \, dt$.

9.5 The Parabola

3. This parabola opens to the right, with V: $(0, 0)$. $4p = 8$, $p = 2$, so the focus is F: $(2, 0)$.

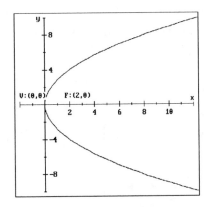

7. $x^2 = -4y$. This parabola opens down, with V: $(0, 0)$. $4p = -4$, $p = -1$, so the focus is F: $(0, -1)$.

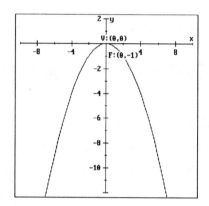

11. $y^2 = 5x$. This parabola opens to the right, with V: $(0, 0)$. $4p = 5$, $p = \frac{5}{4}$, so the focus is F: $(\frac{5}{4}, 0)$.

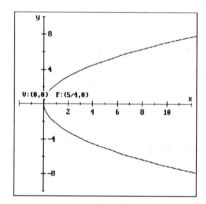

15.   This parabola is translated and opens to the right, with V: $(-2, 1)$. $4p = 2$, $p = \frac{1}{2}$, so the focus is F: $(-\frac{3}{2}, 1)$.

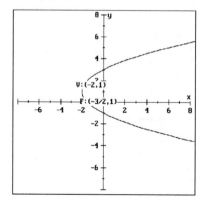

19. Completing the square: $y^2 - 3y + \frac{9}{4} = -4x - 1 + \frac{9}{4}$,

$(y - \frac{3}{2})^2 = -4(x - \frac{5}{16})$. This parabola opens to the left, with V: $(\frac{5}{16}, \frac{3}{2})$.

$4p = -4$, $p = -1$, so the focus is F: $(-\frac{11}{16}, \frac{3}{2})$.

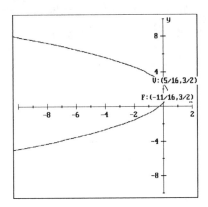

23. Completing the square: $9(x^2 + \frac{2}{3}x) = -18y + 23$,

$9(x + \frac{1}{3})^2 = -18(y - \frac{4}{3})$, $(x + \frac{1}{3})^2 = -2(y - \frac{4}{3})$.

This parabola opens down, with V: $(-\frac{1}{3}, \frac{4}{3})$.

$4p = -2$, $p = -\frac{1}{2}$, so the focus is F: $(-\frac{1}{3}, \frac{5}{6})$.

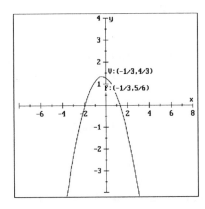

27. With the directrix of $x = 3$, and vertex $(-1, 2)$, the parabola must open to
the left. It is translated, so the general form is $(y - k)^2 = 4p(x - h)$.

$p = $ the distance from the directrix to the vertex $= -4$. The equation is:

$(y - 2)^2 = -16(x + 1)$.

31. With vertex at $(-3, 2)$ and passing through $(-2, -1)$, and axis parallel to the
y-axis, the parabola opens down and has general form: $(x - h)^2 = 4p(y - k)$.

Knowing the vertex: $(x + 3)^2 = 4p(y - 2)$. Use the given point to find p:

$(-2 + 3)^2 = 4p(-1 - 2)$, $4p = -\frac{1}{3}$. The equation is:

$(x + 3)^2 = -\frac{1}{3}(y - 2)$.

35.   This parabola has focus at the pole, and has $p = -9$, and opens up (due to the negative value of p). The vertex will be $(0, -\frac{9}{2})$, x intercepts: ± 9.

Finding the rectangular equation: $r + r\sin\theta = -9$, $\sqrt{x^2 + y^2} + y = -9$, $\sqrt{x^2 + y^2} = -(y + 9)$, $x^2 + y^2 = y^2 + 18y + 81$, $x^2 = 18(y + \frac{9}{2})$.

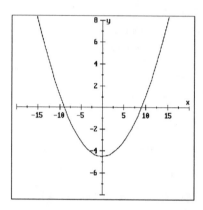

39.   Vertex at $(4, 0)$ and focus at the pole means it opens left, and has $p = 8$.

The general form: $r = \dfrac{8}{1 + \cos\theta}$

43. $y^2 = 4x$ by substitution of polar values is: $(r\sin\theta)^2 = 4r\cos\theta$.

$r\sin^2\theta = 4\cos\theta$, $r = \dfrac{4\cos\theta}{\sin^2\theta}$

47.   Minimize the function giving the distance from the point (x_1, y_1) on the parabola to the point $(2, 0)$. $d^2 = (x_1 - 2)^2 + y_1{}^2$. Minimizing d^2 also minimizes d, so substitute of $y_1{}^2$ and set the derivative $= 0$:

$2(x_1 - 2) + 4 = 0$, $x_1 = 0$, $y_1 = 0$. The origin is the nearest point.

51.   Equate the distance formulas:

$$\sqrt{(x - 4)^2 + (y - 3)^2} = \sqrt{(x + 2)^2 + (y - 1)^2}$$

$x^2 - 8x + 16 + y^2 - 6y + 9 = x^2 + 4x + 4 + y^2 - 2y + 1$

$-12x - 4y = -20$, $3x + y = 5$.

55.   Placing the parabola opening upward with its vertex at the origin gives general equation $x^2 = 4py$. $(6, 4)$ is on the parabola so $36 = 4p(4)$, $p = \frac{9}{4}$.

The focus is $\frac{9}{4}$ m from the vertex.

57.   Analysis: Using the locus definition, the distance from a point on the parabola to the directrix, $(x + c)$, must be the same as the distance to the focus, (use the distance formula).

59. Analysis: This parabola opens to the right with vertex at the origin, focus
at $(c, 0)$, and directrix $x = -c$. Let the intersection of the focal chord with
the parabola be (c, y_1). Find the distance from this point to the directrix.
By symmetry the focal chord is twice this length.

61. Analysis: Distance from the vertex to the focus is c. For any other point,
(x_1, y_1), on $x^2 = 4cy$ what will be its distance to the focus, $(0, c)$?

63. Analysis: This is an isosceles triangle with base of $4c$. To find the height,
find the slope of the tangent at the chord's extremity, and its intersection
with the y-axis.

65. Analysis: **a.** and **b.** were done in #63. **c.** Use the distance formula.

    d. Since $\triangle FPQ$ is isosceles, $\angle FPQ = \angle FQP$ which is a corresponding
angle to $\angle LPT$.

9.6 Conic Sections: The Ellipse and the Hyperbola

SURVIVAL HINT

In an ellipse the major axis is greater than either the minor axis or the distance from the
center to the focus, so $a^2 = b^2 + c^2$. However in a hyperbola the distance from the
center to the focus is greater than either semi-axis, so $a^2 + b^2 = c^2$. Remember
these relationships and use the correct formula.

7. Hyperbola with center at $(0, 0)$, $a = 5$, $b = 3$, $c = \sqrt{34}$

V: $(-5, 0)$
   $(5, 0)$
F: $(-\sqrt{34}, 0)$
   $(\sqrt{34}, 0)$

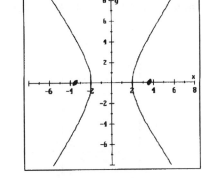

11. $\dfrac{(x-1)^2}{64} + \dfrac{y^2}{16} = 1$

Ellipse with center at $(1, 0)$, $a = 8$, $b = 4$, $c = 4\sqrt{3}$

V: $(-7, 0)$

$(9, 0)$

F: $(1 - 4\sqrt{3}, 0)$

$(1 + 4\sqrt{3}, 0)$

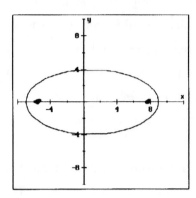

15.   Ellipse with center at $(1, -3)$, $a = 4$, $b = 2$, $c = 2\sqrt{3}$

V: $(1, 1)$

$(1.-7)$

F: $(1, -3 + 2\sqrt{3})$

$(1, -3 - 2\sqrt{3})$

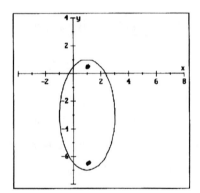

19.   Completing the square: $4(x^2 - 2x + 1) - 9(y^2 - 6y + 9) = 41 + 4 - 81,$

$9(y - 3)^2 - 4(x - 1)^2 = 36,$ $\dfrac{(y - 3)^2}{4} - \dfrac{(x - 1)^2}{9} = 1$

Hyperbola with center at $(1, 3)$, $a = 3$, $b = 2$, $c = \sqrt{13}$

V: $(1, 5)$

$(1, 1)$

F: $(1, 3 + \sqrt{13})$

$(1, 3 - \sqrt{13})$

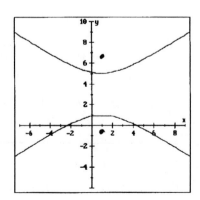

23. Completing the square: $9x^2 + 4(y^2 - 2y + 1) = 32 + 4,$

$9x^2 + 4(y-1)^2 = 36, \dfrac{x^2}{4} + \dfrac{(y-1)^2}{9} = 1$

Ellipse with center at $(0, 1)$, $a = 3$, $b = 2$, $c = \sqrt{5}$

V: $(0, 4)$

$(0, -2)$

F: $(0, 1 + \sqrt{5})$

$(0, 1 - \sqrt{5})$

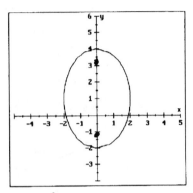

27. $c = 2$, $b = 2$, $a = \sqrt{8}$. $\dfrac{x^2}{8} + \dfrac{y^2}{4} = 1$

31. $a = 4$, $b = 3$. $\dfrac{x^2}{16} + \dfrac{y^2}{9} = 1$

35. $c = 6$, $b = 2$, If the conic is a hyperbola then $a^2 + b^2 = c^2$,

$a^2 = 32$, $a = \pm 4\sqrt{2}$,

$\dfrac{y^2}{4} - \dfrac{x^2}{32} = 1$, a hyperbola opening up and down.

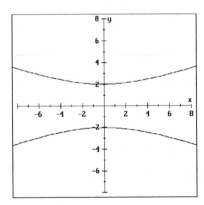

SURVIVAL HINT

All the conics have similar polar form equations, (see p. 609, 620, and 621).

Once the polar equation is in standard form the eccentricity can be determined

and the conic identified. Once identified it is easily graphed by finding the intercepts.

39.   Dividing by 6: $r = \dfrac{\frac{2}{3}}{1 + \frac{1}{6}\cos\theta}$. $\epsilon = \frac{1}{6}$, $\epsilon p = \frac{2}{3}$, $p = 4$

$\epsilon < 1$ is an ellipse. $f(0) = \frac{4}{7}$, $f(\frac{\pi}{2}) = \frac{2}{3}$, $f(\pi) = \frac{4}{5}$, $f(\frac{3\pi}{2}) = \frac{2}{3}$

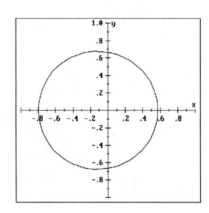

43.   Using implicit differentiation: $10x + 8yy' = 0$; using the point: $y' = \frac{5}{6}$.

$y - 3 = \frac{5}{6}(x + 2)$, $y = \frac{5}{6}x + \frac{14}{3}$ or $5x - 6y + 28 = 0$

47.   Vertical slices will give disks. Using symmetry: $V = 2\pi \displaystyle\int_0^a y^2\,dx$,

$\dfrac{x^2}{a^2} + \dfrac{y^2}{b^2} = 1$, $\dfrac{y^2}{b^2} = \dfrac{a^2 - x^2}{a^2}$, $y^2 = \dfrac{b^2}{a^2}(a^2 - x^2)$,

$V = \dfrac{2\pi b^2}{a^2}\displaystyle\int_0^a (a^2 - x^2)\,dx = \dfrac{2\pi b^2}{a^2}\left(a^2 x - \dfrac{x^3}{3}\right)\Big|_0^a = \dfrac{4}{3}\pi ab^2$

51.   Perpendicular asymptotes, and vertices on the x-axis, imply that $m = \pm 1$.
$a = b = 9$. The equation is: $\dfrac{x^2}{81} - \dfrac{y^2}{81} = 1$.

55. $\dfrac{x^2}{4} - \dfrac{y^2}{9} = 1$. The vertex is at $(2, 0)$. $y^2 = \dfrac{9}{4}(x^2 - 4)$, $y = \dfrac{3}{2}\sqrt{x^2 - 4}$.
Finding the first quadrant area and doubling:

$A = 2\displaystyle\int_2^4 y\,dx = 3\displaystyle\int_2^4 \sqrt{x^2 - 2^2}\,dx = 3\left(\dfrac{x\sqrt{x^2 - 4}}{2} - 2\ln\left(x + \sqrt{x^2 - 4}\right)\right)\Big|_2^4$

(by formula #203 or trigonometric substitution)

$= 3[2\sqrt{12} - 2\ln(4 + \sqrt{12}) + 2\ln 2] = 12\sqrt{3} + 6\ln\dfrac{2}{4 + 2\sqrt{3}}$

$= 6[2\sqrt{3} - \ln(2 + \sqrt{3})] \approx 12.88$

59. Put the sun (focus) at the origin. The general equation is: $r = \dfrac{\epsilon p}{1 - \epsilon \cos \theta}$.
The least distance, l, is the intercept on the negative x-axis, $r(\pi)$, and the greatest
distance, g, is the intercept on the positive x-axis, $r(0)$. The major axis, $2a$,
is equal to $r(0) + r(\pi)$. $2a = \dfrac{\epsilon d}{1 + \epsilon} + \dfrac{\epsilon d}{1 - \epsilon} = \dfrac{2\epsilon d}{1 - \epsilon^2}$.
$\epsilon d = a(1 - \epsilon^2) \approx 9.3 \times 10^7(1 - 0.017^2) \approx 9.2973 \times 10^7$
$l = r(\pi) = \dfrac{\epsilon d}{1 + \epsilon} \approx 9.1419 \times 10^7 \text{mi}.$
$g = r(0) = \dfrac{\epsilon d}{1 - \epsilon} \approx 9.4581 \times 10^7 \text{mi}.$
To check: $l + g = 18.6 \times 10^7 \text{mi}.$ which is the major axis.

61. Analysis: If the signals from the second and third stations arrive at the
same time, the plane is on the perpendicular bisector of the line joining them.
The locus of all points where the distance to $(0, 0)$ minus the distance to $(4, 0)$
is 2 will determine a hyperbola. The plane is at the intersection of the hyperbola
and the line. Solve simultaneously to find the exact location.

63. Analysis: For any point, $P(x, y)$; $\overline{PF_1} - \overline{PF_2} = 2a$. Use the distance formula.

65. Analysis:

    a. Use implicit differentiation to find y', then use the point-slope equation
of a line.

    b. The vertex of the ellipse will be at either $(a, 0)$ or $(0, a)$. Use these points
in the equation from part **a.**

67. Analysis: For the standard rectangular form of a hyperbola with center at
the origin, the equation of an asymptote is $y = \dfrac{b}{a} x$. Any other line parallel to
the asymptote will be $y = \dfrac{b}{a} x + d$. Solve the equation of the hyperbola and line
simultaneously. $[y^2 = (\dfrac{b}{a} x + d)^2]$.

69. Analysis: In addition to the hint also see Sect. 9.5 problem 65. Make use of the
fact that both triangles are right triangles.

Chapter 9 Review

PRACTICE PROBLEMS

27..   **a.**   Parabola. Second degree in y, first degree in x

    b.   Hyperbola. $x^2 - y^2 = 1$

    c.   Circle. Second degree in x and y with equal coefficients

    d.   Circle. $r = 2$, a constant

    e.   Line. First degree in x and y

    f.   Parabola. Second degree in x, first degree in y

    g.   4-leafed rose. Argument of 2θ means 4 leaves

    h.   Cardioid. $r = 5 + 5 \cos \theta$

28.   Translated parabola, opening right, with vertex: $(3, 2)$. $p = 4$

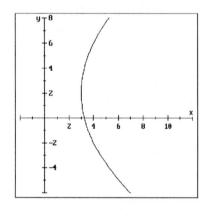

29.   Circle, symmetric to the positive x-axis, with diameter of 2

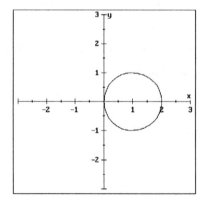

30.   Completing the square: $2(x^2 - 2x + 1) + 3(y^2 + 4y + 4) = -8 + 2 + 12$,

$\dfrac{(x - 1)^2}{3} + \dfrac{(y + 2)^2}{2} = 1$. A translated ellipse with center at $(1, -2)$,

$a = \sqrt{3}, \ b = \sqrt{2}, \ c = \sqrt{5}$

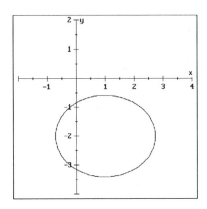

31.   Completing the square: $9(x^2 - 2x + 1) - (y^2 - 12y + 36) = 36 + 9 - 36$.

$\dfrac{(x - 1)^2}{1} - \dfrac{(y - 6)^2}{9} = 1$. A translated hyperbola with center at $(1, 6)$,

$a = 3, \ b = 1, \ c = \sqrt{10}$

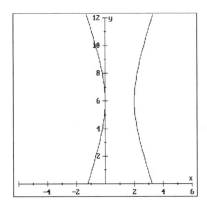

32.   Lemniscate with the leaves on the x-axis, with $r = \pm 1$

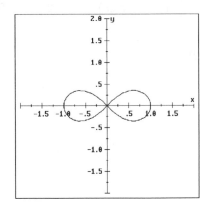

33.   Limaçon, with $r(0) = 2$, $r(\frac{\pi}{2}) = 4$, $r(\pi) = 6$, $r(\frac{3\pi}{2}) = 4$

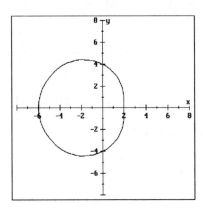

34.

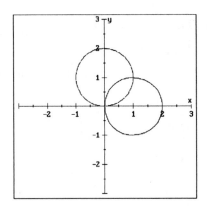

Solving simultaneously, or by symmetry, we see that the intersection is at $(\sqrt{2}, \frac{\pi}{4})$.

We can ignore the scalar factor, a, and introduce it in our final answer.

$$A = 2(\tfrac{1}{2}) \int_0^{\frac{\pi}{4}} r^2 \, d\theta = \int_0^{\frac{\pi}{4}} 4\sin^2\theta \, d\theta. \ \text{ Or } \int_{\frac{\pi}{4}}^{\frac{\pi}{2}} 4\cos^2\theta \, d\theta. \ \text{ We will use the first.}$$

$$A = 4\int_0^{\frac{\pi}{4}} \frac{1 - \cos 2\theta}{2} \, d\theta = 2\int_0^{\frac{\pi}{4}} (1 - \cos 2\theta) \, d\theta = 2\left(\theta - \frac{\sin 2\theta}{2}\right)\Big|_0^{\frac{\pi}{4}}$$

$$= 2(\tfrac{\pi}{4} - \tfrac{1}{2}) = \tfrac{a^2}{2}(\pi - 2) \text{ sq. units}$$

Chapter 10
Vectors in the Plane and in Space

10.1 Vectors in the Plane

1.

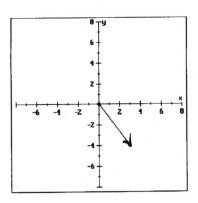

5. $\langle 7 - 3, 2 - (-1) \rangle = \langle 4, 3 \rangle$

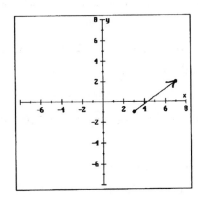

9. $\mathbf{v} = \langle 1 - (-1), -2 - (-2) \rangle = 2\mathbf{i}$

$\|\mathbf{v}\| = \sqrt{(2)^2 + (0)^2} = 2$

13. $\|\mathbf{v}\| = \sqrt{2}; \ \mathbf{u} = \dfrac{\sqrt{2}}{2}(\mathbf{i} + \mathbf{j})$

17. $s\langle -3, 4 \rangle + t\langle 1, -1 \rangle = \langle 6, 0 \rangle; \ -3s + t = 6 \text{ and } 4s - t = 0$

Adding: $s = 6, t = 24$

21. $2\mathbf{u} + 3\mathbf{v} - \mathbf{w} = (6 + 12 - 1)\mathbf{i} + (-8 - 9 - 1)\mathbf{j} = 17\mathbf{i} - 18\mathbf{j}$

25.   For two vectors to be equal their components must be equal:

$x - y - 1 = 0$, and $2x + 3y - 12 = 0$. Triple the first and add to the

second: $5x = 15$, $x = 3$, $y = 2$

29. $\mathbf{i} = \cos 30°$, $\mathbf{j} = \sin 30°$; $\quad \frac{\sqrt{3}}{2}\mathbf{i} + \frac{1}{2}\mathbf{j}$

33. $\mathbf{u} + \mathbf{v} = 5\mathbf{i} + \mathbf{j}$. $\|\mathbf{u} + \mathbf{v}\| = \sqrt{26}$. unit vector $= \frac{5}{\sqrt{26}}\mathbf{i} + \frac{1}{\sqrt{26}}\mathbf{j}$

37.   **a.**   The midpoint of $\overline{PQ} = \frac{1}{2}\langle 12, 6\rangle = \langle 6, 3\rangle$. If the tail of this vector is at

P, its head will be at: $(3, -5)$.

b. $\frac{5}{6}\langle 12, 6\rangle = \langle 10, 5\rangle$. This vector, with tail at P, will end at $(7, -3)$.

41.   Not necessarily equal. Equal magnitudes say nothing about their direction.

45.   Since we do not know the components of $\mathbf{u}$ and $\mathbf{v}$, we can guarantee the sum

is 0 if each of the vectors is 0. $a\mathbf{u} + b\mathbf{u} - b\mathbf{v} + c\mathbf{u} + c\mathbf{v} = \mathbf{0}$,

$(a + b + c)\mathbf{u} = \mathbf{0}$ and $(-b + c)\mathbf{v} = \mathbf{0}$. So $b = c$, and $a + 2b = 0$,

$a = -2b$. One solution: $a = 2$, $b = -1$, $c = -1$.

Parametrically: $a = 2t$, $b = -t$, $c = -t$.

49.   Vertical: $60 \sin 40° \approx 38.57$ ft/s. Horizontal: $60 \cos 40° \approx 45.96$ ft/s.

51.   Analysis: $\mathbf{F}_4 = -(\mathbf{F}_1 + \mathbf{F}_2 + \mathbf{F}_3)$.

53.   Analysis: Draw a triangle and label the vertices P, Q, and R. Let $\mathbf{u}$ be the

vector from P to the midpoint of $\overline{QR}$, $\mathbf{v}$ from Q to the midpoint of $\overline{PR}$, and

$\mathbf{w}$ from R to the midpoint of $\overline{PQ}$. Let $P = (x_1, y_1)$, $Q = (x_2, y_2)$ and

$R = (x_3, y_3)$. Write each of the three vectors in standard form and add them.

Note that $(x_1 + x_2 + x_3)\mathbf{i} = 0$, since you return to the starting point.

55.   Analysis:

a.   Write as a vector triangle: $\mathbf{CN} + \mathbf{AC} = \frac{1}{2}\mathbf{AB}$.

b.   Substitute from **a.**, r and s are scalar multiples.

Then substitute into the given equation, and solve the resulting

vector equation for the scalars r and s.

c.   The selection of the medians was arbitrary. Label the points with

subscripted x and y values, and find the components for the vector

that is $\frac{2}{3}\mathbf{CN}$.

57.   Analysis: Write out each step using the corresponding properties of

the real numbers. Proofs are similar to the one shown for associativity

of vector addition.

10.2 Quadric Surfaces and Graphing in Three Dimensions

SURVIVAL HINT

When using the distance formula it is immaterial as to which point you consider the first or the second, since $(x_1 - x_2)^2 = (x_2 - x_1)^2$. It is usually easier to think in terms of the change, Δx, from one point to the other. $D = \sqrt{(\Delta x)^2 + (\Delta y)^2 + (\Delta z)^2}$

3. $D = \sqrt{\Delta x^2 + \Delta y^2 + \Delta z^2} = \sqrt{6^2 + 11^2 + (-15)^2} = \sqrt{382}$

7. $(x - 0)^2 + (y - 4)^2 + (z + 5)^2 = 9$

11. Completing the square:

$(x^2 - 6x + 9) + (y^2 + 2y + 1) + (z^2 - 2z + 1) = -10 + 9 + 1 + 1$

$(x - 3)^2 + (y + 1)^2 + (z - 1)^2 = 1$. The center is at $(3, -1, 1)$ and the radius is 1.

SURVIVAL HINT

In identifying quadrics, let each variable, one at a time, equal zero and identify
the resulting second degree conic in the coordinate plane. Two or more conics of the
same type give the quadric the "oid" name and the remaining conic describes the type.
For instance, if two coordinate planes have a parabola, and the third an ellipse,
the surface is an elliptic paraboloid.

15. If $x = 0$ we have a hyperbola, if $y = 0$ we get an ellipse, if $z = 0$ the resulting conic is a hyperbola. So this is a hyperboloid. If the equation of a hyperboloid is equal to a positive constant and has one negative sign it is a hyperboloid of one sheet, if it has two negative signs it is a hyperboloid of two sheets.

This is a hyperboloid of one sheet and could be either E or F. The third trace, when $y = 0$, is an ellipse in the xz—plane, so it must be E.

19. This is an elliptic paraboloid, so it is G, H, or K. It has parabolas in the xy and xz planes, so it is G.

23. $\overline{AB} = \sqrt{4^2 + 2^2 + 4^2} = 6$

$\overline{AC} = \sqrt{2^2 + 4^2 + 4^2} = 6$

$\overline{BC} = \sqrt{6^2 + 2^2 + 0^2} = 2\sqrt{10}$. The triangle is isosceles, but since

$6^2 + 6^2 \neq 40$, it is not a right triangle.

31.   All three variables are first degree, so this is a plane. The x-intercept is 10, the y-intercept is 5, and the z-intercept is 2.

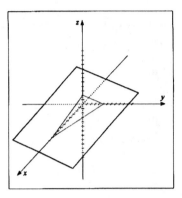

35.   This is a plane parallel to the yz-plane. For all values of y or z, $x = 1$.

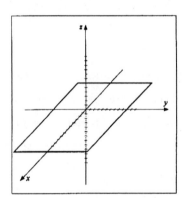

39.   This is an oblique plane parallel to the y-axis. The intercepts in the xz-plane are both equal to 1.

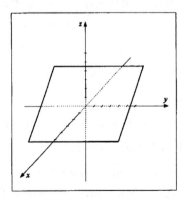

43.   Two of the axis traces are hyperbolas and the other is an ellipse. This is a

hyperboloid of one sheet. It has an ellipse in the xy-plane.

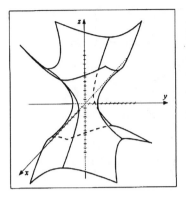

47.   This looks like a hyperboloid of one sheet, but since $x^2 + 2y^2 - 9z^2 = 0$,

it is the degenerate case. Each of the axis traces is a pair of lines, so this is

an elliptic cone.

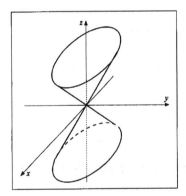

51.   Analysis: Find the midpoint, M, as $\frac{1}{2}$ **BC** from B. Find P as $\frac{2}{3}$ **AM** from A.

10.3 The Dot Product

SURVIVAL HINT

The motivation for the definition of the dot product comes from the problem of finding

the angle between two vectors. If we had done theorem 10.3 before defining the dot

product, the $\cos \theta$ would equal an expression whose numerator was $a_1 b_1 + a_2 b_2 + a_3 b_3$

which we then call, for convenience sake, the dot product. (Note: the • is larger than it

should be to draw attention to it, and my keyboard has nothing more appropriate.)

3.   When finding the standard representation of a vector it is important to pay attention to which is the first and second points. $PQ \neq QP$.

$PQ = (-1 - 1)\mathbf{i} + [1 - (-1)]\mathbf{j} + (4 - 3)\mathbf{k} = -2\mathbf{i} + 2\mathbf{j} + \mathbf{k}$.

$\|PQ\| = \sqrt{(-2)^2 + (2)^2 + (1)^2} = 3$

7. $\mathbf{v} \cdot \mathbf{w} = 3(2) + (-2)(-1) + 4(-6) = -16$

11. $\mathbf{v} \cdot \mathbf{w} = 0$, which means that the vectors are orthogonal.

15. $\sqrt{1^2 + 1^2 + 1^2} = \sqrt{3}$

19. $2\mathbf{u} - 3\mathbf{v} = [2(1) - 3(2)]\mathbf{i} + [2(-2) - 3(4)]\mathbf{j} + [2(2) - 3(-1)]\mathbf{k}$

$= -4\mathbf{i} - 16\mathbf{j} + 7\mathbf{k}$

23. There is a scalar multiple of $s = 2$ for $\mathbf{i}$ and $\mathbf{k}$, but not for $\mathbf{j}$. The vectors are not parallel.

27. $\mathbf{v} + \mathbf{w} = \langle 4, -1, 0 \rangle$, $\mathbf{v} - \mathbf{w} = \langle 2, -3, 2 \rangle$

$(\mathbf{v} + \mathbf{w}) \cdot (\mathbf{v} - \mathbf{w}) = 4(2) + (-1)(-3) + (0)(2) = 11$

31. $\cos \theta = \dfrac{\mathbf{v} \cdot \mathbf{w}}{\|\mathbf{v}\| \, \|\mathbf{w}\|} = \dfrac{1 - 1 + 1}{\sqrt{3} \, \sqrt{3}} = \dfrac{1}{3}$, $\theta \approx 1.23$ or $70.5°$.

35. The scalar projection: $\left| \dfrac{\mathbf{v} \cdot \mathbf{w}}{\|\mathbf{w}\|} \right| = \dfrac{2}{2} = 1$

The vector projection: $\left(\dfrac{\mathbf{v} \cdot \mathbf{w}}{\mathbf{w} \cdot \mathbf{w}} \right) \mathbf{w} = \dfrac{2}{4} \mathbf{w} = \dfrac{1}{2}(2)\mathbf{k} = \mathbf{k}$

39. For $\mathbf{u}$ to be orthogonal to both $\mathbf{v}$ and $\mathbf{w}$, we need $\mathbf{u} \cdot \mathbf{v} = 0$ and also $\mathbf{u} \cdot \mathbf{w} = 0$.

This gives us a system of equations to solve:

$1u_1 + 0u_2 + 1u_3 = 0$, and $1u_1 + 0u_2 - 2u_3 = 0$. Subtracting: $3u_3 = 0$.

$u_3 = 0$, which means $u_1 = 0$ also. u_2, since it has a coefficient of 0, could be any value. Let $u_2 = \pm 1$. $\mathbf{u} = \pm \mathbf{j}$.

43. $\mathbf{u} = -\mathbf{v} = s(-2\mathbf{i} - 3\mathbf{j} + 2\mathbf{k})$.

Using the unit vector: $\mathbf{u} = -\dfrac{1}{\sqrt{17}}(2\mathbf{k} + 3\mathbf{j} - 2\mathbf{k})$.

47. The two vectors, call them $\mathbf{u}$ and $\mathbf{v}$, must have a dot product of 0.

$\mathbf{u} \cdot \mathbf{v} = 6 - 2a - 2a = 0$. $4a = 6$, $a = \dfrac{3}{2}$

51. **a.** $8 - 3 - 1 = 4$

b. $\cos \theta = \dfrac{\mathbf{v} \cdot \mathbf{w}}{\|\mathbf{v}\| \, \|\mathbf{w}\|} = \dfrac{4}{\sqrt{18} \, \sqrt{14}} = \dfrac{2\sqrt{7}}{21}$

c.   We need $\mathbf{v} \cdot (\mathbf{v} - s\mathbf{w}) = 0$

$(4\mathbf{i} - \mathbf{j} + \mathbf{k}) \cdot [(4 - 2s)\mathbf{i} + (-1 - 3s)\mathbf{j} + (1 + s)\mathbf{k}] = 0$,

$16 - 8s + 1 + 3s + 1 + s = 0$, $4s = 18$, $s = \dfrac{9}{2}$

51. (con't.)

    d.   We need $(s\mathbf{v} + \mathbf{w}) \bullet \mathbf{w} = 0$.

$$[(4s + 2)\mathbf{i} + (-s + 3)\mathbf{j} + (s - 1)\mathbf{k}] \bullet (2\mathbf{i} + 3\mathbf{j} - \mathbf{k}) = 0,$$

$$8s + 4 - 3s + 9 - s + 1 = 0, \quad 4s = -14, \quad s = -\frac{7}{2}$$

55.   The work done is $\mathbf{F} \bullet \mathbf{PQ}$. $\mathbf{PQ} = [4 - (-3)]\mathbf{i} + [9 - (-5)]\mathbf{j} + (11 - 4)\mathbf{k}$

$$= 7\mathbf{i} + 14\mathbf{j} + 7\mathbf{k}. \quad W = \frac{6}{7}(7) + (-\frac{2}{7})(14) + \frac{6}{7}(7) = 8 \text{ units}$$

59.   The wind's component on the path of the boat is $1000 \cos 60° = 500$.

$$W = \mathbf{F} \bullet \mathbf{B} = 500\,(50) = 25,000 \text{ ft-lbs}$$

63.   Analysis: The area of an equilateral triangle with side s, is given by $\dfrac{s^2\sqrt{3}}{4}$.

The angle between the two sides is $\frac{\pi}{3}$. If one vector is $\langle s, 0 \rangle$, then the other

is $\left\langle \frac{1}{2}, \frac{\sqrt{3}}{2} \right\rangle$.

65.   Analysis: $\mathbf{v} \bullet \mathbf{v} = \|\mathbf{v}\|^2 = 0$ means that $\mathbf{v} = 0$. If $\mathbf{v} \bullet \mathbf{w} = 0$, what values

can $\mathbf{w}$ have?

67.   Analysis:

    a.   If $\cos \theta = \dfrac{\mathbf{u} \bullet \mathbf{v}}{\|\mathbf{u}\| \, \|\mathbf{v}\|}$, then $\|\mathbf{u} \bullet \mathbf{v}\| = \|\mathbf{u}\| \, \|\mathbf{v}\| \cos \theta$.

    b.   When is $\cos \theta = 1$?

    c.   Make use of the inequality property of real numbers, sometimes called the

triangular inequality: $|a| + |b| \geq |a + b|$.

10.4 The Cross Product

SURVIVAL HINT

Remember that the dot product of two vectors is always a real number, and the cross product of two vectors is always another vector (which is normal to the plane determined by the two vectors and equal in magnitude to the area of the parallelogram determined by them).

3. $\begin{vmatrix} \mathbf{i} & \mathbf{j} & \mathbf{k} \\ 3 & 0 & 2 \\ 2 & 1 & 0 \end{vmatrix} = -2\mathbf{i} + 4\mathbf{j} + 3\mathbf{k}$

7. $\begin{vmatrix} \mathbf{i} & \mathbf{j} & \mathbf{k} \\ 3 & -1 & 2 \\ 2 & 3 & -4 \end{vmatrix} = -2\mathbf{i} + 16\mathbf{j} + 11\mathbf{k}$

11. $\sin \theta \ = \ \dfrac{\|\mathbf{v} \times \mathbf{w}\|}{\|\mathbf{v}\| \, \|\mathbf{w}\|} \ = \ \dfrac{\sqrt{3}}{\sqrt{2} \, \sqrt{2}} \ = \ \dfrac{\sqrt{3}}{2}$

15. $\sin \theta \ = \ \dfrac{\|\mathbf{v} \times \mathbf{w}\|}{\|\mathbf{v}\| \, \|\mathbf{w}\|} \ = \ \dfrac{3\sqrt{6}}{\sqrt{14} \, \sqrt{77}} \ = \ \dfrac{3\sqrt{3}}{7\sqrt{11}} \ = \ \dfrac{3\sqrt{33}}{77}$

19. $\begin{vmatrix} \mathbf{i} & \mathbf{j} & \mathbf{k} \\ 1 & 1 & 1 \\ 3 & 12 & -4 \end{vmatrix} \ = \ -16\mathbf{i} + 7\mathbf{j} + 9\mathbf{k}, \quad \mathbf{u} \ = \ \dfrac{-16\mathbf{i} + 7\mathbf{j} + 9\mathbf{k}}{\sqrt{386}}$

23. $\left| \begin{vmatrix} \mathbf{i} & \mathbf{j} & \mathbf{k} \\ 4 & -1 & 1 \\ 2 & 3 & -1 \end{vmatrix} \right| \ = \ |-2\mathbf{i} + 6\mathbf{j} + 14\mathbf{k}| \ = \ \sqrt{236} \ = \ 2\sqrt{59}$

27. Area of the triangle $= \frac{1}{2}$ Area of the parallelogram $= \dfrac{1}{2} \left| \begin{vmatrix} \mathbf{i} & \mathbf{j} & \mathbf{k} \\ 1 & 1 & -2 \\ 2 & -1 & -1 \end{vmatrix} \right|$

$$= \ \tfrac{1}{2} \, |-3\mathbf{i} - 3\mathbf{j} - 3\mathbf{k}| \ = \ \dfrac{3\sqrt{3}}{2}$$

31. **a.** The cross product of two vectors is a vector, so we have the dot product of two vectors, which is a scalar.

   b. The cross product of two vectors is a vector, so we then have the cross product of those vectors, which will be another vector.

35. $V \ = \ |(\mathbf{u} \times \mathbf{v}) \bullet \mathbf{w}|; \quad \mathbf{u} \times \mathbf{v} \ = \ \begin{vmatrix} \mathbf{i} & \mathbf{j} & \mathbf{k} \\ 2 & 1 & -1 \\ 3 & 0 & 1 \end{vmatrix} \ = \ \langle 1, -5, -3 \rangle.$

$|\langle 1, -5, -3 \rangle \bullet \langle 0, 1, 1 \rangle| \ = \ 8.$

41. If the vector orthogonal to $\mathbf{u}$ and $\mathbf{v}$ is parallel to the vector orthogonal to $\mathbf{u}$ and $\mathbf{w}$, then $\mathbf{u}, \mathbf{v}$ and $\mathbf{w}$ will all be parallel to the same plane. We need the cross products of the vectors to be parallel.

$$\begin{vmatrix} \mathbf{i} & \mathbf{j} & \mathbf{k} \\ 1 & 1 & 0 \\ 2 & -1 & 1 \end{vmatrix} \ = \ \langle 1, -1, -3 \rangle, \qquad \begin{vmatrix} \mathbf{i} & \mathbf{j} & \mathbf{k} \\ 1 & 1 & 0 \\ 1 & 1 & t \end{vmatrix} \ = \ \langle t, -t \rangle,$$

41. (con't.)

$$\begin{vmatrix} \mathbf{i} & \mathbf{j} & \mathbf{k} \\ 2 & -1 & 1 \\ 1 & 1 & t \end{vmatrix} = \langle -1 - t, 1 - 2t, 3 \rangle.$$ These three vectors can be parallel,

each a scalar multiple of the other, only if $t = 0$.

45. For $\mathbf{v} \times \mathbf{w} = \mathbf{w}$, $\begin{vmatrix} \mathbf{i} & \mathbf{j} & \mathbf{k} \\ v_1 & v_2 & v_3 \\ w_1 & w_2 & w_3 \end{vmatrix} = \langle w_1, w_2, w_3 \rangle$,

$\langle v_2 w_3 - v_3 w_2, \ v_3 w_1 - v_1 w_3, \ v_1 w_2 - v_2 w_1 \rangle = \langle w_1, w_2, w_3 \rangle$.

$v_2 w_3 - v_3 w_2 = w_1, \quad v_3 w_1 - v_1 w_3 = w_2, \quad v_1 w_2 - v_2 w_1 = w_3$.

Multiplying the first equation by w_1, the second by w_2 and adding:

$v_2 w_1 w_3 - v_1 w_2 w_3 = w_1^2 + w_2^2$. Multiplying the third equation by w_3

and then adding: $0 = w_1^2 + w_2^2 + w_3^2$. This can be true only if $\mathbf{w} = 0$.

49. $(\mathbf{u} \times \mathbf{v}) \bullet \mathbf{w} = \begin{vmatrix} \mathbf{i} & \mathbf{j} & \mathbf{k} \\ u_1 & u_2 & u_3 \\ v_1 & v_2 & v_3 \end{vmatrix} \bullet (w_1 \mathbf{i} + w_2 \mathbf{j} + w_3 \mathbf{k})$

$= \left(\begin{vmatrix} u_2 & u_3 \\ v_2 & v_3 \end{vmatrix} \mathbf{i} - \begin{vmatrix} u_1 & u_3 \\ v_1 & v_3 \end{vmatrix} \mathbf{j} + \begin{vmatrix} u_1 & u_2 \\ v_1 & v_2 \end{vmatrix} \mathbf{k} \right) \bullet (w_1 \mathbf{i} + w_2 \mathbf{j} + w_3 \mathbf{k})$

$= \begin{vmatrix} u_2 & u_3 \\ v_2 & v_3 \end{vmatrix} w_1 - \begin{vmatrix} u_1 & u_3 \\ v_1 & v_3 \end{vmatrix} w_2 + \begin{vmatrix} u_1 & u_2 \\ v_1 & v_2 \end{vmatrix} w_3$

$= \begin{vmatrix} u_1 & u_2 & u_3 \\ v_1 & v_2 & v_3 \\ w_1 & w_2 & w_3 \end{vmatrix}$, when expanded about the third row.

53. Analysis: The area of the triangular base is $\frac{1}{2}$ the area of the parallelogram

determined by the vectors $\mathbf{AB}$ and $\mathbf{AC}$, which is the magnitude of the cross

product. The dot product with $\mathbf{BD}$ has the same effect as discussed in the triple

scalar product.

55. Analysis: $\mathbf{u} \times \mathbf{v} = \mathbf{w}$ means $\mathbf{w}$ is orthogonal to $\mathbf{u}$ and $\mathbf{v}$. $\mathbf{u} \bullet \mathbf{v} = 0$ means

$\mathbf{u}$ and $\mathbf{v}$ are perpendicular. The three vectors are mutually perpendicular,

like the axes of a coordinate system. So $\mathbf{w} \times \mathbf{u}$ will be parallel to $\mathbf{v}$, and $\mathbf{v} \times \mathbf{w}$

will be parallel to $\mathbf{u}$.

57.   Analysis: Recall that $\cos \theta \;=\; \dfrac{\mathbf{v} \bullet \mathbf{w}}{\|\mathbf{v}\|\,\|\mathbf{w}\|}$, and $\sin \theta \;=\; \dfrac{\|\mathbf{v} \times \mathbf{w}\|}{\|\mathbf{v}\|\,\|\mathbf{w}\|}$.

59.   Analysis: Dot products are commutative, and the resulting expressions both represent the same parallelepiped.

61.   Analysis: You can simplify your work here by letting $\mathbf{w} \times \mathbf{z} \;=\; \mathbf{q}$. Both sides represent vectors so they are equal if their $\mathbf{i}$, $\mathbf{j}$ and $\mathbf{k}$ coefficients are equal.

63.   Analysis: Note the pattern that $\mathbf{u} \bullet (\mathbf{v} \times \mathbf{i}) \;=\; \mathbf{u} \bullet (v_3\mathbf{j} \;-\; v_2\mathbf{k}) \;=\; u_2 v_3 \;-\; u_3 v_2$

$$= \;\begin{vmatrix} u_2 & u_3 \\ v_2 & v_3 \end{vmatrix}\; ,\;\; \text{which is the coefficient of } \mathbf{i} \text{ in the expansion of } \mathbf{u} \times \mathbf{v}.$$

10.5 Lines and Planes in Space

SURVIVAL HINT

You should be able to represent a line in either parametric or symmetric form.

You should be able to represent a plane in either point-normal or standard form.

3.   For standard form, distribute the coefficients and combine the constants:

$4x \;+\; 4 \;-\; 2y \;-\; 2 \;+\; 6z \;-\; 12 \;=\; 0, \quad 4x \;-\; 2y \;+\; 6z \;=\; 10$ or

$2x \;-\; y \;+\; 3z \;=\; 5$

7.   The given data: $(x_0,\, y_0,\, z_0) = (1,\, -1,\, -2)$ and $A\mathbf{i} \;+\; B\mathbf{j} \;+\; C\mathbf{k} \;=\; 3\mathbf{i} \;-\; 2\mathbf{j} \;+\; 5\mathbf{k}$

Parametric equations: $x \;=\; 1 \;+\; 3t, \;\; y \;=\; -1 \;-\; 2t, \;\; z \;=\; -2 \;+\; 5t$

Symmetric equation: $\dfrac{x - 1}{3} \;=\; \dfrac{y + 1}{-2} \;=\; \dfrac{z + 2}{5}$

11.   From the given symmetric equation we know the direction numbers are $[1,\, -3,\, -5]$. We are also given a point on the line: $(1,\, -3,\, 6)$.

Parametric equations: $x \;=\; 1 \;+\; t, \;\; y = -3 \;-\; 3t, \;\; z \;=\; 6 \;-\; 5t$

Symmetric equation: $\dfrac{x - 1}{1} \;=\; \dfrac{y + 3}{-3} \;=\; \dfrac{z - 6}{-5}$

15.   In order for a line to be parallel to both the xy-plane and the yz-plane it must be parallel to the y-axis. It will have direction numbers $[0,\, 1,\, 0]$.

Parametric equations: $x \;=\; 3, \;\; y \;=\; -1 \;+\; t, \;\; z \;=\; 0$

Symmetric equation: None. Symmetric form requires all direction numbers to be non-zero. A and C are zero in this case.

19.   When $x = 0$, $t = 3$: $(0, 4, 9)$

When $y = 0$, $t = -1$: $(8, 0, -3)$

When $z = 0$, $t = 0$: $(6, 1, 0)$

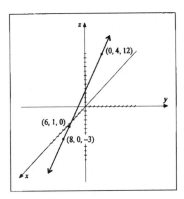

23.   The two lines have the same direction numbers, $[3, -4, -7]$, but pass through different points. They are parallel.

27. $\|\mathbf{v}\| = \sqrt{2^2 + (-3)^2 + (-5)^2} = \sqrt{38}$. $a_1 = 2$, $a_2 = -3$, $a_3 = -5$.

$\cos \alpha = \dfrac{2}{\sqrt{38}}$, $\alpha \approx 1.24$ or $71°$. $\cos \beta = \dfrac{-3}{\sqrt{38}}$, $\beta \approx 2.08$ or $119°$.

$\cos \gamma = \dfrac{-5}{\sqrt{38}}$, $\gamma \approx 2.52$ or $144°$.

31. $\|\mathbf{v}\| = \sqrt{1^2 + (-3)^2 + 9^2} = \sqrt{91}$. $a_1 = 1$, $a_2 = -3$, $a_3 = 9$.

$\cos \alpha = \dfrac{1}{\sqrt{91}}$, $\alpha \approx 1.47$ or $84°$. $\cos \beta = \dfrac{-3}{\sqrt{91}}$, $\beta \approx 1.89$ or $108°$.

$\cos \gamma = \dfrac{9}{\sqrt{91}}$, $\gamma \approx 0.34$ or $19°$.

35.   Using the point-normal equation of a plane,

$A(x - x_0) + B(y - y_0) + C(z - z_0) = 0$, where the point is $(0, -3, 0)$,

and the attitude numbers are $[0, -2, 3]$:

$-2(y + 3) + 3(z) = 0$, $2y - 3z = -6$

39.   The direction numbers are $[4, 2, 1]$, so the direction of the line is

$4\mathbf{i} + 2\mathbf{j} + \mathbf{k} = \mathbf{v}$. $\|\mathbf{v}\| = \sqrt{4^2 + 2^2 + 1^2} = \sqrt{21}$. The two unit

vectors will be: $\pm\dfrac{1}{\sqrt{21}}(4\mathbf{i} + 2\mathbf{j} + \mathbf{k})$.

43.   The line through $(0, 0, 1)$ and $(2, 1, -1)$ contains the vector $2\mathbf{i} + \mathbf{j} - 2\mathbf{k}$.

$\langle 2, 1, -2 \rangle \bullet \langle 3, -4, 1 \rangle = 6 - 4 - 2 = 0$; so they are orthogonal.

47. The normal to the first line has direction numbers $[1, 1, -1]$, and the normal to the second line has direction numbers $[2, -1, 3]$. The cross product of the two normals must be parallel to the line of intersection.

$$\mathbf{N_1} \times \mathbf{N_2} = \begin{vmatrix} \mathbf{i} & \mathbf{j} & \mathbf{k} \\ 1 & 1 & -1 \\ 2 & -1 & 3 \end{vmatrix} = \langle 2, -5, -3 \rangle.$$ These are the direction numbers

for our line. Now we need any point on the line. If $z = 0$, $x + y = 4$ and $2x - y = 1$; $x = \frac{5}{3}$, $y = \frac{7}{3}$. So we have a point: $(\frac{5}{3}, \frac{7}{3}, 0)$. The equation of

the line is: $\dfrac{x - \frac{5}{3}}{2} = \dfrac{y - \frac{7}{3}}{-5} = \dfrac{z - 0}{-3}$.

51. An orthogonal to the given line will have attitude numbers $[3, 5, 2]$. It must also pass through $(2, 1, -1)$. It will have equation:

$$3(x - 2) + 5(y - 1) + 2(z + 1) = 0, \text{ or } 3x + 5y + 2z = 9.$$

55. $\mathbf{N_1} = 2\mathbf{i} + \mathbf{j} - 4\mathbf{k}$, $\mathbf{N_2} = \mathbf{i} - \mathbf{j} + \mathbf{k}$. Let θ be the angle between the normals.

$$\cos\theta = \frac{|\mathbf{N_1} \bullet \mathbf{N_2}|}{\|\mathbf{N_1}\| \, \|\mathbf{N_2}\|} = \frac{|2 - 1 - 4|}{\sqrt{21}\sqrt{3}} = \frac{\sqrt{7}}{7}. \ \theta \approx 1.18 \text{ or } 68°$$

59. The normal to the first plane has direction numbers $[3, 1, -1]$, and the normal to the second plane has direction numbers $[1, -6, -2]$. The cross product of these normals must be parallel to the line of intersection.

$$\begin{vmatrix} \mathbf{i} & \mathbf{j} & \mathbf{k} \\ 3 & 1 & -1 \\ 1 & -6 & -2 \end{vmatrix} = \langle -8, 5, -19 \rangle.$$

Now to find a point; if $x = 0$, $y - z = 5$ and $-6y - 2z = 10$. Multiplying the first equation by -2 and adding: $-8y = 0$, $y = 0$. Substituting into $y - z = 5$, $z = -5$.

So we have a point: $(0, 0, -5)$. The line of intersection is:

$$\frac{x}{-8} = \frac{y}{5} = \frac{z + 5}{-19}$$

63. Analysis: The normal to the first line is $\langle a_1, b_1, c_1 \rangle$, and the normal to the second line is $\langle a_2, b_2, c_2 \rangle$. The dot product of these normals is 0. What does that say about the original lines?

65. Analysis: This is called the intercept form for the equation of a plane. Test it for each of its intercepts, $(a, 0, 0)$, $(0, b, 0)$ and $(0, 0, c)$, to verify that it meets the requirements.

10.6 Vector Methods for Measuring Distance in $\mathbf{R}^3$

SURVIVAL HINT

You will need to remember two formulas for the distance from a point to a line in $\mathbf{R}^3$.
One when the line is in standard form; theorem 10.10, and one when the line is in vector
form; theorem 10.11.

3.   The distance from the point $(x_0,\ y_0)$ to the line $Ax\ +\ By\ +\ C\ =\ 0$ is

given by $d\ =\ \left|\dfrac{Ax_0\ +\ By_0\ +\ C}{\sqrt{A^2\ +\ B^2}}\right|$. $(x_0,\ y_0)\ =\ (4,\ -3)$ and

$Ax\ +\ By\ +\ C\ =\ 0$ is given by $12x\ +\ 5y\ -\ 2\ =\ 0$.

$d\ =\ \left|\dfrac{12(4)\ +\ 5(-3)\ +\ (-2)}{\sqrt{12^2\ +\ 5^2}}\right|\ =\ \dfrac{31}{13}$

7.   The distance from the point $(x_0,\ y_0,\ z_0)$ to the plane $Ax\ +\ By\ +\ Cz\ +\ D\ =\ 0$

is given by: $d\ =\ \left|\dfrac{Ax_0\ +\ By_0\ +\ Cz_0\ +\ D}{\sqrt{A^2\ +\ B^2\ +\ C^2}}\right|$. For the given values:

$d\ =\ \left|\dfrac{1(1)\ +\ 1(0)\ +\ (-1)(-1)\ -\ 1}{\sqrt{1^2\ +\ 1^2\ +\ (-1)^2}}\right|\ =\ \dfrac{1}{\sqrt{3}}\ =\ \dfrac{\sqrt{3}}{3}$

11.   The point $(x_0,\ y_0,\ z_0)\ =\ (a,\ -\ a,\ 2a)$, the line $Ax\ +\ By\ +\ Cz\ +\ D\ =\ 0$
is $2ax\ -\ y\ +\ az\ -\ 4a\ =\ 0$.

$d\ =\ \left|\dfrac{Ax_0\ +\ By_0\ +\ Cz_0\ +\ D}{\sqrt{A^2\ +\ B^2\ +\ C^2}}\right|\ =\ \left|\dfrac{2a(a)\ +\ (-\ 1)(-\ a)\ +\ a(2a)\ -\ 4a}{\sqrt{(2a)^2\ +\ (-\ 1)^2\ +\ a^2}}\right|$

$=\ \left|\dfrac{4a^2\ -\ 3a}{\sqrt{5a^2\ +\ 1}}\right|$.

15.   The plane containing the point $(-\ 3,\ 5,\ 1)$ and having normal $\langle 3,\ 1,\ 5\rangle$ has
equation $3(x\ +\ 3)\ +\ 1(y\ -\ 5)\ +\ 5(z\ -\ 1)\ =\ 0$, or $3x\ +\ y\ +\ 5z\ -\ 1\ =\ 0$.
The distance to this plane from the point $(-\ 1,\ 2,\ 1)$ is given by:

$d\ =\ \left|\dfrac{Ax_0\ +\ By_0\ +\ Cz_0\ +\ D}{\sqrt{A^2\ +\ B^2\ +\ C^2}}\right|\ =\ \left|\dfrac{-\ 3\ +\ 2\ +\ 5\ -\ 1}{\sqrt{3^2\ +\ 1^2\ +\ 5^2}}\right|\ =\ \dfrac{3}{\sqrt{35}}\ =\ \dfrac{3\ \sqrt{35}}{35}$

19. The distance from the point $(1, -2, 2)$ to the line $\frac{x}{1} = \frac{2y}{1} = \frac{z}{-1}$ is given by:
$d = \frac{||\mathbf{v} \times \mathbf{QP}||}{||\mathbf{v}||}$, where $\mathbf{v}$ is a vector in the direction of the line, P is the given point, and Q is any point on the line. We can see that $(0, 0, 0)$ is on the line, so
$\mathbf{QP} = \langle 1, -2, 2 \rangle$, $\mathbf{v} = \langle 1, \frac{1}{2}, -1 \rangle$.

$$\mathbf{v} \times \mathbf{QP} = \begin{vmatrix} \mathbf{i} & \mathbf{j} & \mathbf{k} \\ 1 & \frac{1}{2} & -1 \\ 1 & -2 & 2 \end{vmatrix} = \langle -1, -3, -\frac{5}{2} \rangle. \quad ||\mathbf{v} \times \mathbf{QP}|| = \frac{\sqrt{65}}{2}$$

$$d = \frac{\frac{\sqrt{65}}{2}}{||\mathbf{v}||} = \frac{\frac{\sqrt{65}}{2}}{\frac{3}{2}} = \frac{\sqrt{65}}{3}$$

23. We have the center of the sphere, but we need the radius, which is the distance from the center to the tangent plane. We use the formula for the distance from a point to a plane. $d = \left| \frac{Ax_0 + By_0 + Cz_0 + D}{\sqrt{A^2 + B^2 + C^2}} \right|$

$$= \left| \frac{2(-2) + 3(3) + (-6)(7) - 5}{\sqrt{2^2 + 3^2 + (-6)^2}} \right| = 6$$

So the equation of the sphere is $(x + 2)^2 + (y - 3)^2 + (z - 7)^2 = 36$

27. We need the distance from the point to the line to equal 5.

$$d = \frac{||\mathbf{v} \times \mathbf{QP}||}{||\mathbf{v}||} = \left| \frac{Ax_0 + By_0 + Cz_0 + D}{\sqrt{A^2 + B^2 + C^2}} \right| = 5$$

$\mathbf{v} = \langle 4, -1, 3 \rangle$, and $Q(1, -1, 0)$ is on the line, $\mathbf{QP} = \langle x - 1, y + 1, z \rangle$

$$\mathbf{v} \times \mathbf{QP} = \begin{vmatrix} \mathbf{i} & \mathbf{j} & \mathbf{k} \\ 4 & -1 & 3 \\ x-1 & y+1 & z \end{vmatrix} = \langle -3(y + 1) - z, 3(x - 1) - 4z, 4(y + 1) + (x - 1) \rangle$$

$$\frac{||\mathbf{v} \times \mathbf{QP}||}{||\mathbf{v}||} = \frac{\sqrt{(3y + z + 3)^2 + (3x - 4z - 3)^2 + (x + 4y + 3)^2}}{\sqrt{26}} = 5$$

$$(3y + z + 3)^2 + (3x - 4z - 3)^2 + (x + 4y + 3)^2 = 650$$

31.   The distance must be measured perpendicular to both lines, therefore, it will be

in the direction of their cross product.

$$\mathbf{v}_1 \times \mathbf{v}_2 = \begin{vmatrix} \mathbf{i} & \mathbf{j} & \mathbf{k} \\ 1 & -2 & 0 \\ 1 & -1 & 1 \end{vmatrix} = \langle -2, -1, 1 \rangle. \text{ Now if } \mathbf{P}_1\mathbf{P}_2 \text{ is a vector joining}$$

a point of line one to a point of line two, then the length of the common normal

is the component of $\mathbf{P}_1\mathbf{P}_2$ in the direction of the normal: $d = \dfrac{\mathbf{n} \bullet \mathbf{P}_1\mathbf{P}_2}{||\mathbf{n}||}$.

$(-1, 0, 3)$ is a point on line one, and $(0, -1, 2)$ is a point on the second line.

$$\mathbf{P}_1\mathbf{P}_2 = \langle -1, -1, -1 \rangle. \quad d = \frac{2 + 1 - 1}{\sqrt{(-2)^2 + (-1)^2 + 1)^2}} = \frac{2}{\sqrt{6}} = \frac{\sqrt{6}}{3}$$

33.   Analysis:

a. $|D_1 - D_2|$ represents the distance between the planes in the z direction.

The perpendicular distance between the planes is the component of some

$\mathbf{P}_1\mathbf{P}_2$ in the direction of the common normal. See last half of #31.

b.   In this case $(0, 0, 1)$ is on the first plane and $(0, 0, 2)$ is on the second

plane. They are 1 unit apart in the z direction. They are actually less

than 1 unit apart in the direction normal to both. Notice that if we use

the points $(2, 0, 0)$ and $(4, 0, 0)$ they are 2 units apart in the x

direction.

Chapter 10 Review

PRACTICE PROBLEMS

34.   **a.** $[2(2) + 3(3)]\mathbf{i} + [2(-3) + 3(-2)]\mathbf{j} + [2(1) + 3(0)]\mathbf{k}$

$= 13\mathbf{i} - 12\mathbf{j} + 2\mathbf{k}$

b. $\left(\sqrt{2^2 + (-3)^2 + 1^2} \right)^2 - \left(\sqrt{3^2 + (-2)^2} \right)^2 = 14 - 13 = 1$

c. $\left(\dfrac{\mathbf{v} \bullet \mathbf{w}}{\mathbf{w} \bullet \mathbf{w}} \right)\mathbf{w} = \frac{12}{13}(3\mathbf{i} - 2\mathbf{j})$

d. $\left| \dfrac{\mathbf{w} \bullet \mathbf{v}}{||\mathbf{v}||} \right| = \dfrac{12}{\sqrt{14}} = \dfrac{6\sqrt{14}}{7}$

35. **a.** $2(0) + (-5)(1) + 1(-3) = -5 - 3 = -8$

b. $\begin{vmatrix} \mathbf{i} & \mathbf{j} & \mathbf{k} \\ 2 & -5 & 1 \\ 0 & 1 & -3 \end{vmatrix} = 14\mathbf{i} + 6\mathbf{j} + 2\mathbf{k}$

36. **a.** $\begin{vmatrix} \mathbf{i} & \mathbf{j} & \mathbf{k} \\ 2 & -3 & 1 \\ 1 & 1 & -2 \end{vmatrix} \bullet \langle 3, 0, 5 \rangle = \langle 5, 5, 5 \rangle \bullet \langle 3, 0, 5 \rangle = 40$

b.   Not possible to take the cross product of a scalar and a vector.

c. $\begin{vmatrix} \mathbf{i} & \mathbf{j} & \mathbf{k} \\ 2 & -3 & 1 \\ 1 & 1 & -2 \end{vmatrix} \times \langle 3, 0, 5 \rangle = \langle 5, 5, 5 \rangle \times \langle 3, 0, 5 \rangle$

$= \begin{vmatrix} \mathbf{i} & \mathbf{j} & \mathbf{k} \\ 5 & 5 & 5 \\ 3 & 0 & 5 \end{vmatrix} = 25\mathbf{i} - 10\mathbf{j} - 15\mathbf{k}$

d.   Not possible to take the dot product of a scalar and a vector.

37.   Using Q and the vector $\mathbf{PQ} = \langle -1, 6, -4 \rangle$: $\dfrac{x}{-1} = \dfrac{y+2}{6} = \dfrac{z-1}{-4}$

38.   Using point P and the direction numbers $[2, 0, 3]$:

$2(x - 1) + 0(y - 1) + 3(z - 3) = 0$, or $2x + 3z = 11$

39.   The direction numbers for each plane give a normal vector. The cross product of these vectors give the direction of the line of intersection. We can find a point by looking at intercepts.

$\mathbf{N}_1 \times \mathbf{N}_2 = \begin{vmatrix} \mathbf{i} & \mathbf{j} & \mathbf{k} \\ 2 & 3 & 1 \\ 0 & 1 & -3 \end{vmatrix} = \langle -10, 6, 2 \rangle$. If $z = 0$ in the second plane,

$y = 5$. Use these values in the first plane to find $x = -\dfrac{13}{2}$. We now have a point on the line of intersection and its direction. The line is:

$\dfrac{x + \frac{13}{2}}{-10} = \dfrac{y - 5}{6} = \dfrac{z}{2}$ or $\dfrac{x + \frac{13}{2}}{5} = \dfrac{y - 5}{-3} = \dfrac{z}{-1}$

40. Use P as the point, and find the direction numbers by finding $\mathbf{PQ} \times \mathbf{PR}$.

$$\mathbf{PQ} \times \mathbf{PR} = \begin{vmatrix} \mathbf{i} & \mathbf{j} & \mathbf{k} \\ 1 & -5 & 6 \\ 3 & -2 & -1 \end{vmatrix} = \langle 17, 19, 13 \rangle$$

The plane is: $17x + 19(y - 2) + 13(z + 1) = 0$, or $17x + 19y + 13z = 25$.

41. The direction numbers for the line: $[2, -3, 3]$. Using the point $(1, 2, -1)$:

$$\frac{x - 1}{2} = \frac{y - 2}{-3} = \frac{z + 1}{3}$$

42. $\cos \alpha = \dfrac{-2}{\sqrt{(-2)^2 + 3^2 + 1^2}} = \dfrac{-2}{\sqrt{14}}$; so $\alpha \approx 2.13$ or $122°$

$\cos \beta = \dfrac{3}{\sqrt{14}}$; so $\beta \approx 0.64$ or $37°$

$\cos \gamma = \dfrac{1}{\sqrt{14}}$; so $\gamma \approx 1.30$ or $74°$

43. **a.** The direction numbers are not scalar multiples, so they are not parallel.

If $z = 3$, $t = \frac{3}{2}$, $x = 0 \neq -2$. They are skew.

b. The direction numbers are not scalar multiples, so they are not parallel.

To determine if they intersect we need to solve a system of equations.

1st line: $x = 7 + 5t_1$, $y = 6 + 4t_1$, $z = 8 + 5t_1$.

2nd line: $x = 8 + 6t_2$, $y = 6 + 4t_2$, $z = 9 + 6t_2$.

Equating x_1 and x_2: $7 + 5t_1 = 8 + 6t_2$, $5t_1 - 6t_2 = 1$.

From the y equations we see that $t_1 = t_2$, so $5t^1 - 6t_1 = 1$,

$t_1 = -1 = t_2$. Both equations contain the point $(2, 2, 3)$.

44. **a.** $V = (\mathbf{u} \times \mathbf{v}) \bullet \mathbf{w} = \begin{vmatrix} \mathbf{i} & \mathbf{j} & \mathbf{k} \\ 2 & 1 & 0 \\ 1 & -1 & -1 \end{vmatrix} \bullet \langle 3, 0, 1 \rangle$

$= |\langle -1, 2, -3 \rangle \bullet \langle 3, 0, 1 \rangle| = 6$ cubic units

b. Volume of a tetrahedron $= \frac{1}{6}|(\mathbf{u} \times \mathbf{v}) \bullet \mathbf{w}|$ (see 10.4 #53)

$2 = \frac{1}{6} A^2 |(\mathbf{u} \times \mathbf{v}) \bullet \mathbf{w}|$, $12 = A^2 \begin{vmatrix} \mathbf{i} & \mathbf{j} & \mathbf{k} \\ 2 & 1 & 0 \\ 1 & -1 & -1 \end{vmatrix} \bullet (3\mathbf{i} + \mathbf{k})$

$12 = A^2(6)$, $A^2 = 2$, $A = \sqrt{2}$

45. The distance from a point to a plane is given by $d = \left| \dfrac{Ax_0 + By_0 + Cz_0 + D}{\sqrt{A^2 + B^2 + C^2}} \right|$

$$= \left| \frac{2(-1) + 5(1) + -1(4) - 3}{\sqrt{2^2 + 5^2 + (-1)^2}} \right| = \frac{4}{\sqrt{30}} = \frac{2\sqrt{30}}{15}$$

46. The distance must be measured perpendicular to both lines. That will be in the direction of their cross product.

$$\mathbf{v}_1 \times \mathbf{v}_2 = \begin{vmatrix} \mathbf{i} & \mathbf{j} & \mathbf{k} \\ 1 & 2 & 3 \\ -1 & 1 & 1 \end{vmatrix} = -\mathbf{i} - 4\mathbf{j} + 3\mathbf{k}. \text{ Now if } \mathbf{P}_1\mathbf{P}_2 \text{ is a vector joining}$$

a point on the first line to a point on the second line, then the length of the common normal is the component of $\mathbf{P}_1\mathbf{P}_2$ in the direction of the normal.

$d = \dfrac{|\mathbf{n} \bullet \mathbf{P}_1\mathbf{P}_2|}{\|\mathbf{n}\|}$. $(0, 0, -1)$ is a point on the first line and $(1, 2, 0)$ is a point on the second line, so $\mathbf{P}_1\mathbf{P}_2 = \mathbf{i} + 2\mathbf{j} + \mathbf{k}$.

$$d = \frac{|\langle -1, -4, 3 \rangle \bullet \langle 1, 2, 1 \rangle|}{\sqrt{(-1)^2 + (-4)^2 + 3^2}} = \frac{6}{\sqrt{26}} = \frac{3\sqrt{26}}{13}$$

47. The distance from a point to a line is given by: $d = \dfrac{\|\mathbf{v} \times \mathbf{QP}\|}{\|\mathbf{v}\|}$, where $\mathbf{v}$ is a vector in the direction of the line, P is the given point, and Q is any point on

the line. $\mathbf{v} \times \mathbf{QP} = \begin{vmatrix} \mathbf{i} & \mathbf{j} & \mathbf{k} \\ 3 & 5 & -1 \\ -2 & -5 & -1 \end{vmatrix} = -10\mathbf{i} + 5\mathbf{j} - 5\mathbf{k}$

$\|\mathbf{v} \times \mathbf{QP}\| = \sqrt{(-10)^2 + 5^2 + (-5)^2} = 5\sqrt{6}.$

$\|\mathbf{v}\| = \sqrt{3^2 + 5^2 + (-1)^2} = \sqrt{35}.$ $d = \dfrac{5\sqrt{6}}{\sqrt{35}} = \dfrac{\sqrt{210}}{7}$

48. The path of the plane is represented by the sum of the vector of the plane, $\langle 0, -200 \rangle$, and the vector of the wind, $\langle 25\sqrt{2}, 25\sqrt{2}, \rangle$.

Path: $\langle 25\sqrt{2}, 25\sqrt{2} - 200 \rangle$. The magnitude of this vector is the groundspeed:

$\sqrt{(25\sqrt{2})^2 + (25\sqrt{2} - 200)^2} \approx 168.4$ mph.

49. The work performed on the sled is the horizontal component of the force times the displacement. In vector terms: $W = \mathbf{F} \bullet \mathbf{PQ} = \langle \frac{3\sqrt{3}}{2}, \frac{3}{2} \rangle \bullet \langle 50, 0 \rangle$

$= 75\sqrt{3} \approx 130$ ft-lbs.

Chapter 11
Vector Calculus

11.1 Introduction to Vector Functions

SURVIVAL HINT

Generally speaking the operations on vector functions are exactly what you would expect or want them to be. They are the same as the scalar functions applied to the component fuctions. Be careful with the cross product however, since it is non-commutative.

3. The domain is the intersection of the domains of the component functions.
    f_1 and f_2 are defined on all reals, but f_3 is not defined for $t = \frac{\pi}{2} + k\pi$.
    The domain is all reals except $\frac{\pi}{2} + k\pi$.

7. The domain of all component functions is the reals, except for f_1 which
    has a domain of $t > 0$. The domain of $\mathbf{F}(t) - \mathbf{G}(t)$ is $t > 0$.

11. Since z is always 0, this is a circle in the xy plane. It begins at $(0, 1)$,
    has a radius of 1, and traces in a counter-clockwise direction.

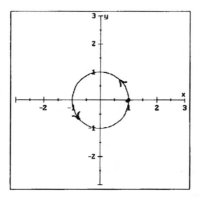

15. This is a helix (spiral) with radius of 1, beginning at (1, 0, 0), and tracing in a counter-clockwise direction.

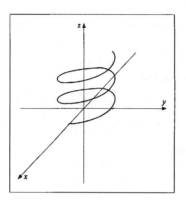

19. If $x = t$, $y = x^2 + 1$ (a parabola), and $z = x^2$ (also a parabola).

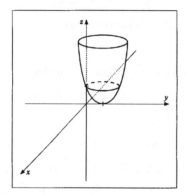

23. $x = 2 \sin t$, $y = 2 \sin t$, $z = \sqrt{8} \cos t$.

$$x^2 + y^2 + z^2 = 4 \sin^2 t + 4 \sin^2 t + 8 \cos^2 t$$
$$= 8(\sin^2 t + \cos^2 t) = 8$$

Which is a sphere with center at the origin and a radius of $2\sqrt{2}$.

27. $\mathbf{F}(t) = 2t\mathbf{i} + (1 - t)\mathbf{j} + \sin t \, \mathbf{k}$

31. $[2(2t) - 3(1 - t)]\mathbf{i} + [2(-5) - 3(0)]\mathbf{j} + [2(t^2) - 3(\frac{1}{t})]\mathbf{k}$

   $= (7t - 3)\mathbf{i} - 10\mathbf{j} + (2t^2 - \frac{3}{t})\mathbf{k}$

35. $(1 - t)\sin t$ (One vector has $\mathbf{j} = 0$ and the other has $\mathbf{k} = 0$)

39. $[2e^t(2t) + t(1 - t) + 10 \sin t]\mathbf{i} + [2e^t(-5) + 10e^t]\mathbf{j} + [2e^t t^2 + 1]\mathbf{k}$

   $= (4te^t - t^2 + t + 10 \sin t)\mathbf{i} + [2e^t t^2 + 1]\mathbf{k}$.

43. $3\mathbf{i} + e^2\mathbf{j}$

47. $\mathbf{i} + e\mathbf{k}$

SURVIVAL HINT

The vector function is discontinuous at a value of t for which any one of its component

functions is discontinuous.

51. Continuous for all reals, since each component function is everywhere continuous.

55. Continuous for all real values; $t \neq 0$.

59. Analysis: Write out the expression as real-valued expressions, then use properties

of limits and properties of reals to rearrange the terms.

$$\lim_{t \to t_0} [\mathbf{F}(t) + \mathbf{G}(t)] = \lim_{t \to t_0} [f_1(t)\mathbf{i} + f_2(t)\mathbf{j} + f_3(t)\mathbf{k} + g_1(t)\mathbf{i} + g_2(t)\mathbf{j} + g_3(t)\mathbf{k}]$$

$$= \lim_{t \to t_0} [f_1(t)\mathbf{i} + f_2(t)\mathbf{j} \ u + f_3(t)\mathbf{k}] + \lim_{t \to t_0} [g_1(t)\mathbf{i} + g_2(t)\mathbf{j} + g_3(t)\mathbf{k}], \ \ldots$$

61. Analysis: Write out the expression as real-valued expressions, then use properties

of limits and properties of reals to rearrange the terms.

11.2 Differentiation and Integration of Vector Functions

SURVIVAL HINT

The rules for the derivatives of the product (scalar, dot, or cross) of vector functions are

easy to remember as they all follow the same pattern as that of two real valued functions:

the product rule. Again, be careful with the cross product, as it is non-commutative.

1. $\mathbf{F}'(t) = \mathbf{i} + 2t\mathbf{j} + (1 + 3t^2)\mathbf{k}$

5. $\mathbf{F}'(t) = 2t \, \mathbf{i} - t^{-2}\mathbf{j} + 2e^{2t}\mathbf{k}, \ \ \mathbf{F}''(t) = 2\mathbf{i} + 2t^{-3}\mathbf{j} + 4e^{2t}\mathbf{k}$

9. $f(x) = 2x^2 - 3x^3 - 3x^2 = -3x^3 - x^2, \ \ f'(x) = -9x^2 - 2x$

13. $\mathbf{R}(t) = \langle t, t^2, 2t \rangle, \ \ \mathbf{V}(t) = \mathbf{R}'(t) = \langle 1, 2t, 2 \rangle,$

$\mathbf{A}(t) = \mathbf{V}'(t) = \mathbf{R}''(t) = \langle 0, 2, 0 \rangle.$ Speed $= \|\mathbf{V}\|$.

$\|\mathbf{V}(1)\| = \sqrt{1^2 + 2^2 + 2^2} = 3.$ Direction of motion is the tangent vector.

$\mathbf{R}(1) = (1, 1, 2), \ \mathbf{V}(1) = \langle 1, 2, 2 \rangle.$ Direction of motion at $t = 1$ is the line

$\dfrac{x - 1}{1} = \dfrac{y - 1}{2} = \dfrac{z - 2}{2}$, or as a unit vector $\dfrac{\mathbf{V}}{\|\mathbf{V}\|} = \frac{1}{3}\mathbf{i} + \frac{2}{3}\mathbf{j} + \frac{2}{3}\mathbf{k}$

17. $\mathbf{R}(t) = \langle e^t, e^{-t}, e^{2t}\rangle$, $\mathbf{V}(t) = \mathbf{R}'(t) = \langle e^t, -e^{-t}, 2e^{2t}\rangle$,

$\mathbf{A}(t) = \mathbf{V}'(t) = \mathbf{R}''(t) = \langle e^t, e^{-t}, 4e^{2t}\rangle$. Speed $= \|\mathbf{V}\|$.

$\|\mathbf{V}(\ln 2)\| = \sqrt{2^2 + (\frac{1}{2})^2 + 8^2} = \frac{\sqrt{273}}{2}$. Direction of motion is the tangent vector.

$\mathbf{R}(\ln 2) = (2, \frac{1}{2}, 4)$, $\mathbf{V}(\ln 2) = \langle 2, -\frac{1}{2}, 8\rangle$. Direction of motion at $t = \ln 2$

is the line $\dfrac{x - 2}{2} = \dfrac{y - \frac{1}{2}}{-\frac{1}{2}} = \dfrac{z - 4}{8}$, or $\dfrac{x - 2}{2} = \dfrac{1 - 2y}{1} = \dfrac{z - 4}{8}$,

or as a unit vector $\dfrac{\mathbf{V}}{\|\mathbf{V}\|} = \dfrac{2\sqrt{273}}{273}(2\mathbf{i} - \frac{1}{2}\mathbf{j} + 8\mathbf{k})$

21. $\mathbf{F}(t) = \dfrac{1}{1 + 2t}\langle t, t^2, t^3\rangle$, a scalar function times a vector function. Using the scalar multiple rule: $\mathbf{F}'(t) = h'\mathbf{F} + h\mathbf{F}'$

$= \dfrac{-2}{(1 + 2t)^2}\langle t, t^2, t^3\rangle + \dfrac{1}{1 + 2t}\langle 1, 2t, 3t^2\rangle$.

$\mathbf{F}(0) = \langle 0, 0, 0\rangle$, $\mathbf{F}'(0) = \langle 1, 0, 0\rangle$

$\mathbf{F}(2) = \langle\frac{2}{5}, \frac{4}{5}, \frac{8}{5}\rangle$, $\mathbf{F}'(2) = \langle\frac{1}{25}, \frac{12}{25}, \frac{44}{25}\rangle$. Tangent: $\dfrac{x - \frac{2}{5}}{\frac{1}{25}} = \dfrac{y - \frac{4}{5}}{\frac{12}{25}} = \dfrac{z - \frac{8}{5}}{\frac{44}{25}}$,

or $\dfrac{25x - 10}{1} = \dfrac{25y - 20}{12} = \dfrac{25z - 40}{44}$, or as a unit vector $\frac{1}{25}(\mathbf{i} + 12\mathbf{j} + 44\mathbf{k})$

25. Find the antiderivative of each component function: $\dfrac{t^2}{2}\mathbf{i} - \dfrac{e^{3t}}{3}\mathbf{j} + 3t\mathbf{k} + \mathbf{C}$.

29. Use integration by parts on the **i** component.

$\dfrac{t^2}{2}(\ln|t| - \frac{1}{2})\mathbf{i} - \cos(1 - t)\mathbf{j} + \dfrac{t^2}{2}\mathbf{k} + C$

33. Use the linearity rule. Since all the coefficients are constants, their derivatives are 0. $\dfrac{d}{dt}(\mathbf{v} + t\mathbf{w}) = \mathbf{w}$.

37. $3\mathbf{F} - 2\mathbf{G} = \langle 9 + 3t^2 - 2\sin(2 - t), -3\cos 3t, 3t^{-1} + 2e^{2t}\rangle$

$(3\mathbf{F} - 2\mathbf{G})' = \langle 6t + 2\cos(2 - t), 9\sin 3t, -\dfrac{3}{t^2} + 4e^{2t}\rangle$

$\mathbf{F}' = \langle 2t, 3\sin 3t, -\dfrac{1}{t^2}\rangle$. $\mathbf{G}' = \langle-\cos(2 - t), 0, -2e^{2t}\rangle$

$3\mathbf{F}' - 2\mathbf{G}' = \langle 6t + 2\cos(2 - t), 9\sin 3t, -\dfrac{3}{t^2} + 4e^{2t}\rangle$

41. Integrate each component. For the **i** component set up $u^{1/2}\,du$ form, for the **j** component use $\tan^{-1}t$.

$\dfrac{1}{3}(1 + t^2)^{3/2}\mathbf{i} + (\tan^{-1}t)\mathbf{j}\Big|_0^1 = \dfrac{2\sqrt{2}}{3}\mathbf{i} + \dfrac{\pi}{4}\mathbf{j} - \dfrac{1}{3}\mathbf{i} - 0\mathbf{j} = \dfrac{2\sqrt{2} - 1}{3}\mathbf{i} + \dfrac{\pi}{4}\mathbf{j}$

45. Analysis: Write out $\mathbf{F}(t)\bullet\mathbf{F}'(t)$ as a scalar expression, and do the same for the right side of the equation. The radicals should cancel and leave the desired result.

47.   Analysis: Let $\mathbf{F}(t) = u_1(t)\mathbf{i} + u_2(t)\mathbf{j} + u_3(t)\mathbf{k}$ and

$\mathbf{G}(t) = v_1(t)\mathbf{i} + v_2(t)\mathbf{j} + v_3(t)\mathbf{k}.$

Since these are scalar functions the right hand side of the equation can be

commutated and associated to equal the left hand side.

49.   The obvious proof makes note of the fact that $\|\mathbf{F}(t)\|$ is a scalar, so the derivative

of the quotient is represented by the first term on the right. The second term

is 0 because a vector and its derivative are orthogonal, so their dot product is 0.

The not-so-obvious approach would be to use the quotient rule for the derivative,

and then show that $\|\mathbf{F}\|' = \dfrac{\mathbf{F} \bullet \mathbf{F}'}{\|\mathbf{F}\|}.$

11.3 Modeling Ballistics and Planetary Motion

SURVIVAL HINT

This is basically the same material as was covered in Section 2.4, except that the path of

the projectile has been broken into its horizontal and vertical components. This allows us

to write a vector function, which gives the position, velocity, and acceleration at any time

for the parameter t. A review of some of the problems in Section 2.4 might be useful.

3. $T_f = \dfrac{2}{g} V_0 \sin \alpha = \dfrac{2}{9.8} (850) \sin 48.5° \approx 129.92$ sec

$R = \dfrac{V_0{}^2}{g} \sin 2\alpha = \dfrac{850^2}{9.8} \sin 2(48.5°) \approx 73{,}174.96$ m

7. $T_f = \dfrac{2}{g} V_0 \sin \alpha = \dfrac{2}{32} (100) \sin 14.11° \approx 1.5237$ sec

$R = \dfrac{V_0{}^2}{g} \sin 2\alpha = \dfrac{100^2}{32} \sin 2(14.11°) \approx 147.77$ ft

11. $\theta = 2t,\ r = \sin 2t.$ $V = \dfrac{dr}{dt}\mathbf{u}_r + r\dfrac{d\theta}{dt}\mathbf{u}_\theta = (2 \cos 2t)\mathbf{u}_r + (2 \sin 2t)\mathbf{u}_\theta.$

$A = \left\{\dfrac{d^2r}{dt^2} - r\left(\dfrac{d\theta}{dt}\right)^2\right\}\mathbf{u}_r + \left\{r\dfrac{d^2\theta}{dt^2} + 2\dfrac{dr}{dt}\dfrac{d\theta}{dt}\right\}\mathbf{u}_\theta$

$= [-4 \sin 2t - (\sin 2t)(4)]\mathbf{u}_r + [(\sin 2t)(0) + 2(2)(2 \cos 2t)]\mathbf{u}_\theta$

$= (-8 \sin 2t)\,\mathbf{u}_r + (8 \cos 2t)\,\mathbf{u}_\theta$

15. $\alpha = 45°$, so we can use the formula for maximum range: $R_m = \dfrac{V_0{}^2}{g}.$

$2000 = \dfrac{V_0{}^2}{9.8},\ V_0 \approx 140$ m/sec

19. The maximum height is reached when $y'(t) = 0$. We can use this equation to find the time at which this occurs, then use that time in $y(t)$ to find the height.

$y(t) = -16t^2 + (V_0 \sin \alpha) t + s_0 = -16t^2 + 45t + 4$

$y'(t) = -32t + 45 = 0$ when $t = \frac{45}{32}$ sec

$y(\frac{45}{32}) = -16 \left(\frac{45}{32}\right)^2 + 45(\frac{45}{32}) + 4 \approx 35.64$ ft

The ball will land when $y = 0$. We can use $y(t) = 0$ to find the time of flight, and then use that time to find $x(t)$. $-16t^2 + 45t + 4 = 0$ when $t \approx 2.8987$ sec

$x(t) = (v_0 \cos \alpha) t = \frac{\sqrt{3}}{2}(90) t = 45\sqrt{3}\, t.$ $x(2.8987) \approx 225.93$ ft

To find the distance to the fence we will find t for $y(t) = 5$, then use that time in $x(t)$. $-16t^2 + 45t + 4 = 5$ when $t \approx 2.79$ sec. $x(2.79) = 45\sqrt{3}\,(2.79)$

≈ 217.47 ft

23. Since $\alpha = 45°$ we can use the maximum range formula to find the distance to the green. We can then let $x(t)$ equal that distance in order to find t.

$R_m = \frac{V_0^{\,2}}{g} = \frac{125^2}{32}.$ $x(t) = (v_0 \cos \alpha) t,$ $125 \frac{\sqrt{2}}{2} t = \frac{125^2}{32},$ $t \approx 5.52$ sec

27. $\mathbf{V} = \frac{dr}{dt}\mathbf{u}_r + r\frac{d\theta}{dt}\mathbf{u}_\theta = (2\cos t)\mathbf{u}_r + (3 + 2\sin t)3t^2\mathbf{u}_\theta.$

$\mathbf{A} = \left\{\frac{d^2r}{dt^2} - r\left(\frac{d\theta}{dt}\right)^2\right\}\mathbf{u}_r + \left\{r\frac{d^2\theta}{dt^2} + 2\frac{dr}{dt}\frac{d\theta}{dt}\right\}\mathbf{u}_\theta$

$= \left\{-2\sin t - (3 + 2\sin t)(3t^2)^2\right\}\mathbf{u}_r + \left\{(3 + 2\sin t)(6t) + 2(2\cos t)(3t^2)\right\}\mathbf{u}_\theta$

$= \left\{-2\sin t - 27t^4 - 18t^4\sin t\right\}\mathbf{u}_r + \left\{18t + 12t\sin t + 12t^2\cos t\right\}\mathbf{u}_\theta$

31. $\mathbf{r}(t) = (a\cos \omega t)\mathbf{i} + (a\sin \omega t)\mathbf{j}.$

$\mathbf{v}(t) = \mathbf{r}'(t) = (-\omega a\sin \omega t)\mathbf{i} + (\omega a\cos \omega t)\mathbf{j}.$

$\mathbf{a}(t) = \mathbf{v}'(t) = \mathbf{r}''(t) = (-\omega^2 a\cos \omega t)\mathbf{i} + (-\omega^2 a\sin \omega t)\mathbf{j}.$

$||\mathbf{a}(t)|| = \sqrt{(-\omega^2 a\cos \omega t)^2 + (-\omega^2 a\sin \omega t)^2} = \omega^2 |a|$

35. Analysis: Distribute the mass, m, to each term of the acceleration formula, and and redefine the terms in brackets as indicated.

37. Analysis: Proofs are often done by working backward from the desired result. Let $mr^2 \frac{d\theta}{dt} = C$, m is a constant, so $r^2 \frac{d\theta}{dt} = \frac{C}{m}$. Now differentiate, using the product rule on the left, and see what results.

39. Analysis: Write $\mathbf{R}'$ as $\mathbf{V}$, write $\mathbf{V}$ as polar formulas in terms of $\sin \theta$ and $\cos \theta$. Differentiate using the product rule, and then find the magnitude.

11.4 Tangent and Normal Vectors; Curvature

3. $\mathbf{T}(t) = \dfrac{\mathbf{R}'(t)}{\|\mathbf{R}'(t)\|} = \dfrac{e^t(\cos t - \sin t)\mathbf{i} + e^t(\cos t + \sin t)\mathbf{j}}{\sqrt{e^{2t}(\cos t - \sin t)^2 + e^{2t}(\cos t + \sin t)^2}}$

$\qquad\qquad = \dfrac{e^t(\cos t - \sin t)\mathbf{i} + e^t(\cos t + \sin t)\mathbf{j}}{e^t\sqrt{2}}$

$\qquad\qquad = \dfrac{\sqrt{2}}{2}[(\cos t - \sin t)\mathbf{i} + (\cos t + \sin t)\mathbf{j}].$

$\quad\ \mathbf{N}(t) = \dfrac{\mathbf{T}'(t)}{\|\mathbf{T}'(t)\|} = \dfrac{\sqrt{2}}{2}\dfrac{(-\sin t - \cos t)\mathbf{i} + (-\sin t + \cos t)\mathbf{j}}{\sqrt{\frac{1}{2}(-\sin t - \cos t)^2 + \frac{1}{2}(-\sin t + \cos t)^2}}$

$\qquad\qquad = -\dfrac{\sqrt{2}}{2}[(\sin t + \cos t)\mathbf{i} + (\sin t - \cos t)\mathbf{j}]$

7. $\mathbf{T}(t) = \dfrac{\mathbf{R}'(t)}{\|\mathbf{R}'(t)\|} = \dfrac{\frac{1}{t}\mathbf{i} + 2t\mathbf{k}}{\sqrt{\frac{1}{t^2} + 4t^2}} = \dfrac{\mathbf{i} + 2t^2\mathbf{k}}{\sqrt{1 + 4t^4}}.$

$\quad\ \mathbf{T}'(t) = \dfrac{\sqrt{1 + 4t^4}\,4t\mathbf{k} - \dfrac{(\mathbf{i} + 2t^2\mathbf{k})8t^3}{\sqrt{1 + 4t^4}}}{1 + 4t^4} = \dfrac{-8t^3\mathbf{i} + 4t\mathbf{k}}{(1 + 4t^4)^{3/2}}$

$\quad\ \|\mathbf{T}'(t)\| = \sqrt{\dfrac{(-8t^3)^2 + (4t)^2}{(1 + 4t^4)^3}} = \dfrac{4t\sqrt{4t^4 + 1}}{(1 + 4t^4)^{3/2}} = \dfrac{4t}{1 + 4t^4}.$

$\quad\ \mathbf{N}(t) = \dfrac{\mathbf{T}'(t)}{\|\mathbf{T}'(t)\|} = \dfrac{\dfrac{-8t^3\mathbf{i} + 4t\mathbf{k}}{(1 + 4t^4)^{3/2}}}{\dfrac{4t}{1 + 4t^4}} = \dfrac{-2t^2\mathbf{i} + \mathbf{k}}{\sqrt{1 + 4t^4}}.$

SURVIVAL HINT

The arc length formula is easily recalled if you think of it as the sum (definite integral) of lines in $\mathbb{R}^3$ (space diagonals); $\sqrt{(\Delta x)^2 + (\Delta y)^2 + (\Delta z)^2}$, where the delta change is found by using the derivative.

11. $\mathbf{R}'(t) = 3\mathbf{i} - (3\sin t)\mathbf{j} + (3\cos t)\mathbf{k}.$

$\quad\ \|\mathbf{R}'(t)\| = \sqrt{3^2 + 9\sin^2 t + 9\cos^2 t} = 3\sqrt{2}.$

$\quad\ \text{Arc Length} = \displaystyle\int_0^{\frac{\pi}{2}} 3\sqrt{2}\,dt = \dfrac{3\sqrt{2}\,\pi}{2}$

SURVIVAL HINT

There are so many different formulas for curvature, dependent upon the form of the given information (p. 837), that you should find out which ones your instructor expects you to know.

15. $\kappa = 0$, as this is a straight line. $\rho = \frac{1}{\kappa} = \infty$. Notice that a straight line is the circumference of a circle with infinite radius.

19. Let $x = t$, then $y = at^2 + bt$.

$\mathbf{F}(t) = t\mathbf{i} + (at^2 + bt)\mathbf{j}$. $\mathbf{V}(t) = \mathbf{i} + (2at + b)\mathbf{j}$. $\mathbf{A}(t) = 2a\mathbf{j}$.

$\|\mathbf{V}\| = \sqrt{1^2 + (2at + b)^2}$

$$\mathbf{V} \times \mathbf{A} = \begin{vmatrix} \mathbf{i} & \mathbf{j} & \mathbf{k} \\ 1 & 2at+b & 0 \\ 0 & 2a & 0 \end{vmatrix} = 2a\mathbf{k}$$

$\kappa = \dfrac{\|\mathbf{V} \times \mathbf{A}\|}{\|\mathbf{V}\|^3} = \dfrac{2\,|a|}{[1 + (2at + b)^2]^{3/2}}$. $\kappa(c) = \dfrac{2\,|a|}{[1 + (2ac + b)^2]^{3/2}}$.

$\rho = \dfrac{1}{\kappa}(c) = \dfrac{[1 + (2ac + b)^2]^{3/2}}{2|a|}$

23. $\mathbf{F}(t) = t\mathbf{i} + (\sin t)\mathbf{j}$. $\mathbf{V}(t) = \mathbf{i} + (\cos t)\mathbf{j}$. $\mathbf{A}(t) = (-\sin t)\mathbf{j}$

$\|\mathbf{V}\| = \sqrt{1 + \cos^2 t}$

$$\mathbf{V} \times \mathbf{A} = \begin{vmatrix} \mathbf{i} & \mathbf{j} & \mathbf{k} \\ 1 & \cos t & 0 \\ 0 & -\sin t & 0 \end{vmatrix} = (-\sin t)\mathbf{k}.$$

$\kappa = \dfrac{\|\mathbf{V} \times \mathbf{A}\|}{\|\mathbf{V}\|^3} = \dfrac{|\sin t|}{(1 + \cos^2 t)^{3/2}}$; $\kappa(\frac{\pi}{2}) = \frac{1}{1} = 1$; $\rho = \frac{1}{\kappa}(\frac{\pi}{2}) = 1$

27. If $\mathbf{u}$ and $\mathbf{v}$ are constant vectors, $\mathbf{R}(t) = \mathbf{u} + \mathbf{v}t$, then $\mathbf{V}(t) = \mathbf{v}$, and $\mathbf{A}(t) = \mathbf{0}$; $\mathbf{V} \times \mathbf{A}$ will be 0, so κ will be 0.

31.   **a.** $T = \dfrac{R'(t)}{||R'(t)||} = \dfrac{(\cos t)i + (-\sin t)j + k}{\sqrt{\cos^2 t + \sin^2 t + 1}}$

$= \dfrac{\sqrt{2}}{2}[(\cos t)i + (-\sin t)j + k].$

$T(\pi) = \dfrac{\sqrt{2}}{2}[-i + k]$

b. $\kappa = \dfrac{||R' \times R''||}{||R'||^3}.$ $R' = (\cos t)i - (\sin t)j + k.$

$||R'|| = \sqrt{\cos^2 t + \sin^2 t + 1} = \sqrt{2}.$

$R'' = (-\sin t)i + (-\cos t)j.$

$||R' \times R''|| = \begin{vmatrix} i & j & k \\ \cos t & -\sin t & 1 \\ -\sin t & -\cos t & 0 \end{vmatrix} = ||(\cos t)i - (\sin t)j - k|| = \sqrt{2}$

$\kappa = \dfrac{\sqrt{2}}{2\sqrt{2}} = \dfrac{1}{2}$

c. $S = \displaystyle\int_0^\pi ||R'|| \, dt = \int_0^\pi \sqrt{2} \, dt = \sqrt{2}\,\pi$

35.   Let $x = t$, then $y = t^6 - 3t^2$. $F = ti + (t^6 - 3t^2)j$.

To find the extrema: $y' = 6x^5 - 6x = 6x(x^2 + 1)(x + 1)(x - 1) = 0$

when $x = 0, 1, -1$.   (Which are also the t values.)

$V = \langle 1, 6t^5 - 6t \rangle,$ $A = \langle 0, 30t^4 - 6 \rangle.$

$V \times A = \begin{vmatrix} i & j & k \\ 1 & 6t^5 - 6t & 0 \\ 0 & 30t^4 - 6 & 0 \end{vmatrix} = (30t^4 - 6)k$

$\kappa = \dfrac{||V \times A||}{||V||^3} = \dfrac{|30t^4 - 6|}{[1 + (6t^5 - 6t)^2]^{3/2}}$

$\dfrac{1}{\kappa} = \dfrac{[1 + (6t^5 - 6t)^2]^{3/2}}{|30t^4 - 6|}$

$\dfrac{1}{\kappa}(0) = \dfrac{1}{6}.$ $\dfrac{1}{\kappa}(1) = \dfrac{1}{24}.$ $\dfrac{1}{\kappa}(-1) = \dfrac{1}{24}$

39. $\mathbf{R}(t) = t\mathbf{i} + t^2\mathbf{j} + t^3\mathbf{k};\ \mathbf{V}(t) = \mathbf{i} + 2t\mathbf{j} + 3t^2\mathbf{k};\ \mathbf{A}(t) = 2\mathbf{j} + 6t\mathbf{k}$

$$\mathbf{V} \times \mathbf{A} = \begin{vmatrix} \mathbf{i} & \mathbf{j} & \mathbf{k} \\ 1 & 2t & 3t^2 \\ 0 & 2 & 6t \end{vmatrix} = 6t^2\mathbf{i} - 6t\mathbf{j} + 2\mathbf{k}$$

$$\|\mathbf{V} \times \mathbf{A}\| = \sqrt{36t^4 + 36t^2 + 4} = 2\sqrt{9t^4 + 9t^2 + 1}$$

$$\kappa = \frac{\|\mathbf{V} \times \mathbf{A}\|}{\|\mathbf{V}\|^3} = \frac{2\sqrt{9t^4 + 9t^2 + 1}}{\left(\sqrt{1 + 4t^2 + 9t^4}\right)^3}$$

43. $y = x^{-1},\ y' = -x^{-2},\ y'' = 2x^{-3}$

$$\kappa = \frac{|x'y'' - y'x''|}{[(x')^2 + (y')^2]^{3/2}} = \frac{|1(2x^{-3}) - (-x^{-2})(0)|}{[1^2 + (-x^{-2})^2]^{3/2}} = \frac{2|x^{-3}|}{(1 + x^{-4})^{3/2}}$$

$$= \frac{\dfrac{2}{|x^3|}}{\dfrac{(x^4 + 1)^{3/2}}{x^6}} = \frac{2|x^3|}{(x^4 + 1)^{3/2}}$$

49.   Analysis: In addition to the Hint you might look at part of the proof of Theorem 11.8 in the next section.
Also make use of: $\dfrac{d\mathbf{T}}{dt} = \dfrac{ds}{dt}\dfrac{d\mathbf{T}}{ds} = \kappa\dfrac{ds}{dt}\left(\dfrac{\frac{d\mathbf{T}}{ds}}{\kappa}\right) = \kappa\dfrac{ds}{dt}\mathbf{N}$

51.   Analysis: Find $\mathbf{R}'$ and $\mathbf{R}''$, then use of the formula in Problem 49 will yield the desired result.

53.   Analysis: Substitution of the identity given in the Hint, into the formula of Problem 49, gives the result directly.

55.   Analysis:

a.   The derivative of $\mathbf{B} \bullet \mathbf{T}$ with respect to s gives $\dfrac{d\mathbf{B}}{ds} \bullet \mathbf{T} + \mathbf{B} \bullet \dfrac{d\mathbf{T}}{ds}$.
Show that the first term is 0, then the factors are orthogonal.

b.   Same procedure as **a**.

c.   Make use of the fact that $\mathbf{N} = \dfrac{\mathbf{T}'}{\|\mathbf{T}'\|}$.

11.5 Tangential and Normal Components of Acceleration

SURVIVAL HINT

Most of these problems involve finding several different derivatives, and the sums and products of several components. It is essential that your work be well organized and each step be properly labeled.

3. $\mathbf{R}(t) = (t \sin t)\mathbf{i} + (t \cos t)\mathbf{j}, \quad \mathbf{V}(t) = (\sin t + t \cos t)\mathbf{i} + (\cos t - t \sin t)\mathbf{j},$

$\mathbf{A}(t) = (2 \cos t - t \sin t)\mathbf{i} + (-2 \sin t - t \cos t)\mathbf{j}$

$\dfrac{ds}{dt} = \|\mathbf{V}\| = \sqrt{\sin^2 t + t^2\cos^2 t + \cos^2 t + t^2\sin^2 t} = \sqrt{1 + t^2}$

$A_T = \dfrac{d^2 s}{dt^2} = \dfrac{t}{\sqrt{1 + t^2}}$

$A_N = \sqrt{\|\mathbf{A}\|^2 - A_T^2} = \sqrt{t^2 + 4 - \dfrac{t^2}{1 + t^2}}$

$\quad = \dfrac{\sqrt{(t^2 + 4)(1 + t^2) - t^2}}{\sqrt{1 + t^2}} = \dfrac{\sqrt{t^4 + 4t^2 + 4}}{\sqrt{1 + t^2}} = \dfrac{t^2 + 2}{\sqrt{t^2 + 1}}$

7. $\mathbf{R}(t) = (4 \cos t)\mathbf{i} + (\sin t)\mathbf{k}, \quad \mathbf{V}(t) = (-4 \sin t)\mathbf{i} + (\cos t)\mathbf{k},$

$\mathbf{A}(t) = (-4 \cos t)\mathbf{i} + (-\sin t)\mathbf{k}. \quad \|\mathbf{A}\| = \sqrt{16 \cos^2 t + \sin^2 t}.$

$\dfrac{ds}{dt} = \|\mathbf{V}\| = \sqrt{16 \sin^2 t + \cos^2 t}$

$A_T = \dfrac{d^2 s}{dt^2} = \dfrac{-32 \sin t \cos t + 2 \cos t \sin t}{2\sqrt{16 \sin^2 t + \cos^2 t}} = \dfrac{-15 \sin t \cos t}{\sqrt{16 \sin^2 t + \cos^2 t}}.$

$A_N = \sqrt{\|\mathbf{A}\|^2 - A_T^2} = \sqrt{16 \cos^2 t + \sin^2 t - \dfrac{225 \sin^2 t \cos^2 t}{16 \sin^2 t + \cos^2 t}}$

$\quad = \sqrt{\dfrac{16 \sin^4 t + 16 \cos^4 t + 32\sin^2 t \cos^2 t}{16 \sin^2 t + \cos^2 t}}$

$\quad = \dfrac{4 \sqrt{\sin^4 t + 2 \sin^2 t \cos^2 t + \cos^4 t}}{\sqrt{16 \sin^2 t + \cos^2 t}} = \dfrac{4}{\sqrt{16 \sin^2 t + \cos^2 t}}$

11. $\mathbf{R}(t) = t\mathbf{i} + 4t^2\mathbf{j}.$ $\mathbf{V} = \mathbf{i} + 8t\mathbf{j}.$ $\mathbf{A} = 8\mathbf{j}.$ $\|\mathbf{A}\| = 8$

$\dfrac{ds}{dt} = \|\mathbf{V}\| = \sqrt{1 + 64t^2}.$ If $\dfrac{ds}{dt} = 20,$ $1 + 64t^2 = 400,$ $t = \dfrac{\sqrt{399}}{8}$

$A_T = \dfrac{d^2s}{dt^2} = \dfrac{64\,t}{\sqrt{1 + 64t^2}}.$ $A_N = \sqrt{\|\mathbf{A}\|^2 - A_T{}^2} = \sqrt{64 - \dfrac{(64\,t)^2}{1 + 64\,t^2}}$

$\qquad = \dfrac{8}{\sqrt{1 + 64t^2}}.$ $A_T\left(\dfrac{\sqrt{399}}{8}\right) = \dfrac{8\sqrt{399}}{20} \approx 7.98999$

$A_N\left(\dfrac{\sqrt{399}}{8}\right) = \dfrac{8}{20} = 0.4$

15. **a.** We wish to find the force normal to the path of the car. $F_N = m\kappa\left(\dfrac{ds}{dt}\right)^2.$

$\qquad m = \dfrac{3500}{32.2},$ $\kappa = \dfrac{1}{200},$ $\dfrac{ds}{dt} = 55 \text{ mph} = \dfrac{242}{3} \text{ fps}$

$\qquad F_N = \dfrac{3500}{32.2}\left(\dfrac{1}{200}\right)\left(\dfrac{242}{3}\right)^2 \approx 3536.5 \text{ lbs}$

b. $\theta = \tan^{-1}\left(\dfrac{3500}{3536.5}\right) \approx 44.7°$

19. First we will compute all the necessary components for the formulas:

$\mathbf{R}(t) = t\mathbf{i} + 2t\mathbf{j} + t^2\mathbf{k},$ $\mathbf{R}' = \mathbf{i} + 2\mathbf{j} + 2t\mathbf{k},$ $\mathbf{R}'' = 2\mathbf{k}$

$\|\mathbf{R}'\| = \sqrt{1 + 4 + 4t^2} = \sqrt{5 + 4t^2}$

$\mathbf{R}' \bullet \mathbf{R}'' = 4t$

$\mathbf{R}' \times \mathbf{R}'' = \begin{vmatrix} \mathbf{i} & \mathbf{j} & \mathbf{k} \\ 1 & 2 & 2t \\ 0 & 0 & 2 \end{vmatrix} = 4\mathbf{i} - 2\mathbf{j}$

$\|\mathbf{R}' \times \mathbf{R}''\| = \sqrt{16 + 4} = 2\sqrt{5}$

$A_T = \dfrac{\mathbf{R}' \bullet \mathbf{R}''}{\|\mathbf{R}'\|} = \dfrac{4t}{\sqrt{5 + 4t^2}}$

$A_N = \dfrac{\|\mathbf{R}' \times \mathbf{R}''\|}{\|\mathbf{R}'\|} = \dfrac{2\sqrt{5}}{\sqrt{5 + 4t^2}}$

23. Analysis:

a. The motion is circular, so $\mathbf{R} = (\cos \omega t)\mathbf{i} + (\sin \omega t)\mathbf{j}.$

Choose an appropriate formula to find $A_N = \dfrac{W}{g}(60\pi^2\omega^2)$

b. The hint is sufficient. $\omega \approx 4.83 \text{ rev/min}$

25. Analysis: Use the form $\mathbf{R}(x) = x\mathbf{i} + f(x)\mathbf{j}$. Find $\mathbf{V}$ and $\|\mathbf{V}\|$ and use the suggested formulas.

27. Analysis: $\mathbf{R} = [(v_0\cos\alpha)t]\mathbf{i} + [-\frac{1}{2}gt^2 + (v_0\sin\alpha)t]\mathbf{j}$, and since the time of flight is given by $\frac{2}{g}v_0\sin\alpha$, half of this time is when the maximum height occurs.

29. Analysis:

   a. Example 4 establishes that the period, T, is equal to $\frac{2\pi R}{v}$, where R is the distance from the center of the earth $(R_e + R)$, and $v = \sqrt{\frac{GM}{R}}$.
   g in our problem must represent the force of gravity at the height of the satellite, which is the gravitational constant times the mass of the earth. Make the appropriate substitutions.

   b. Solve part **a** for R.

Chapter 11 Review

PRACTICE PROBLEMS

24. This is a helix with radius of 3, climbing in a counter-clockwise direction.

$$s = \int_0^{2\pi} \|\mathbf{R}'\|\, dt.$$

$\mathbf{R} = (3\cos t)\mathbf{i} + (3\sin t)\mathbf{j} + t\mathbf{k}$, $\mathbf{R}' = (-3\sin t)\mathbf{i} + (3\cos t)\mathbf{j} + \mathbf{k}$,
$\|\mathbf{R}'\| = \sqrt{9\sin^2 t + 9\cos^2 t + 1} = \sqrt{10}$.

$$s = \int_0^{2\pi} \sqrt{10}\, dt = 2\pi\sqrt{10}$$

25. $\mathbf{F}' = \dfrac{1}{(1+t)^2}\mathbf{i} + \dfrac{t\cos t - \sin t}{t^2}\mathbf{j} + (-\sin t)\mathbf{k}$.

$\mathbf{F}'' = -\dfrac{2}{(1+t)^3}\mathbf{i} + \dfrac{-t^2\sin t - 2t\cos t + 2\sin t}{t^3}\mathbf{j} - (\cos t)\mathbf{k}$

26.   Finding the cross product:

$$\begin{vmatrix} \mathbf{i} & \mathbf{j} & \mathbf{k} \\ 3t & 0 & 3 \\ 0 & \ln t & -t^2 \end{vmatrix} = (-3 \ln t)\mathbf{i} + (3t^3)\mathbf{j} + (3t \ln t)\mathbf{k}.$$

So our integral is: $\displaystyle\int_1^2 [(-3 \ln t)\mathbf{i} + (3t^3)\mathbf{j} + (3t \ln t)\mathbf{k}]\, dt$

$$= -3(t \ln t - t)\mathbf{i} + \frac{3t^4}{4}\mathbf{j} + 3\left(\frac{t^2}{2}[\ln t - \tfrac{1}{2}]\right)\mathbf{k} \,\Big|_1^2 \quad \text{(formulas 499 \& 502)}$$

$$= (6 - 6 \ln 2)\mathbf{i} + 12\mathbf{j} + (6 \ln 2 - 3)\mathbf{k} - \left(3\mathbf{i} + \tfrac{3}{4}\mathbf{j} - \tfrac{3}{4}\mathbf{k}\right)$$

$$= (3 - 6 \ln 2)\mathbf{i} + \frac{45}{4}\mathbf{j} + (6 \ln 2 - \tfrac{9}{4})\mathbf{k}$$

27.   Antidifferentiation will lead us back to the original $\mathbf{F}$.

$\mathbf{F}'' = e^t \mathbf{i} - t^2 \mathbf{j} + 3\mathbf{k},\ \ \mathbf{F}' = e^t \mathbf{i} - \frac{t^3}{3}\mathbf{j} + 3t\mathbf{k} + C,$ but $\mathbf{F}'(0) = 3\mathbf{k}$ so

$\mathbf{F}' = (e^t - 1)\mathbf{i} - \frac{t^3}{3}\mathbf{j} + (3t + 3)\mathbf{k}.$

$\mathbf{F} = (e^t - t)\mathbf{i} - \frac{t^4}{12}\mathbf{j} + (\frac{3t^2}{2} + 3t)\mathbf{k} + C,$ but $\mathbf{F}(0) = \mathbf{i} - 2\mathbf{j}$ so

$\mathbf{F} = (e^t - t)\mathbf{i} - \left(\frac{t^4}{12} + 2\right)\mathbf{j} + (\frac{3t^2}{2} + 3t)\mathbf{k}$

28. $\mathbf{R} = t\mathbf{i} + 2t\mathbf{j} + te^t\mathbf{k}.$ $\mathbf{V} = \mathbf{R}' = \mathbf{i} + 2\mathbf{j} + e^t(t + 1)\mathbf{k}.$

$\frac{ds}{dt} = \|\mathbf{V}\| = \sqrt{1 + 4 + e^{2t}(t+1)^2} = \sqrt{5 + e^{2t}(t+1)^2}.$

$\mathbf{A} = \mathbf{V}' = e^t(t + 2)\mathbf{k}$

29. $\mathbf{R} = t^2\mathbf{i} + 3t\mathbf{j} - 3t\mathbf{k};\ \ \mathbf{T} = \dfrac{\mathbf{R}'}{\|\mathbf{R}'\|} = \dfrac{2t\mathbf{i} + 3\mathbf{j} - 3\mathbf{k}}{\sqrt{4t^2 + 18}}$

To find $\mathbf{T}'$, treat $\mathbf{T}$ as a scalar times a vector and use the product rule:

$$\mathbf{T}' = \frac{-4t}{(4t^2 + 18)^{3/2}}(2t\mathbf{i} + 3\mathbf{j} - 3\mathbf{k}) + \frac{1}{(4t^2 + 18)^{1/2}}(2\mathbf{i})$$

$$= \frac{36\mathbf{i} - 12t\mathbf{j} + 12t\mathbf{k}}{(4t^2 + 18)^{3/2}}$$

$$\mathbf{N} = \frac{\mathbf{T}'}{\|\mathbf{T}'\|} = \frac{3\mathbf{i} - t\mathbf{j} + t\mathbf{k}}{(2t^2 + 9)^{1/2}} \qquad \mathbf{A} = \mathbf{R}'' = 2\mathbf{i}$$

29. (con't.)

$$\mathbf{R}' \times \mathbf{R}'' = \begin{bmatrix} \mathbf{i} & \mathbf{j} & \mathbf{k} \\ 2t & 3 & -3 \\ 2 & 0 & 0 \end{bmatrix} = -6\mathbf{j} - 6\mathbf{k}$$

And now $\kappa = \dfrac{\|\mathbf{R}' \times \mathbf{R}''\|}{\|\mathbf{R}'\|^3} = \dfrac{\sqrt{36+36}}{(4t^2+18)^{3/2}}$

$$= \dfrac{\sqrt{72}}{(4t^2+18)^{3/2}} = \dfrac{3}{(2t^2+9)^{3/2}}$$

$A_T = \dfrac{d^2s}{dt^2} = \dfrac{4t}{(4t^2+18)^{1/2}}$

$A_N = \dfrac{3}{(2t^2+9)^{3/2}}(4t^2+18) = \dfrac{6\sqrt{2}}{(4t^2+18)^{1/2}}$

To check this, does $A_T{}^2 + A_N{}^2 = \|\mathbf{A}\|^2$?

$$\dfrac{16t^2}{(4t^2+18)} + \dfrac{72}{(4t^2+18)} = 4$$

30. **a.** $\mathbf{R} = [(v_0\cos\alpha)t]\mathbf{i} + [(v_0\sin\alpha)t - \tfrac{1}{2}gt^2 + s_0]\mathbf{j}$, in our case:

$\mathbf{R} = (25\sqrt{3}\,t)\mathbf{i} + (25t - 16t^2)\mathbf{j}$. The maximum height is the value

of the coefficient of $\mathbf{j}$ at half the total time of flight, $t = \frac{25}{32}$ sec (see **b**).

$\mathbf{R}(\frac{25}{32}) = 25(\frac{25}{32}) - 16(\frac{25}{32})^2 \approx 9.76$ ft

b. $T_f = \frac{2}{g}v_0\sin\alpha = \frac{1}{16}(50)(\frac{1}{2}) = \frac{25}{16}$ sec

Range $= \dfrac{v_0{}^2}{g}\sin 2\alpha = \dfrac{50^2}{32}\sin 60° \approx 67.7$ ft

SURVIVAL HINT

The Cumulative Review for Chapters 7-11 can be very valuable to refresh some
of the skills and concepts that you may not have been using often. It is also
a valuable tool in preparing for a final exam. If you do not have the time to actually
do all of the problems, try looking at each one to see if you recall the concept
involved and how to proceed with the solution. If you are confident about your
ability to solve the problem, do not spend the time. If you feel a little uncertain
about the problem, refer back to the appropriate section, review the concepts,
look in your old homework for a similar problem, and then see if you can work it.
Be more concerned about understanding the concept than about getting exactly the
right answer. Do not spend a lot of your time looking for algebraic and
arithemetic errors.

Chapter 12
Partial Differentiation

12.1 Functions of Several Variables

3. $f(x, y, z) = x^2 y e^{2x} + (x + y - z)^2$

a. $f(0, 0, 0) = 0^2 0 e^0 + (0 + 0 - 0)^2 = 0$

b. $f(1, -1, 1) = 1^2(-1)e^2 + (1 - 1 - 1)^2 = -e^2 + 1$

c. $f(-1, 1, -1) = (-1)^2(1)e^{-2} + (-1 + 1 + 1)^2 = e^{-2} + 1$

d. $f(x, x, x) = x^2 x e^{2x} + (x + x - x)^2 = x^3 e^{2x} + x^2$

$\dfrac{d}{dx} f(x, x, x) = 2x^3 e^{2x} + 3x^2 e^{2x} + 2x$

e. $f(1, y, 1) = 1^2 y e^2 + (1 + y - 1)^2 = e^2 y + y^2$

$\dfrac{d}{dy} f(1, y, 1) = e^2 + 2y$

f. $f(1, 1, z^2) = 1^2 1 e^2 + (1 + 1 - z^2)^2 = e^2 + (2 - z^2)^2$

$\dfrac{d}{dz} f(1, 1, z^2) = -4z(2 - z^2) = 4z(z^2 - 2)$

9. The radicand must be positive so the domain is $x - y > 0$ or $x > y$

Since the radical is always positive, the range will be $f(x, y) > 0$

13. The radicand must be non-negative so the domain is $u \sin v \geq 0$

Since the radical is always non-negative, the range is $f(u, v) \geq 0$

17. The radicand in the denominator must be positive, so the domain is

$9 - x^2 - y^2 > 0$, $x^2 + y^2 < 9$, which is the interior of a circle centered

at the origin, with a radius of 3. Since the radical is always non-negative, the

range is $f(x, y) > 0$

21. $g(x, y) = x^2 - y$

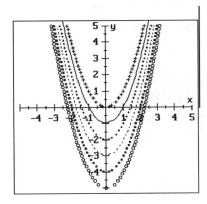

SURVIVAL HINT

The C trace of $y = f(x)$ is a point on a line. The C level curve for $z = f(x, y)$ is a curve or line in a plane, which is a slice of $\mathbb{R}^3$. The C level surface for $w = f(x, y, z)$ is a surface in a solid, which is a slice of $\mathbb{R}^4$. Since we are trapped in a three-dimensional world we can not visualize $\mathbb{R}^4$, but the mathematics is not restricted to three variables.

25. For all values of y we have $x^2 + z^2 = 1$. All trace planes parallel to the xz-plane will be circles with the y-axis as center and a radius of 1.

The graph is a cylinder.

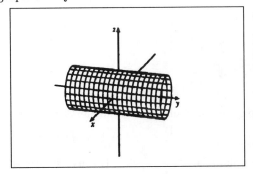

29. This function can be written as $2x^2 + 2y^2 - z = 1$, which is a paraboloid. If $x = 0$ we have a parabola in the yz-plane, if $y = 0$ we have a parabola in the xz-plane, and if $z = 0$ we have a circle in the xy-plane. The z-axis is the axis of the paraboloid and its vertex is at $(0, 0, -1)$.

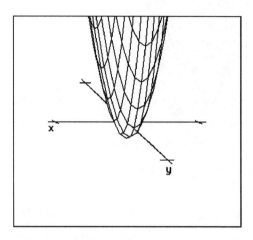

33. If $x = 0$ the trace in the yz-plane is $y^2 + z^2 = -1$, an empty set. If $x \geq 3$ then the trace in a plane parallel to the yz-plane is a circle. If $y = 0$ the trace in the xz-plane is $\frac{x^2}{9} - z^2 = 1$, a hyperbola. If $z = 0$ the trace in the xy-plane is $\frac{x^2}{9} - y^2 = 1$, a hyperbola. The surface is a circular hyperboloid of two sheets with vertices at $(\pm 3, 0, 0)$.

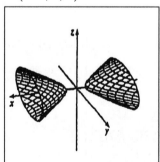

37. If $x = 0$ the trace in the yz-plane is $z^2 - \frac{y^2}{4} = 1$, a hyperbola. If $y = 0$ the trace in the xz-plane is $z^2 - \frac{x^2}{9} = 1$, a hyperbola. If $z = 0$ the trace in the xy-plane is $\frac{x^2}{9} + \frac{y^2}{4} = -1$, an empty set. If $z \geq 1$ the trace is an ellipse. The surface is an elliptical hyperboloid of two sheets with vertices at $(0, 0, \pm 1)$.

37. (con't.)

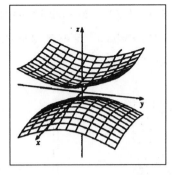

41.   The c-level curves are circles centered on the z-axis, which could be B, E, or F.

The traces in the xz and the yz-planes are always positive. It is E.

45. $z = x$ is a line in the xz-plane, and these values hold for all values of y.

This is a plane passing through the y-axis.

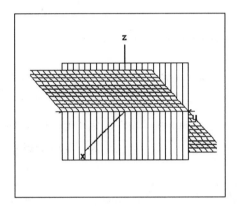

49.   If $x = 0$ the trace in the yz-plane is $y = -z$, a straight line. If $y = 0$ the

trace in the xz-plane is $z = x^2$, a parabola. If $z = 0$ the trace in the xy-plane

is $y = x^2$, a parabola.

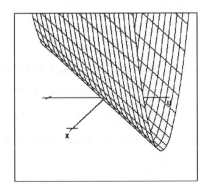

53.   If $x = 0$ the trace in the yz-plane is $z = \sqrt{y}$, a half-parabola. If $y = 0$ the

trace in the xz-plane is $z = \sqrt{x}$, a half-parabola. If $z = 0$ the trace in the

xy-plane is $y = -x$, a straight line.

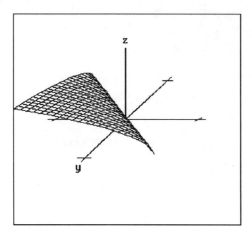

57.   **a.** $f(x, y) = Ax^a y^b$. $f(2x, 2y) = A(2x)^a(2y)^b = 2^{a+b}Ax^a y^b$, and since

$a + b > 1$ the production more than doubles.

b.   In this case $a + b < 1$ so the production less than doubles.

c.   In this case $a + b = 1$ so the production exactly doubles.

61.   If $PV = kT$, then $P = \dfrac{kT}{V}$. If $T = 0$, the trace in the PV plane is $P = 0$.

But if $T > 0$, the traces in planes parallel to the PV plane are hyperbolas.

If $V > 0$, the traces in planes parallel to the PT plane are straight lines.

If $P > 0$, the traces in planes parallel to the VT plane are straight lines.

12.2 Limits and Continuity

3. $\displaystyle\lim_{(x,\, y)\to(-1,\, 0)} (xy^2 + x^3y + 5) = (-1)0^2 + (-1)^3 0 + 5 = 5$

7. $\displaystyle\lim_{(x,\, y)\to(1,\, 0)} e^{xy} = e^0 = 1$

SURVIVAL HINT

It might be a good idea to review the concepts and definitions of limit and continuity as presented in Sections 1.7 and 1.8; since the concepts and definitions for functions of several variables are a direct extension from $\mathbb{R}^2$ to $\mathbb{R}^3$. If you remember and understand the ϵ-δ definition of limit and the three part definition of continuity, then the multi-variable definitions are the obvious extensions.

11. $\displaystyle\lim_{(x,\ y)\to(0,\ 0)} \frac{x^2 - 2xy + y^2}{x - y} = \lim_{(x,\ y)\to(0,\ 0)} (x - y) = 0$

Notice that the function is not defined at $(0, 0)$. This is precisely the situation in which limits become valuable.

15. $\displaystyle\lim_{(x,\ y)\to(0,\ 0)} \frac{\sin(x + y)}{x + y}$ is of the form $\displaystyle\lim_{x\to 0}\frac{\sin x}{x}$ and therefore $= 1$

19. $\displaystyle\lim_{(x,\ y)\to(2,\ 1)} \frac{x^2 - 4y^2}{x - 2y} = \lim_{(x,\ y)\to(2,\ 1)} (x + 2y) = 2 + 2(1) = 4$

23. $\displaystyle\lim_{(x,\ y)\to(0,\ 0)} \frac{x + y}{x - y}$ along the line $x = 0$ is $\displaystyle\lim_{y\to 0}\frac{y}{(-y)} = \lim_{y\to 0}(-1) = -1$

$\displaystyle\lim_{(x,\ y)\to(0,\ 0)} \frac{x + y}{x - y}$ along the line $y = 0$ is $\displaystyle\lim_{x\to 0}\frac{x}{x} = \lim_{x\to 0}(1) = 1$

Since we have different limits as we approach from different directions, the limit does not exist.

27. $\displaystyle\lim_{(x,\ y)\to(0,\ 0)} (\sin x - \cos y) = \sin 0 - \cos 0 = -1$

31. $\displaystyle\lim_{(x,\ y)\to(0,\ 0)} \frac{x - y^2}{x^2 + y^2}$ If we substitute the limiting values into the function

we get $\frac{0}{0}$ which could be any value. But if we approach the origin along the line $x = 0$ we have $\displaystyle\lim_{y\to 0}\frac{(-y^2)}{y^2} = -1$. The limit does not exist.

35. **a.** As we approach the origin along any straight line path, $y = kx$,

$$\lim_{(x,\ y)\to(0,\ 0)} \frac{xy^3}{x^2 + y^6} = \lim_{(x,\ y)\to(0,\ 0)} \frac{k^3 x^4}{x^2 + k^6 x^6}$$

$$= \lim_{(x,\ y)\to(0,\ 0)} \frac{k^3 x^2}{1 + k^6 x^4} = 0$$

35. (con't.)

b.   As we approach the origin along the path $x = y^3$

$$\lim_{(x,\,y)\to(0,\,0)} \frac{xy^3}{x^2 + y^6} = \lim_{(x,\,y)\to(0,\,0)} \frac{y^6}{y^6 + y^6}$$

$$= \lim_{(x,\,y)\to(0,\,0)} \frac{1}{2} = \frac{1}{2}$$

Since different paths of approach have different limits, the limit does not exist, so the function can not be continuous.

39. $$\lim_{(x,\,y)\to(0,\,0)} \frac{3x^3 - 3y^3}{x^2 - y^2} = \lim_{(x,\,y)\to(0,\,0)} \frac{3(x - y)(x^2 + xy + y^2)}{(x + y)(x - y)}$$

$$= \lim_{(x,\,y)\to(0,\,0)} \frac{3(x^2 + xy + y^2)}{x + y}$$

along the line $x = y$: $$= \lim_{(x,\,y)\to(0,\,0)} \frac{9x^2}{2x}$$

$$= \lim_{(x,\,y)\to(0,\,0)} \frac{9}{2}x = 0$$

along the line $y = 0$: $$= \lim_{(x,\,y)\to(0,\,0)} \frac{3x^2}{x}$$

$$= \lim_{(x,\,y)\to(0,\,0)} 3x = 0$$

along the line $x = 0$: $$= \lim_{(x,\,y)\to(0,\,0)} \frac{3y^2}{y}$$

$$= \lim_{(x,\,y)\to(0,\,0)} 3y = 0$$

If the function is to be continuous $f(0, 0)$ must equal $\lim_{(x,\,y)\to(0,\,0)} f(x, y) = 0$. (However, we have not established that the limit really is 0 since we have not approached the origin along *every* path.) Nevertheless, if the limit exists there, $B = 0$.

43.   Analysis: We need to be able to show that for any $\epsilon > 0$ there exists a $\delta > 0$ such that $|f(x, y) - L| < \epsilon$ whenever $\sqrt{(x - x_0)^2 + (y - y_0)^2} < \delta$. In this case we need to transform the absolute value inequality $|x + y^2| < \epsilon$ into $|x^2 + y^2| < g(\epsilon)$, where $g(\epsilon)$ is the resulting transformation of ϵ, and then let δ be $\leq g(\epsilon)$. Remember that for values less than 1, $x^2 < x$.

45.   Analysis: Transform $\left| \dfrac{x^2 - y^2}{x + y} - 2 \right| < \epsilon$ into $\sqrt{(x - 1)^2 + (y + 1)^2} < g(\epsilon)$ where $g(\epsilon)$ is the resulting transformation of ϵ, and then let δ be $\leq g(\epsilon)$.

Begin by factoring and reducing. Make use of the identity $|a + b| \leq |a| + |b|$.

47. Analysis: If $f(a, b)$ is continuous, and $f(a, b) = L > 0$, then
$|f(a, b) - L| \leq |f(a, b) + L| \leq |L + L| = 2L$. Complete the proof to find
an appropriate δ neighborhood of (a, b).

49. Analysis: For any given ϵ, let $|f(x, y) - L| \leq \frac{\epsilon}{2}$ and $|g(x, y) - M| \leq \frac{\epsilon}{2}$.
Now add and find the appropriate δ.

12.3 Partial Derivatives

SURVIVAL HINT

When you hold one variable fixed and find the derivative with respect to the other, you
are reducing the surface in $\mathbb{R}^3$ to the line in $\mathbb{R}^2$. The first and second derivatives give the
slope and concavity as you move along the curve determined by the trace through a given
point. The mixed partial indicates how the slope would change if you moved to another
trace. The analogy of moving in various directions from a point on a hillside is useful.

3. $f(x, y) = x^3 + x^2 y + xy^2 + y^3$

For f_x treat y as a constant. For f_y treat x as a constant.

$f_x = 3x^2 + 2xy + y^2, \ f_y = x^2 + 2xy + 3y^2,$

$f_{xx} = 6x + 2y, \ f_{yx} = 2x + 2y$

7. $f(x, y) = \ln(2x + 3y)$

$f_x = \dfrac{2}{2x + 3y}, \ f_y = \dfrac{3}{2x + 3y}, \ f_{xx} = \dfrac{-4}{(2x + 3y)^2}, \ f_{yx} = \dfrac{-6}{(2x + 3y)^2}$

11. $f(x, y) = (3x^2 + y^4)^{1/2}$

$f_x = \dfrac{3x}{(3x^2 + y^4)^{1/2}}, \ f_y = \dfrac{2y^3}{(3x^2 + y^4)^{1/2}}$

15. $f(x, y) = \sin^{-1} xy \ \left(\text{Recall } f'(\sin^{-1} u) = \dfrac{1}{\sqrt{1 - u^2}} du\right)$

$f_x = \dfrac{y}{\sqrt{1 - x^2 y^2}}, \ f_y = \dfrac{x}{\sqrt{1 - x^2 y^2}}$

19. $f(x, y, z) = \dfrac{x + y^2}{z}$

$f_x = \dfrac{1}{z}, \ f_y = \dfrac{2y}{z}, \ f_z = -\dfrac{x + y^2}{z^2}$

23. Remember that z is assumed to be a function of x and y.

Differentiating with respect to x: $\dfrac{2x}{9} + \dfrac{2z}{2}\dfrac{\partial z}{\partial x} = 0$, $z\dfrac{\partial z}{\partial x} = -\dfrac{2x}{9}$, $\dfrac{\partial z}{\partial x} = -\dfrac{2x}{9z}$

Differentiating with respect to y: $-\dfrac{2y}{4} + \dfrac{2z}{2}\dfrac{\partial z}{\partial y} = 0$, $\dfrac{\partial z}{\partial y} = \dfrac{y}{2z}$

27. Differentiating with respect to x: $\dfrac{1}{2\sqrt{x}} + (\cos xz)\left(z + x\dfrac{\partial z}{\partial x}\right) = 0$,

$(x\cos xz)\dfrac{\partial z}{\partial x} = -\left(\dfrac{1}{2\sqrt{x}} + z\cos xz\right)$, $\dfrac{\partial z}{\partial x} = -\dfrac{\dfrac{1}{2\sqrt{x}} + z\cos xz}{x\cos xz}$

$\dfrac{\partial z}{\partial x} = -\dfrac{1}{2x^{3/2}\cos xz} - \dfrac{z}{x}$

Differentiating with respect to y: $2y + (\cos xz)\left(z + x\dfrac{\partial z}{\partial y}\right) = 0$,

$(x\cos xz)\dfrac{\partial z}{\partial y} = -2y$, $\dfrac{\partial z}{\partial y} = -\dfrac{2y}{x\cos xz}$

31. Find $\dfrac{dy}{dx}$ for $x^2 + xy^2 + y^3 = 3$ by implicit differentiation, then evaluate it at the point $(1, 1, 3)$.

$2x + x(2y)\dfrac{dy}{dx} + y^2 + 3y^2\dfrac{dy}{dx} = 0$, $(2xy + 3y^2)\dfrac{dy}{dx} = -(2x + y^2)$,

$\dfrac{dy}{dx} = -\dfrac{2x + y^2}{2xy + 3y^2}$. $\dfrac{dy}{dx}(1, 1, 3) = -\dfrac{2 + 1}{2 + 3} = -\dfrac{3}{5}$

35. $f(x, y) = e^x\sin y$ is the product of two functions that are continuously differentiable, so it will be continuously differentiable. It is given that the second derivatives are continuous, so the only remaining requirement for the function to be harmonic is to show $f_{xx} + f_{yy} = 0$.

$f_x = e^x\sin y$ ($\sin y$ is a constant), $f_{xx} = e^x\sin y$

$f_y = e^x\cos y$ (e^x is a constant), $f_{yy} = -e^x\sin y$. $f_{xx} + f_{yy} = 0$

39. $f(x, y) = \cos xy^2$. $f_x = -y^2\sin xy^2$,

$f_{xy} = -[y^2(2xy)\cos xy^2 + 2y\sin xy^2] = -2xy^3\cos xy^2 - 2y\sin xy^2$

$f_y = -(2xy)\sin xy^2$,

$f_{yx} = -[(2y)(\sin xy^2) + (2xy)(\cos xy^2)(y^2)] = -2xy^3\cos xy^2 - 2y\sin xy^2$

They are the same.

43. **a.** $C_m = \sigma(T - t)(-0.67m^{-1.67}) = -0.67\sigma(T - t)m^{-1.67}$

b. $C_T = \sigma m^{-0.67}$ (Write $C = \sigma Tm^{-0.67} - \sigma tm^{-0.67}$, where the second term is a constant when differentiating with respect to T.)

c. $C_t = -\sigma m^{-0.67}$

47. **a.** We are asked to find $\frac{\partial T}{\partial y}(2, 1)$. Moving parallel to the y-axis means y

is changing, and x is remaining constant. $\frac{\partial T}{\partial y} = 4xy + 1$, $\frac{\partial T}{\partial y}(2, 1) = 9$

  b. In this case x is changing and y is constant.

$\frac{\partial T}{\partial x} = 3x^2 + 2y^2$. $\frac{\partial T}{\partial x}(2, 1) = 3(2)^2 + 2(1)^2 = 14$

51. Analysis:

  a. Find $\frac{\partial C}{\partial t}$ and $\frac{\partial^2 C}{\partial x^2}$ and substitute these values into the diffusion equation

to determine the relationship between a, b, and δ.

  b. Find $\frac{\partial C}{\partial t}$ and $\frac{\partial^2 C}{\partial x^2}$ and substitute these values into the diffusion equation

to verify that the coefficient of $\frac{\partial^2 C}{\partial x^2}$ is δ.

12.4 Tangent Planes, Approximations, and Differentiability

SURVIVAL HINT

The equation of a tangent plane can be remembered as an extension of the point-slope

formula for a line in $\mathbb{R}^2$: $y - y_0 = m(x - x_0)$.   For the plane in $\mathbb{R}^3$:

$z - z_0 = m(x - x_0) + n(y - y_0)$ where m and n are the slopes in the x and y

directions, found with the partial derivatives.

3. The equation of the tangent plane is given by:

$z - z_0 = f_x(x_0, y_0)(x - x_0) + f_y(x_0, y_0)(y - y_0)$. In this case

$f_x = 2x + y \cos xy$, $f_x(x_0, y_0) = 2(0) + 2 \cos 0 = 2$

$f_y = 2y + x \cos xy$, $f_y(x_0, y_0) = 2(2) + 0 \cos 0 = 4$

The tangent plane is: $z - 4 = 2(x - 0) + 4(y - 2)$ or

$2x + 4y - z - 4 = 0$

SURVIVAL HINT

The total differential can be remembered as an $\mathbb{R}^3$ extension of the concepts of the

differential in $\mathbb{R}^2$; $dy = f'(x) dx$. This says the change along the tangent line is the rate

of change, $f'(x)$, multiplied by the change in the x direction, Δx, or in the limiting case,

dx. In $\mathbb{R}^3$ the change in z along the tangent plane, dz or df, is the rate of change in the x

direction, $\frac{\partial z}{\partial x}$, multiplied by the change in the x direction, dx, plus the same computation

for change of z in the y direction: $df = \frac{\partial z}{\partial x} dx + \frac{\partial z}{\partial y} dy$.

7. The total differential is given by: $df = f_x\, dx + f_y\, dy$.

$df = 10xy^3\, dx + 15x^2y^2\, dy$

11. $df = -\dfrac{y}{x^2}\, dx + \dfrac{1}{x}\, dy$

15. $df = 9x^2\, dx - 8y^3\, dy + 5\, dz$

19. By Theorem 12.3 $f(x,\, y)$ is differentiable if both of the partials are continuous.

$f_x = y^3 + 3y^2,\ f_y = 3xy^2 + 6xy$ both of which are everywhere continuous.

23. Find $f(1,\, 2) + df$ where $dx = 0.01$ and $dy = 0.03$. $f(1,\, 2) = 35$.

$df = f_x\, dx + f_y\, dy = 12x^3(0.01) + 8y^3(0.03) = 0.12 + 1.92 = 2.04$

$f(1.01,\, 2.03) \approx 35 + 2.04 = 37.04$ Actual $f(1.01,\, 2.03) = 37.085\,445\,65$

27. Find $f(1,\, 1) + df$ where $dx = 0.01$ and $dy = -0.02$. $f(1,\, 1) = e$

$df = f_x\, dx + f_y\, dy = ye^{xy}(0.01) + xe^{xy}(-0.02) = 0.01e - 0.02e = -0.01e$

$f(1.01,\, 0.98) \approx e - 0.01e = 0.99e \approx 2.6911$ Actual $f(1.01,\, 0.98) \approx 2.690\,70$

31. a. If f is differentiable there exists ϵ_1 and ϵ_2 that tend to 0 as Δx and Δy tend to 0, where $\Delta f = f_x\, dx + f_y\, dy + \epsilon_1 \Delta x + \epsilon_2 \Delta y$. If x and y are "sufficiently close" to 0, then $f(x,\, y) \approx f(0,\, 0) + \Delta f$.

b. For $f(x,\, y) = \dfrac{1}{1 + x - y}$, $f(0,\, 0) = 1$, $f_x = -\dfrac{1}{(1 + x - y)^2}$,

$f_x(0,\, 0) = -1$, $f_y = \dfrac{1}{(1 + x - y)^2}$, $f_y(0,\, 0) = 1$,

$f(x,\, y) \approx f(0,\, 0) + x f_x(0,\, 0) + y f_y(0,\, 0)$

$\approx 1 + x(-1) + y(1) = 1 - x + y$

c. $f(0,\, 0) = \dfrac{1}{2}$, $f_x = -\dfrac{2(x + 1)}{[(x + 1)^2 + (y + 1)^2]^2}$, $f_x(0,\, 0) = -\dfrac{1}{2}$.

$f_y = -\dfrac{2(y + 1)}{[(x + 1)^2 + (y + 1)^2]^2}$, $f_y(0,\, 0) = -\dfrac{1}{2}$.

$f(x,\, y) \approx f(0,\, 0) + x f_x(0,\, 0) + y f_y(0,\, 0)$

$= \dfrac{1}{2} - \dfrac{1}{2}x - \dfrac{1}{2}y = \dfrac{1}{2}(1 - x - y)$

35. Revenue is the number of units sold times the unit price.

a. $R(x,\, y) = x(4{,}000 - 500x) + y(3{,}000 - 450y)$

$= -500x^2 + 4{,}000x - 450y^2 + 3{,}000y$

35. (con't.)

    b.   Since the changes are "small" with respect to the current prices, the total differential will give a decent approximation of the change.

$$R = x\,p(x) + y\,q(y).\quad dR = x\,dp + p(x)\,dx + y\,dq + q(y)\,dy$$

To find x and y, substitute the given prices into $p(x)$ and $q(y)$:

$$500 = 4{,}000 - 500x,\quad 500x = 3{,}500,\quad x = 7$$

$$750 = 3{,}000 - 450y,\quad 450y = 2{,}250.\quad y = 5$$

Also: $dp = -500\,dx,\quad dx = \dfrac{20}{-500} = -0.04$

$$dq = -450\,dy,\quad dy = \dfrac{18}{-450} = -0.04$$

$$dR = 7(20) + (500)(-0.04) + 5(18) + (750)(-0.04)$$

$$= 140 - 20 + 90 - 30 = \$180$$

39. $F(x, y) = \dfrac{1.786\,xy}{1.798x + y}.\quad f_x = \dfrac{(1.798x + y)(1.786y) - (1.786\,xy)(1.798)}{(1.798x + y)^2}$

$$f_y = \dfrac{(1.798x + y)(1.786x) - 1.786\,xy(1)}{(1.798x + y)^2}.\quad x = 5,\ y = 4,\ dx = 0.1,\ dy = 0.04$$

$$dF = f_x\,dx + f_y\,dy = \dfrac{(28.586)(0.1) + (80.2807)(0.04)}{168.7401} \approx 0.03597$$

43. $T = 2\pi\sqrt{\dfrac{L}{g}}.\quad L = 4.03,\ g = 32.2,\ dL = -0.03,\ dg = -0.2.$

$$f_L = \dfrac{2\pi}{2\sqrt{\tfrac{L}{g}}}\dfrac{1}{g} = \dfrac{\pi}{\sqrt{Lg}}.\quad f_g = \dfrac{2\pi}{2\sqrt{\tfrac{L}{g}}}\left(-\dfrac{L}{g^2}\right) = -\dfrac{\pi\sqrt{L}}{g^{3/2}}.$$

$$dT = f_L\,dL + f_g\,dg = \dfrac{\pi}{\sqrt{129.766}}(-0.03) + \left(-\dfrac{\pi\sqrt{4.03}}{32.2\sqrt{32.2}}\right)(-0.2)$$

$$\approx -0.0013704$$

45.   Analysis: For differentiability $\displaystyle\lim_{(x,\,y)\to(0,\,0)} f(x, y) = f(0, 0)$ along any path. Try approaching $(0, 0)$ along $y = 0$.

47.   Analysis: The altitude, h, perpendicular to b, equals $a\sin\theta$. The area of the triangle, $A = \frac{1}{2}\,ab\sin\theta.\ dA = f_a\,da + f_b\,db$ (θ is a constant). The percentage change in A is $\dfrac{dA}{A}$.

12.5 Chain Rules

SURVIVAL HINT

Check to see if the given parameters restrict the domain of $f(x, y,)$. In Problem #5, x and y must both be positive, so $f(x, y)$ has only first octant values.

5. $f(x, y) = (4 + y^2)x$, $x = e^{2t}$, $y = e^{3t}$

    a. Substituting: $z = (4 + e^{6t})e^{2t} = 4e^{2t} + e^{8t}$

        $\dfrac{dz}{dt} = 8e^{2t} + 8e^{8t} = 8e^{2t}(1 + e^{6t})$

    b. $\dfrac{dz}{dt} = \dfrac{\partial z}{\partial x}\dfrac{dx}{dt} + \dfrac{\partial z}{\partial y}\dfrac{dy}{dt} = (4 + y^2)2e^{2t} + (2xy)3e^{3t}$

        In terms of t: $(4 + e^{6t})2e^{2t} + (2e^{2t}e^{3t})3e^{3t} = 8e^{2t}(1 + e^{6t})$

9. $F(x, y) = x^2 + y^2$, $x = u\sin v$, $y = u - 2v$

    a. Substituting: $z = u^2\sin^2 v + u^2 - 4uv + 4v^2$

        $\dfrac{\partial z}{\partial u} = 2u\sin^2 v + 2u - 4v$, $\dfrac{\partial z}{\partial v} = 2u^2\sin v \cos v - 4u + 8v$

    b. $\dfrac{\partial z}{\partial u} = \dfrac{\partial z}{\partial x}\dfrac{\partial x}{\partial u} + \dfrac{\partial z}{\partial y}\dfrac{\partial y}{\partial u} = 2x\sin v + 2y(1) = 2u\sin^2 v + 2(u - 2v)$

        $\dfrac{\partial z}{\partial v} = \dfrac{\partial z}{\partial x}\dfrac{\partial x}{\partial v} + \dfrac{\partial z}{\partial y}\dfrac{\partial y}{\partial v} = 2x\,u\cos v + 2y(-2)$

            $= 2u^2\sin v\cos v - 4(u - 2v)$

13. $\dfrac{\partial w}{\partial s} = \dfrac{\partial w}{\partial x}\dfrac{\partial x}{\partial s} + \dfrac{\partial w}{\partial y}\dfrac{\partial y}{\partial s} + \dfrac{\partial w}{\partial z}\dfrac{\partial z}{\partial s}$

    $\dfrac{\partial w}{\partial t} = \dfrac{\partial w}{\partial x}\dfrac{\partial x}{\partial t} + \dfrac{\partial w}{\partial y}\dfrac{\partial y}{\partial t} + \dfrac{\partial w}{\partial z}\dfrac{\partial z}{\partial t}$

17. $\dfrac{\partial w}{\partial t} = \dfrac{\partial w}{\partial x}\dfrac{\partial x}{\partial t} + \dfrac{\partial w}{\partial y}\dfrac{\partial y}{\partial t} + \dfrac{\partial w}{\partial z}\dfrac{\partial z}{\partial t}$

    $= [yz\cos(xyz)](-3) + [xz\cos(xyz)](-e^{1-t}) + [xy\cos(xyz)](4)$

    $= \cos(xyz)[-3yz - e^{1-t}xz + 4xy]$

21. $\dfrac{\partial w}{\partial r} = \dfrac{\partial w}{\partial x}\dfrac{\partial x}{\partial r} + \dfrac{\partial w}{\partial y}\dfrac{\partial y}{\partial r} + \dfrac{\partial w}{\partial z}\dfrac{\partial z}{\partial r}$

    $= \dfrac{1}{2 - z}(2s) + \dfrac{1}{2 - z}(t\cos rt) + \left[\dfrac{x + y}{(2 - z)^2}\right](0)$

    $= \dfrac{2s + t\cos rt}{2 - z}$

    $\dfrac{\partial w}{\partial t} = \dfrac{\partial w}{\partial x}\dfrac{\partial x}{\partial t} + \dfrac{\partial w}{\partial y}\dfrac{\partial y}{\partial t} + \dfrac{\partial w}{\partial z}\dfrac{\partial z}{\partial t}$

    $= \dfrac{1}{2 - z}(0) + \dfrac{1}{2 - z}(r\cos rt) + \left[\dfrac{x + y}{(2 - z)^2}\right](2st)$

    $= \dfrac{(2 - z)(r\cos rt) + 2st(x + y)}{(2 - z)^2}$

25. **a.** $\quad - x^{-2} - z^{-2} \dfrac{\partial z}{\partial x} = 0$

$$\frac{\partial z}{\partial x} = -\frac{z^2}{x^2}, \quad \frac{\partial z}{\partial y} = -\frac{z^2}{y^2}$$

$$\frac{\partial^2 z}{\partial x \partial y} = -\frac{1}{x^2}\left(2z\frac{\partial z}{\partial y}\right) = \frac{2z^3}{x^2 y^2}$$

b. Differentiating $\dfrac{\partial z}{\partial x}$ again with respect to x: (quotient rule)

$$\frac{\partial^2 z}{\partial x^2} = -\frac{x^2\left(2z\frac{\partial z}{\partial x}\right) - z^2(2x)}{x^4} = \frac{2z^2(x+z)}{x^4}$$

c. Differentiating $\dfrac{\partial z}{\partial y}$ again with respect to y:

$$\frac{\partial^2 z}{\partial y^2} = -\frac{y^2\left(2z\frac{\partial z}{\partial y}\right) - z^2(2y)}{y^4} = \frac{2z^2(y+z)}{y^4}$$

29. $\quad \dfrac{\partial z}{\partial u} = \dfrac{\partial z}{\partial x}\cdot\dfrac{\partial x}{\partial u} + \dfrac{\partial z}{\partial y}\cdot\dfrac{\partial y}{\partial u} = \dfrac{\partial z}{\partial x}\cdot a + \dfrac{\partial z}{\partial y}\cdot 0 = a\dfrac{\partial z}{\partial x}.$

$$\frac{\partial z}{\partial v} = \frac{\partial z}{\partial x}\cdot\frac{\partial x}{\partial v} + \frac{\partial z}{\partial y}\cdot\frac{\partial y}{\partial v} = \frac{\partial z}{\partial x}\cdot 0 + \frac{\partial z}{\partial y}\cdot b = b\frac{\partial z}{\partial y}.$$

$$\frac{\partial^2 z}{\partial u^2} = \frac{\partial}{\partial u}\left(\frac{\partial z}{\partial u}\right) = \frac{\partial}{\partial u}\left(\frac{\partial z}{\partial x}\cdot\frac{\partial x}{\partial u} + \frac{\partial z}{\partial y}\cdot\frac{\partial y}{\partial u}\right). \text{ Using the product rule:}$$

$$= \frac{\partial}{\partial u}\left(\frac{\partial z}{\partial x}\right)\cdot\frac{\partial x}{\partial u} + \frac{\partial z}{\partial x}\cdot\frac{\partial}{\partial u}\left(\frac{\partial x}{\partial u}\right) + \frac{\partial}{\partial u}\left(\frac{\partial z}{\partial y}\right)\cdot\frac{\partial y}{\partial u} + \frac{\partial z}{\partial y}\cdot\frac{\partial}{\partial u}\left(\frac{\partial y}{\partial u}\right)$$

Now the chain rule must again be used on $\dfrac{\partial}{\partial u}\left(\dfrac{\partial z}{\partial x}\right)$ and $\dfrac{\partial}{\partial u}\left(\dfrac{\partial z}{\partial y}\right)$,

$$= \left\{\frac{\partial}{\partial x}\left(\frac{\partial z}{\partial x}\right)\cdot\frac{\partial x}{\partial u} + \frac{\partial}{\partial y}\left(\frac{\partial z}{\partial x}\right)\cdot\frac{\partial y}{\partial u}\right\}\cdot\frac{\partial x}{\partial u} + \frac{\partial z}{\partial x}\cdot\frac{\partial}{\partial u}\left(\frac{\partial x}{\partial u}\right) +$$

$$\left\{\frac{\partial}{\partial x}\left(\frac{\partial z}{\partial y}\right)\cdot\frac{\partial x}{\partial u} + \frac{\partial}{\partial y}\left(\frac{\partial z}{\partial y}\right)\cdot\frac{\partial y}{\partial u}\right\}\cdot\frac{\partial y}{\partial u} + \frac{\partial z}{\partial y}\cdot\frac{\partial}{\partial u}\left(\frac{\partial y}{\partial u}\right)$$

For the given functions: $\quad \dfrac{\partial^2 z}{\partial u^2} = \dfrac{\partial^2 z}{\partial x^2}a^2 \quad$ and in like manner $\dfrac{\partial^2 z}{\partial v^2} = \dfrac{\partial^2 z}{\partial y^2}b^2$.

33. If we let $w = uv^2$ then $\dfrac{\partial z}{\partial u} = \dfrac{\partial z}{\partial w}\cdot\dfrac{\partial w}{\partial u} = \dfrac{\partial z}{\partial w}v^2 \quad$ and $\dfrac{\partial z}{\partial v} = \dfrac{\partial z}{\partial w}\cdot\dfrac{\partial w}{\partial v} = \dfrac{\partial z}{\partial w}2uv$

Since $z = f(w)$, equating the $\dfrac{\partial z}{\partial w}$ we have: $\quad \dfrac{\dfrac{\partial z}{\partial u}}{v^2} = \dfrac{\dfrac{\partial z}{\partial v}}{2uv}$ and $2uv\dfrac{\partial z}{\partial u} = v^2\dfrac{\partial z}{\partial v}$

or $\quad 2v\dfrac{\partial z}{\partial u} - v\dfrac{\partial z}{\partial v} = 0.$

37. Let $z = \frac{r - s}{s}$, so now $w = f(z)$. $\frac{\partial w}{\partial r} = \frac{\partial w}{\partial z} \cdot \frac{\partial z}{\partial r} = \frac{\partial w}{\partial z} \frac{1}{s}$
and $\frac{\partial w}{\partial s} = \frac{\partial w}{\partial z} \cdot \frac{\partial z}{\partial s} = \frac{\partial w}{\partial z}\left(-\frac{r}{s^2}\right)$.

Since $w = f(z)$, equating the $\frac{\partial w}{\partial z}$ we have:

$s\frac{\partial w}{\partial r} = \left(-\frac{s^2}{r}\right)\frac{\partial w}{\partial s}$ or $r\frac{\partial w}{\partial r} + s\frac{\partial w}{\partial s} = 0$.

41. Let $u = x^2 + y^2$, then $z = f(u)$

  a. $\frac{\partial z}{\partial x} = f'(u)\frac{\partial u}{\partial x} = 2x\,f'(u)$

$\frac{\partial^2 z}{\partial x^2} = 2\,f'(u) + 2x\,f''(u)\frac{\partial u}{\partial x} = 2\,f'(u) + 4x^2\,f''(u)$

$= 2\,f'(x^2 + y^2) + 4x^2\,f''(x^2 + y^2)$

  b. $\frac{\partial z}{\partial y} = f'(u)\frac{\partial u}{\partial y} = 2y\,f'(u)$

$\frac{\partial^2 z}{\partial y^2} = 2\,f'(u) + 2y\,f''(u)\frac{\partial u}{\partial y} = 2\,f'(u) + 4y^2\,f''(u)$

$= 2\,f'(x^2 + y^2) + 4y^2\,f''(x^2 + y^2)$

  c. $\frac{\partial z}{\partial x} = 2x\,f'(u)$, $\frac{\partial^2 z}{\partial x \partial y} = 0[f'(u)] + 2x\left(f''(u)\frac{\partial u}{\partial y}\right)$

$= 4xy\,f''(x^2 + y^2)$

45. Analysis: If $F(x, y) = C$ then $dF = F_x\,dx + F_y\,dy = 0$.

Find F_x and F_y and substitute.

47. Analysis: Take the partial with respect to y for each equation:

$x\frac{\partial u}{\partial y} + y\frac{\partial v}{\partial y} + v - u\frac{\partial v}{\partial y} - v\frac{\partial u}{\partial y} = 0$, $(x - v)\frac{\partial u}{\partial y} + (y - u)\frac{\partial v}{\partial y} = -v$

$y\frac{\partial u}{\partial y} + u - x\frac{\partial v}{\partial y} + u\frac{\partial v}{\partial y} + v\frac{\partial u}{\partial y} = 0$, $(y + v)\frac{\partial u}{\partial y} + (-x + u)\frac{\partial v}{\partial y} = -u$

Now solve these two equations simultaneously. Cramer's rule would be a good method.

49. Analysis: If $F(x, y, z) = 0$ and $G(x, y, z) = 0$ then

$dF = \frac{\partial F}{\partial x}\,dx + \frac{\partial F}{\partial y}\,dy + \frac{\partial F}{\partial z}\,dz = 0$ and $\frac{\partial G}{\partial x}\,dx + \frac{\partial G}{\partial y}\,dy + \frac{\partial G}{\partial z}\,dz = 0$

We want only the dx and dy terms, so eliminate the dz term by multiplication and

subtraction: $F_x G_z dx + F_y G_z dy + F_z G_z dz = 0$

$\underline{F_z G_x dx + F_z G_y dy + F_z G_z dz = 0}$

$(F_x G_z - F_z G_x)dx + (F_y G_z - F_z G_y)dy = 0$

Now find $\frac{dy}{dx}$. In like manner eliminate the dy term and then solve for $\frac{dz}{dx}$.

12.6 Directional Derivatives and the Gradient

1. $\nabla f = f_x \mathbf{i} + f_y \mathbf{j} = (2x - 2y)\mathbf{i} - 2x\mathbf{j}$

5. $\nabla f = f_u \mathbf{i} + f_v \mathbf{j} = e^{3-v}\mathbf{i} - ue^{3-v}\mathbf{j} = e^{3-v}(\mathbf{i} - u\mathbf{j})$

9. $\nabla f = f_x \mathbf{i} + f_y \mathbf{j} + f_z \mathbf{k} = e^{y+3z}\mathbf{i} + x\,e^{y+3z}\mathbf{j} + 3x\,e^{y+3z}\mathbf{k}$

$\quad\quad = e^{y+3z}(\mathbf{i} + x\mathbf{j} + 3x\mathbf{k})$

SURVIVAL HINT

Using the hillside analogy for the directional derivative: If you are standing on the hillside $z = f(x, y)$ at the point P, and walk in the compass direction indicated by the unit vector $\mathbf{u}$, the directional derivative will tell you the instantaneous rate of change in z.

13. $\mathbf{V} = \frac{\sqrt{2}}{2}\mathbf{i} + \frac{\sqrt{2}}{2}\mathbf{j}$, $\nabla f = \frac{2x}{x^2 + 3y}\mathbf{i} + \frac{3}{x^2 + 3y}\mathbf{j}$, $\nabla f(1, 1) = \frac{1}{2}\mathbf{i} + \frac{3}{4}\mathbf{j}$

$\quad\quad D_{\mathbf{u}}f = \nabla f \bullet \mathbf{V} = \frac{\sqrt{2}}{4} + \frac{3\sqrt{2}}{8} = \frac{5\sqrt{2}}{8}$

17. $\mathbf{N} = \nabla f = 2x\,\mathbf{i} + 2y\,\mathbf{j} + 2z\,\mathbf{k}$

$\quad\quad \mathbf{N}(1, -1, 1) = 2\mathbf{i} - 2\mathbf{j} + 2\mathbf{k}$

$\quad\quad \mathbf{N}_u = \frac{\mathbf{N}}{\sqrt{12}} = \frac{\sqrt{3}}{3}(\mathbf{i} - \mathbf{j} + \mathbf{k})$

$\quad\quad$ The tangent plane is: $x - y + z - 3 = 0$

21. Write the function as $\ln x - \ln(y - z) = 0$

$\quad\quad \mathbf{N} = \nabla f = \frac{1}{x}\mathbf{i} - \frac{1}{y - z}\mathbf{j} + \frac{1}{y - z}\mathbf{k}$

$\quad\quad \mathbf{N}(2, 5, 3) = \frac{1}{2}\mathbf{i} - \frac{1}{2}\mathbf{j} + \frac{1}{2}\mathbf{k} = \frac{1}{2}(\mathbf{i} - \mathbf{j} + \mathbf{k})$

$\quad\quad \mathbf{N}_u(2, 5, 3) = \frac{\sqrt{3}}{3}(\mathbf{i} - \mathbf{j} + \mathbf{k})$

$\quad\quad$ The tangent plane is: $x - y + z = 0$

25. $\nabla f = 3x\,\mathbf{i} + 2y\,\mathbf{j}$

$\quad$ **a.** $\nabla f(1, -1) = 3\mathbf{i} - 2\mathbf{j}$, $\|\nabla f\| = \sqrt{13}$

$\quad$ **b.** $\nabla f(1, 1) = 3\mathbf{i} + 2\mathbf{j}$, $\|\nabla f\| = \sqrt{13}$

29. $\nabla f = 2ax\mathbf{i} + 2by\mathbf{j} + 2cz\mathbf{k}$, $\nabla f(a, b, c) = 2(a^2\mathbf{i} + b^2\mathbf{j} + c^2\mathbf{k})$

$\quad\quad \|\nabla f\| = 2\sqrt{a^4 + b^4 + c^4}$

33. $\nabla f = [2(x + y) + 2(z + x)]\mathbf{i} + [2(x + y) + 2(y + z)]\mathbf{j}$

$\quad\quad + [2(y + z) + 2(z + x)]\mathbf{k}$

$\quad\quad \nabla f(2, -1, 2) = 10\mathbf{i} + 4\mathbf{j} + 10\mathbf{k}$, $\|\nabla f\| = \sqrt{216} = 6\sqrt{6}$

37. $\mathbf{N} = \nabla f = \dfrac{2x}{a^2}\mathbf{i} + \dfrac{2y}{b^2}\mathbf{j}, \quad \mathbf{N}(x_0, y_0) = \dfrac{2x_0}{a^2}\mathbf{i} + \dfrac{2y_0}{b^2}\mathbf{j}$

$\|\mathbf{N}\| = \sqrt{\dfrac{4x_0{}^2}{a^4} + \dfrac{4y_0{}^2}{b^4}} = \dfrac{2\sqrt{b^4x_0{}^2 + a^4y_0{}^2}}{a^2b^2}$

$\mathbf{N}_u = \dfrac{\mathbf{N}}{\|\mathbf{N}\|} = \dfrac{b^2x_0\mathbf{i} + a^2y_0\mathbf{j}}{\sqrt{b^4x_0{}^2 + a^4y_0{}^2}}$

41. $\nabla f = e^{x^2y^2}(2xy^2\mathbf{i} + 2x^2y\mathbf{j})$

$\nabla(1, -1) = 2e(\mathbf{i} - \mathbf{j}), \quad \mathbf{V} = \langle 2 - 1, 3 - (-1)\rangle = \langle 1, 4\rangle, \quad \mathbf{u} = \langle \tfrac{1}{\sqrt{17}}, \tfrac{4}{\sqrt{17}}\rangle$

$D_u = \nabla f \bullet \mathbf{u} = -2e\left(\dfrac{3}{\sqrt{17}}\right) = -\dfrac{6e}{\sqrt{17}}$

45. The direction of greatest increase is given by:

$\nabla T = (y + z)\mathbf{i} + (x + z)\mathbf{j} + (x + y)\mathbf{k}.$

$\nabla T(1, 1, 1) = 2(\mathbf{i} + \mathbf{j} + \mathbf{k})$ direction of movement.

As a unit vector: $\dfrac{\sqrt{3}}{3}(\mathbf{i} + \mathbf{j} + \mathbf{k})$

$\|\nabla T\| = 2\sqrt{3}$ maximum rate of temperature change.

49. To find the directional derivative of f in the direction of u we need f_x anf f_y, which we can find with a system of equations.

$(f_x\mathbf{i} + f_y\mathbf{j}) \bullet \left(\dfrac{3\mathbf{i} - 4\mathbf{j}}{5}\right) = 8; \quad 3f_x - 4f_y = 40$

$(f_x\mathbf{i} + f_y\mathbf{j}) \bullet \left(\dfrac{12\mathbf{i} + 5\mathbf{j}}{13}\right) = 1; \quad 12f_x + 5f_y = 13$

Solving simultaneously: $f_x = 4, \ f_y = -7$

Now for $\mathbf{V} = 3\mathbf{i} - 5\mathbf{j}$:

$(f_x\mathbf{i} + f_y\mathbf{j}) \bullet \left(\dfrac{3\mathbf{i} - 5\mathbf{j}}{\sqrt{34}}\right) = (4\mathbf{i} - 7\mathbf{j}) \bullet \left(\dfrac{3\mathbf{i} - 5\mathbf{j}}{\sqrt{34}}\right) = \dfrac{12 + 35}{\sqrt{34}} \approx 8.06$

53. Write the path of the bug as the vector function: $\mathbf{R}(t) = x(t)\mathbf{i} + y(t)\mathbf{j}$

$\mathbf{R}'(t) = \dfrac{dx}{dt}\mathbf{i} + \dfrac{dy}{dt}\mathbf{j} \quad T(x, y) = 1 - ax^2 - by^2, \quad T_x = -2ax, \ T_y = -2by$

$\nabla T = -2ax\mathbf{i} - 2by\mathbf{j}$ so $-2ax = \dfrac{dx}{dt}$ and $-2by = \dfrac{dy}{dt}.$

Separating variables and solving for $x(t)$ and $y(t)$:

$\displaystyle\int (-2a)\, dt = \int \dfrac{dx}{x}$ and $\displaystyle\int (-2b)\, dt = \int \dfrac{dy}{y}$

$\ln x = -2at + C$ and $\ln y = -2bt + C$

$x = e^{-2at+C}$ and $y = e^{-2bt+C}$ Using initial value at $t = 0$ of (x_0, y_0)

$x_0 = e^C$ and $y_0 = e^C$ so $x = x_0e^{-2at}$ and $y = y_0e^{-2bt}$

$\mathbf{R}(t) = x_0e^{-2at}\mathbf{i} + y_0e^{-2bt}\mathbf{j}$

53. (con't.)

If we wish to eliminate the parameter t:

$$t = -\frac{1}{2a}\ln\left(\frac{x}{x_0}\right) = -\frac{1}{2b}\ln\left(\frac{y}{y_0}\right)$$

$$\left(\frac{x}{x_0}\right)^{-\frac{1}{2a}} = \left(\frac{y}{y_0}\right)^{-\frac{1}{2b}} \text{ and } y = y_0\left(\frac{x}{x_0}\right)^{b/a}$$

57. Analysis:

    a. $r = \sqrt{x^2 + y^2 + z^2}$, $\frac{\partial}{\partial x}\left(\frac{1}{r}\right) = -\frac{1}{r^2}\frac{\partial r}{\partial x} = -\frac{1}{r^2}\left(\frac{2x}{2\sqrt{x^2 + y^2 + z^2}}\right)$

        $= -\frac{x}{r^3}$. Likewise for the other partials.

    b. $\nabla V = \frac{\partial V}{\partial x}\mathbf{i} + \frac{\partial V}{\partial y}\mathbf{j} + \frac{\partial V}{\partial z}\mathbf{k}$. Now substitute from **a.**

59. Analysis: $\nabla fg = (fg)_x\mathbf{i} + (fg)_y\mathbf{j} + (fg)_z\mathbf{k}$

              $= (fg_x + f_xg)\mathbf{i} + (fg_y + f_yg)\mathbf{j} + (fg_z + f_zg)\mathbf{k}$

Now commute and associate to complete the proof.

61. Analysis: The directional derivative for the sum $D_s f$ is given by

$$\nabla f \bullet \frac{S}{\|S\|} = \nabla f \bullet \frac{u + v}{\|u + v\|}, \text{ where } \|u + v\| \text{ is a constant.}$$

63. Analysis:

    a. $\nabla r = \frac{x}{r}\mathbf{i} + \frac{y}{r}\mathbf{j} + \frac{z}{r}\mathbf{k}$, show that it has the same direction as $\mathbf{R}$

    b. $\nabla r^n = \nabla(x^2 + y^2 + z^2)^{n/2} = \frac{n}{2}(r)^{n/2 - 1}(2x\mathbf{i} + 2y\mathbf{j} + 2z\mathbf{k})$

                  $= nr^{n-2/2}(x\mathbf{i} + y\mathbf{j} + z\mathbf{k})$

                  $= n(r^{1/2})^{n-2}\mathbf{R}$

12.7 Extrema of Functions of Two Variables

SURVIVAL HINT

Note that, as in section 3.1, the critical points are only *candidates* for extrema. If you are walking in the x direction along a level trail on a hillside, $\frac{\partial f}{\partial x} = 0$ but $\frac{\partial f}{\partial y}$ may go upslope to your right and downslope to your left. The point is not an extreme. If you are at a saddle point *both* partials are zero and the candidate is still not an extreme. The candidate must be tested with the determinant.

7. $f(x, y) = (1 + x^2 + y^2)e^{1 - x^2 - y^2}$

$f_x = (1 + x^2 + y^2)e^{1 - x^2 - y^2}(-2x) + e^{1 - x^2 - y^2}(2x)$

$\quad = e^{1 - x^2 - y^2}(2x)[-1 - x^2 - y^2 + 1]$

$f_y = (1 + x^2 + y^2)e^{1 - x^2 - y^2}(-2y) + e^{1 - x^2 - y^2}(2y)$

$\quad = e^{1 - x^2 - y^2}(2y)[-1 - x^2 - y^2 + 1]$

Setting the partials equal to zero:

$e^{1 - x^2 - y^2}(2x)[-x^2 - y^2] = e^{1 - x^2 - y^2}(2y)[-x^2 - y^2] = 0$

$x = y = 0$. $f(0, 0) = e$. The discriminant test fails, $D = 0$, so examine the function to determine extrema. Let $x^2 + y^2 = \Delta$, $(1 + \Delta)e^{1 - \Delta} < e$.

There is a relative maximum at $(0, 0)$.

11. $f(x, y) = -x^3 + 9x - 4y^2$

$f_x = -3x^2 + 9$, $f_y = -8y$, $f_x = f_y = 0$ at $(\sqrt{3}, 0)$ and $(-\sqrt{3}, 0)$

$f_{xx} = -6x$, $f_{yy} = -8$, $f_{xy} = 0$, $D = 48x$.

$D(\sqrt{3}, 0) > 0$, $f_{xx}(\sqrt{3}, 0) < 0$ so $(\sqrt{3}, 0)$ is a relative maximum.

$D(-\sqrt{3}, 0) < 0$, so $(-\sqrt{3}, 0)$ is a saddle point.

15. $f(x, y) = x^{-1} + y^{-1} + 2xy$

$f_x = -\dfrac{1}{x^2} + 2y$, $f_y = -\dfrac{1}{y^2} + 2x$, $f_x = f_y = 0$ when $y = \dfrac{1}{2x^2}$,

$x = \dfrac{1}{2y^2}$, $y = \dfrac{1}{2\left(\dfrac{1}{2x^2}\right)^2} = 2y^4$, $y^3 = \dfrac{1}{2}$, $y = \dfrac{1}{\sqrt[3]{2}}$, likewise $x = \dfrac{1}{\sqrt[3]{2}}$

$f_{xx} = \dfrac{2}{x^3}$, $f_{yy} = \dfrac{2}{y^3}$, $f_{xy} = 2$, $D = \dfrac{4}{x^3 y^3} - 4$

$D\left(\dfrac{1}{\sqrt[3]{2}}, \dfrac{1}{\sqrt[3]{2}}\right) > 0$, and $f_{xx} > 0$, so $\left(\dfrac{1}{\sqrt[3]{2}}, \dfrac{1}{\sqrt[3]{2}}\right)$ is a relative minimum.

Also consider points where f_x and f_y are undefined: $x = 0$ and/or $y = 0$.
However $f(x, y)$ is undefined at these points also.

19. $f(x, y) = x^2 + y^2 - 6xy + 9x + 5y + 2$

$f_x = 2x - 6y + 9$, $f_y = 2y - 6x + 5$, $f_x = f_y = 0$ when

$2x - 6y = -9$ and $2y - 6x = -5$. Solving this system we get $x = \dfrac{3}{2}$ and

$y = 2$. f_x and f_y are nowhere undefined.

$f_{xx} = 2$, $f_{yy} = 2$, $f_{xy} = -6$, $D = 4 - 36 < 0$.

So $\left(\dfrac{3}{2}, 2\right)$ is a saddle point.

SURVIVAL HINT

In $\mathbb{R}^2$ the extreme value theorem requires a continuous function on a closed interval. Note that the $\mathbb{R}^3$ extension of the extreme value theorem requires that f be continuous on a closed and bounded set S. In $\mathbb{R}^2$ the curve may have corners, cusps, or vertical tangents, as long as it is continuous. Likewise in $\mathbb{R}^3$ the surface may have creases, pointed peaks and vertical tangents if it is continuous. It is sometimes difficult to verify the continuity of the surface. Most of your examples will be sums and products of continuous functions, which are continuous.

23. $f(x, y) = e^{x^2 + 2x + y^2}$

$f_x = (2x + 2)\, e^{x^2 + 2x + y^2}$, $f_y = 2y\, e^{x^2 + 2x + y^2}$, $f_x = f_y = 0$ when

$2x + 2 = 2y = 0$, $x = -1$, $y = 0$.

f_x and f_y are nowhere undefined.

$f_{xx} = (2x + 2)^2 e^{x^2 + 2x + y^2} + 2\, e^{x^2 + 2x + y^2} = e^{x^2 + 2x + y^2}(4x^2 + 8x + 6)$

$f_{yy} = 4y^2\, e^{x^2 + 2x + y^2} + 2\, e^{x^2 + 2x + y^2} = e^{x^2 + 2x + y^2}(4y^2 + 2)$

$f_{xy} = 2y(2x + 2)\, e^{x^2 + 2x + y^2}$

$D(-1, 0) = 2e^{-1}(2e^{-1}) - 0 > 0$, and $f_{xx}(-1, 0) > 0$,

So $(-1, 0)$ is a relative minimum. Checking the boundry points:

If $x^2 + 2x + y^2 = 0$, $f(x, y) = e^{x^2 + 2x + (-x^2 - 2x)} = e^0 = 1$.

$f(-1, 0) = e^{-1}$, so $f(x, y)$ has an absolute maximum of 1 along the boundry and an absolute minimum of e^{-1} at $(-1, 0)$.

SURVIVAL HINT

When doing a least squares regression line it is essential that you carefully organize and label your data.

27. $m = \dfrac{n\sum\limits_{1}^{n} x_k y_k - \left(\sum\limits_{1}^{n} x_k\right)\left(\sum\limits_{1}^{n} y_k\right)}{n\sum\limits_{1}^{n} x_k^2 - \left(\sum\limits_{1}^{n} x_k\right)^2}$ and $b = \dfrac{\sum\limits_{1}^{n} x_k^2 \sum\limits_{1}^{n} y_k - \left(\sum\limits_{1}^{n} x_k\right)\left(\sum\limits_{1}^{n} x_k y_k\right)}{n\sum\limits_{1}^{n} x_k^2 - \left(\sum\limits_{1}^{n} x_k\right)^2}$

Organize the data:

x	y	x^2	xy
0	1	0	0
1	1.6	1	1.6
2.2	3	4.84	6.6
3.1	3.9	9.61	12.09
4	5	16	20
10.3	14.5	31.45	40.29

$m = \dfrac{5(40.29) - (10.3)(14.5)}{5(31.45) - (10.3)^2} = \dfrac{52.1}{51.16} \approx 1.01838$

$b = \dfrac{(31.45)(14.5) - (10.3)(40.29)}{51.16} = \dfrac{41.038}{51.16} \approx 0.80215$

Least squares regression line: $y = 1.02x + 0.80$

31. To find the area of a triangle when three sides are given use Heron's semi-perimeter formula: $A = \sqrt{s(s - a)(s - b)(s - c)}$ where $s = \dfrac{P}{2}$. The values that maximize A^2 will also maximize A.

$A^2 = \dfrac{P}{2}\left(\dfrac{P}{2} - a\right)\left(\dfrac{P}{2} - b\right)\left(\dfrac{P}{2} - c\right)$. To make this a function of two variables substitute $c = P - a - b$.

$16A^2 = P(P - 2a)(P - 2b)(2a + 2b - P)$

$f_a = P(P - 2b)[(P - 2a)2 + (2a + 2b - P)(- 2)]$

$\quad = P(P - 2b)(4P - 8a - 4b)$

$f_b = P(P - 2a)[(P - 2b)2 + (2a + 2b - P)(- 2)]$

$\quad = P(P - 2a)(4P - 8b - 4a)$

For extrema $f_a = f_b = 0$ at $a = \dfrac{P}{2}$, $b = \dfrac{P}{2}$, $c = 0$. $A = 0$, a minimum.

Also, solving the system: $8a + 4b = 4P$, $4a + 8b = 4P$,

$a = \dfrac{P}{3}$, $b = \dfrac{P}{3}$, $c = \dfrac{P}{3}$.

31. (con't.)

Verifying that this is a maximum:

$$f_{aa} = P(P - 2b)(-8), \quad f_{aa}\left(\frac{P}{3}, \frac{P}{3}\right) = P\left(\frac{P}{3}\right)(-8) = -\frac{8}{3}P^2$$

$$f_{bb} = P(P - 2a)(-8), \quad f_{bb}\left(\frac{P}{3}, \frac{P}{3}\right) = P\left(\frac{P}{3}\right)(-8) = -\frac{8}{3}P^2$$

$$f_{ab} = P[(P - 2b)(-4) + (4P - 8a - 4b)(-2)] = P(16a + 16b - 12P)$$

$$f_{ab}\left(\frac{P}{3}, \frac{P}{3}\right) = -\frac{4}{3}P^2. \quad D = \frac{64}{9}P^4 - \frac{16}{9}P^4 > 0 \text{ and } f_{aa} < 0 \text{ so}$$

$a = b = c = \dfrac{P}{3}$ is a maximum.

35. Use $V_0 = xyz$ to write E as a function of two variables:

$$E(x, y) = \frac{k^2}{8m}\left(\frac{1}{x^2} + \frac{1}{y^2} + \frac{1}{\left(\frac{V_0}{xy}\right)^2}\right)$$

Minimize $\dfrac{8m}{k^2}E = \dfrac{1}{x^2} + \dfrac{1}{y^2} + \dfrac{x^2 y^2}{V_0^2}$

$$f_x = -\frac{2}{x^3} + \frac{2xy^2}{V_0^2}, \quad f_y = -\frac{2}{y^3} + \frac{2x^2 y}{V_0^2}$$

$$-\frac{2}{x^3} + \frac{2xy^2}{V_0^2} = 0 \text{ when } x^4 y^2 = V_0^2, \; x^4 = \frac{V_0^2}{y^2}, \; x = \sqrt{\frac{V_0}{y}}$$

Similarily $y = \sqrt{\dfrac{V_0}{x}}$ and $z = \dfrac{V_0}{xy} = \dfrac{V_0}{x\sqrt{\frac{V_0}{x}}} = \sqrt{\dfrac{V_0}{x}} = y$

$V_0 = x^3, \quad x = y = z = \sqrt[3]{V_0}$

Verifying that this is a minimum: $f_{xx} = \dfrac{6}{x^4} + \dfrac{2y^2}{V_0^2}, \; f_{yy} = \dfrac{6}{y^4} + \dfrac{2x^2}{V_0^2},$

$$f_{xy} = \frac{4xy}{V_0^2}. \quad D = \left(\frac{6}{V_0^{4/3}}\right)^2 - \frac{16 V_0^{4/3}}{V_0^4} = \frac{64}{V_0^{8/3}} - \frac{16}{V_0^{8/3}} > 0$$

and $f_{xx} > 0$ so there is a minimum at $(\sqrt[3]{V_0}, \sqrt[3]{V_0}, \sqrt[3]{V_0})$

39. Profit = Revenue − Cost, Revenue = (#sold)(cost per unit).

$$P = (40 - 8x + 5y)100x + (50 + 9x - 7y)100y$$
$$\qquad - 1{,}000(40 - 8x + 5y) - 3{,}000(50 + 9x - 7y)$$

$$\frac{P}{100} = -8x^2 - 7y^2 + 14xy - 150x + 210y - 1900$$

$$P_x = -16x + 14y - 150, \quad P_y = -14y + 14x + 210$$

Solving simultaneously, $x = 30, \; y = 45$. Since $P(0, 0) = 0$,

(30, 45) must be a maximum. Thus, the first type should be priced at

$3,000 and the second type priced at $4,500.

47.   Analysis:

    a.   Verify that $(0, 0)$ is a critical point and then examine the traces in the xz and yz planes to show a saddle point at $(0, 0)$.

    b.   Verify that $D = 0$ then show that $g(x, y) \geq 0$, so $(0, 0)$ is a minimum.

    c.   Verify that $D = 0$ then show that on one side of $y = -x$ $f(x, y)$ increases and it decreases on the other side.

51.   Analysis: Find F_m and F_b for $F(m, b)$ then use Cramer's rule to solve the system of equations.

12.8 Lagrange Multipliers

SURVIVAL HINT

The method of Lagrange multipliers requires the solution of a system of equations. There is no single set of steps to follow in solving a system of equations. Substitution, addition or subtraction of equations, and some cleverness are usually required. Often one equation can be solved for λ, and that value substituted into the other equation involving λ. Solve the resulting equation with the constraint equation to find x and y. These values can then be used to evaluate λ.

1. $f(x, y) = xy$, $g(x, y) = 2x + 2y - 5$

Solve the system: $f_x = \lambda g_x$, $f_y = \lambda g_y$, $2x + 2y = 5$.

$y = 2\lambda$, $x = 2\lambda$, $2x + 2y = 5$.

$2(2\lambda) + 2(2\lambda) = 5$, $\lambda = \frac{5}{8}$, $x = y = \frac{5}{4}$.

$f(\frac{5}{4}, \frac{5}{4}) = \frac{25}{16}$. Testing another constraint point, $(0, \frac{5}{2})$, gives $f(0, \frac{5}{2}) = 0$, a lesser value, so $f(\frac{5}{4}, \frac{5}{4}) = \frac{25}{16}$ is a maximum.

5. $f(x, y) = x^2 + y^2$, $g(x, y) = xy - 1$

Solve the system: $f_x = \lambda g_x$, $f_y = \lambda g_y$, $xy = 1$.

$2x = \lambda y$, $2y = \lambda x$, $xy = 1$.

Multiplying the first two equations: $4xy = \lambda^2 xy$, $4 = \lambda^2$, $\lambda = \pm 2$.

$2x = \pm 2y$, $x = \pm y$. Now the constraint equation gives us two critical points: $(1, 1)$ and $(-1, -1)$. $f(1, 1) = 2$ and $f(-1, -1) = 2$.

Testing another constraint point, $f(2, \frac{1}{2}) = \frac{17}{4}$, a greater value. So f has a minimum value of 2 at $(1, 1)$ and $(-1, -1)$.

9. $f(x, y) = \cos x + \cos y, \; g(x, y) = x - y + \frac{\pi}{4}.$

Solve the system: $-\sin x = 1\lambda, \; -\sin y = -\lambda, \; x - y + \frac{\pi}{4} = 0.$

$\sin x = -\sin y, \; \sin x = \sin(-y), \; x = -y + 2\pi k.$

Using the constraint equation: $-y + 2\pi k - y + \frac{\pi}{4} = 0, \; 2y = 2\pi k + \frac{\pi}{4},$

$y = \frac{\pi}{8} + k\pi.$ In like manner $x - (-x + 2\pi k) + \frac{\pi}{4} = 0, \; x = -\frac{\pi}{8} + k\pi.$

$k = 0: \; f(-\frac{\pi}{8}, \frac{\pi}{8}) \approx 1.84776$

$k = 1: \; f(-\frac{7\pi}{8}, \frac{9\pi}{8}) \approx -1.84776$

$k = 2: \; f(-\frac{15\pi}{8}, \frac{17\pi}{8}) \approx 1.84776$

and so on, f has a period of 2π.

f has a maximum of 1.84776 for $f(-\frac{\pi}{8} + k\pi, \frac{\pi}{8} + k\pi)$ with k even, and a

minimum of -1.84776 for $f(-\frac{\pi}{8} + k\pi, \frac{\pi}{8} + k\pi)$ with k odd.

13. $f(x, y, z) = 2x^2 + 4y^2 + z^2, \; g(x, y, z) = 4x - 8y + 2z - 10$

$4x = 4\lambda, \; 8y = \lambda(-8), \; 2z = \lambda(2), \; 4x - 8y + 2z = 10,$

$4\lambda + 8\lambda + 2\lambda = 10, \; \lambda = \frac{5}{7}.$ Critical point: $(\frac{5}{7}, -\frac{5}{7}, \frac{5}{7}).$

$f(\frac{5}{7}, -\frac{5}{7}, \frac{5}{7}) = \frac{25}{49}(2 + 4 + 1) = \frac{25}{7}.$

Testing another constraint point: $f(\frac{5}{2}, 0, 0) = \frac{25}{2}$, a greater value, so f has

a minimum of $\frac{25}{7}$ at $(\frac{5}{7}, -\frac{5}{7}, \frac{5}{7}).$ By using negative values for any two variables

in the constraint equation, the third variable can be made arbitrarily large,

thus f can be made arbitrarily large and does not have a maximum.

17. Find critical points for $D^2 = f(x, y, z) = x^2 + y^2 + z^2$, and

$g(x, y, z) = Ax + By + Cz. \; 2x = A\lambda, \; 2y = B\lambda, \; 2z = C\lambda.$

$A(\frac{A}{2}\lambda) + B(\frac{B}{2}\lambda) + C(\frac{C}{2}\lambda) = 0, \; \frac{\lambda}{2}(A^2 + B^2 + C^2) = 0.$

But $A^2 + B^2 + C^2 \neq 0$, so $\lambda = 0$ and $x = y = z = 0.$

$f(0, 0, 0) = 0$ is the minimum distance to the plane. The plane

$Ax + By + Cz = 0$ passes through the origin even if A, B, C are non-zero.

21. Maximize $f(x, y, z) = xy^2z$ subject to $x + y + z = 12.$

Solve the system: $y^2z = \lambda, \; 2xyz = \lambda, \; xy^2 = \lambda$ and $x + y + z = 12.$

Equate the first and third equations: $y^2z = xy^2$, so $x = z$, or $y = 0.$

Divide the first and second: $\frac{y^2z}{2xyz} = 1, \; \frac{y}{2x} = 1, \; y = 2x.$

Substitute into the constraint equation: $x + 2x + x = 12, \; x = 3, \; y = 6,$

$z = 3. \; f(3, 6, 3) = 324.$

25.   Maximize $f(x, y, z) = xy$ subject to $2x + 2y = 320$ or $x + y = 160$.

$y = \lambda$, $x = \lambda$, $x + y = 160$, $\lambda = 80 = x = y$.

The maximum area is 6,400 sq yd with a square field 80 by 80 yd.

29.   Maximize $f(x, y) = 50x^{1/2}y^{3/2}$ subject to $x + y = 8$.

$$\frac{25y^{3/2}}{x^{1/2}} = \lambda, \quad 75x^{1/2}y^{1/2} = \lambda, \quad x + y = 8.$$

$$\frac{25y^{3/2}}{x^{1/2}} = 75x^{1/2}y^{1/2}, \quad y = 3x. \quad \text{Using the constraint equation:}$$

$x + 3x = 8$, $x = 2$, $y = 6$.

$2,000 to development and $6,000 to promotion gives the maximum sales of

$f(2, 6) = 50\sqrt{2}\,(6\sqrt{6}) = 600\sqrt{3}$ units.

33.   Maximize $V = xyz$ subject to $2x + 2y + z = 108$.

$yz = 2\lambda$, $xz = 2\lambda$, $xy = \lambda$.

Equating the first and second: $x = y$.

Dividing the second by the third: $\frac{xz}{xy} = 2$, $z = 2y$.

Using the constraint equation: $2y + 2y + 2y = 108$, $y = 18$, $x = 18$,

$z = 36$.   The maximum volume is $V(18, 18, 36) = 11,664$ cu in or 6.75 cu ft.

37. $f(x, y, z) = xy + xz$, $g(x, y, z) = 2x + 3z - 5$, $h(x, y, z) = xy - 4$.

$y + z = 2\lambda + y\mu$, $x = 0\lambda + x\mu$, $x = 3\lambda + 0\mu$.

From equation two: $\mu = 1$. Substitute into equation one: $z = 2\lambda$.

Using the first constraint: $2(3\lambda) + 3(2\lambda) = 5$, $\lambda = \frac{5}{12}$.

$x = \frac{5}{4}$. Substitute into the second constraint: $\frac{5}{4}y = 4$, $y = \frac{16}{5}$

Substitute x into the first constraint: $2(\frac{5}{4}) + 3z = 5$, $z = \frac{5}{6}$

There is a maximum of $f(\frac{5}{4}, \frac{16}{5}, \frac{5}{6}) = \frac{5}{4}(\frac{16}{5}) + \frac{5}{4}(\frac{5}{6}) = \frac{121}{24}$ at $(\frac{5}{4}, \frac{16}{5}, \frac{5}{6})$.

41.   Analysis: Maximize $f(x, y, z) = xyz$ subject to $\frac{x^2}{a^2} + \frac{y^2}{b^2} + \frac{z^2}{c^2} = 1$

$yz = \frac{2x}{a^2}\lambda$, $xz = \frac{2y}{b^2}\lambda$, $xy = \frac{2z}{c^2}\lambda$.

Divide equations one and two: $\frac{yz}{xz} = \frac{xb^2}{ya^2}$, $y^2 = \frac{b^2x^2}{a^2}$.

Divide equations two and three to express z in terms of x and y.

Substitute into the constraint equation and solve for x.

43.   Analysis: The critical points occur where the slopes of $f(x, y)$ and $g(x, y)$

are equal. $\nabla f = f_x\,\mathbf{i} + f_y\,\mathbf{j}$, $\nabla g = p\,\mathbf{i} + q\,\mathbf{j}$. Equate and solve.

45. Analysis: $p = \alpha A x^{\alpha-1} y^{\beta} \lambda$, $q = \beta A x^{\alpha} y^{\beta-1} \lambda$.

Dividing: $\dfrac{p}{q} = \dfrac{\alpha y}{\beta x}$, $y = \dfrac{p\beta x}{q\alpha}$. Substitute this value of y into the constraint

equation and solve for x, remembering that $x^{\alpha+\beta} = x^1$.

Chapter 12 Review

PRACTICE PROBLEMS

32. $f(x, y) = \sin^{-1} xy$. Recall $\dfrac{d}{dx} \sin^{-1} u = \dfrac{1}{\sqrt{1 - u^2}} \dfrac{du}{dx}$

$f_x = \dfrac{y}{\sqrt{1 - x^2 y^2}}$, $f_y = \dfrac{x}{\sqrt{1 - x^2 y^2}}$

$f_{xy} = \dfrac{\sqrt{1 - x^2 y^2}\,(1) - \dfrac{y(-2x^2 y)}{2\sqrt{1 - x^2 y^2}}}{1 - x^2 y^2} = \dfrac{1 - x^2 y^2 + x^2 y^2}{(1 - x^2 y^2)^{3/2}} = \dfrac{1}{(1 - x^2 y^2)^{3/2}}$

$f_{yx} = \dfrac{\sqrt{1 - x^2 y^2}\,(1) - \dfrac{x(-2xy^2)}{2\sqrt{1 - x^2 y^2}}}{1 - x^2 y^2} = \dfrac{1 - x^2 y^2 + x^2 y^2}{(1 - x^2 y^2)^{3/2}} = \dfrac{1}{(1 - x^2 y^2)^{3/2}}$

33. $\dfrac{dw}{dt} = \dfrac{dw}{dx} \cdot \dfrac{dx}{dt} + \dfrac{dw}{dy} \cdot \dfrac{dy}{dt} + \dfrac{dw}{dz} \cdot \dfrac{dz}{dt}$

$= 2xy(t \cos t + \sin t) + (x^2 + 2yz)(-t \sin t + \cos t) + y^2(2)$

$\dfrac{dw}{dt}(\pi) = 2xy(-\pi) + (x^2 + 2yz)(-1) + 2y^2$

$= -2\pi xy + 2y^2 - x^2 - 2yz$

Now if $t = \pi$, $x = 0$, $y = -\pi$, $z = 2\pi$

$\dfrac{dw}{dt}(\pi) = 0 + 2\pi^2 - 0 - 2(-\pi)(2\pi) = 6\pi^2$

34. $f(x, y, z) = xy + yz + xz$ at $(1, 2, -1)$

a. $\nabla f = (y + z)\mathbf{i} + (x + z)\mathbf{j} + (y + x)\mathbf{k} = \mathbf{i} + 3\mathbf{k}$

b. $V_u = \dfrac{-2\mathbf{i} - \mathbf{j}}{\sqrt{5}}$ $df = \nabla f \bullet V_u = \dfrac{-2}{\sqrt{5}} = \dfrac{-2\sqrt{5}}{5}$

c. The directional derivative has its greatest value in the direction of the

gradient: $\dfrac{\mathbf{i} + 3\mathbf{k}}{\sqrt{10}}$. The magnitude is $\|\nabla f\| = \sqrt{10}$

35. $\displaystyle\lim_{(x, y)\to(0, 0)} f(x, y)$ along the line $y = x$ is $\displaystyle\lim_{x\to 0} \dfrac{x^3}{x^3 + x^3} = \dfrac{1}{2}$.

The limit does not equal $f(x, y)$ so the function is not continuous.

36. $f(x, y) = \ln \frac{y}{x} = \ln y - \ln x.$

$f_x = -\frac{1}{x}, \; f_y = \frac{1}{y}, \; f_{yy} = -\frac{1}{y^2}, \; f_{xy} = 0$

37. $f(x, y, z) = x^2 y + y^2 z + z^2 x$

$\dfrac{\partial f}{\partial x} = 2xy + z^2, \; \dfrac{\partial f}{\partial y} = x^2 + 2yz, \; \dfrac{\partial f}{\partial z} = y^2 + 2zx$

$\dfrac{\partial f}{\partial x} + \dfrac{\partial f}{\partial y} + \dfrac{\partial f}{\partial z} = 2xy + z^2 + x^2 + 2yz + y^2 + 2zx$

$$= (x + y + z)^2$$

38. $f(x, y) = x^4 + 2x^2 y^2 + y^4$

$\nabla f = (4x^3 + 4xy^2)\mathbf{i} + (4x^2 y + 4y^3)\mathbf{j}$

$\nabla f(2, -2) = 64(\mathbf{i} - \mathbf{j})$ in the direction of $\frac{2\pi}{3}$ with the positive x-axis:

$V = -\frac{1}{2}\mathbf{i} + \frac{\sqrt{3}}{2}\mathbf{j}. \; \nabla f \bullet V = -32 - 32\sqrt{3} = -32(1 + \sqrt{3}) \approx -87.4$

39. $f_x = e^{-(x^2+y^2)}[2x + (x^2 + y^2)(-2x)] = 2x\,e^{-(x^2+y^2)}(1 - x^2 - y^2)$

$f_y = e^{-(x^2+y^2)}[2y + (x^2 + y^2)(-2y)] = 2y\,e^{-(x^2+y^2)}(1 - x^2 - y^2)$

$f_x = f_y = 0$ if $x = y = 0$ or if $x^2 + y^2 = 1$

40. $f(x, y) = x^2 + 2y^2 + 2x + 3$ subject to $g(x, y) = x^2 + y^2 - 4$

$2x + 2 = 2x\lambda, \; 4y = 2y\lambda,$ and $x^2 + y^2 = 4.$

From the second equation: $\lambda = 2$ or $y = 0.$

If $\lambda = 2, \; 2x + 2 = 4x, \; x = 1,$ and from the constraint equation

$y = \pm\sqrt{3}.$ If $y = 0$ the constraint equation gives $x = \pm 2.$

The set of candidates: $\{(1, \sqrt{3}), (1, -\sqrt{3}), (2, 0), (-2, 0)\}.$

$f(1, \sqrt{3}) = 12, \; f(1, -\sqrt{3}) = 12, \; f(2, 0) = 11, \; f(-2, 0) = 3.$

$f(x, y)$ has a maximum of 12 at $(1, \pm\sqrt{3})$ and a minimum of 3 at $(-2, 0).$

Chapter 13
Multiple Integration

13.1 Double Integration Over Rectangular Regions

3. This is a right triangular prism. The legs of the triangle are each 4, $y = 4$, $z = 4$, and the height of the prism is 3, $x = 3$.

$V = Bh = \frac{1}{2}(4)(4)(3) = 24$ cu units.

SURVIVAL HINT

Be careful to use the x boundaries with dx and y boundaries with dy, for whichever iteration you choose.

$$\int_{x_1}^{x_2} \int_{y_1}^{y_2} f(x,\ y)\ dy\ dx \quad \text{or} \quad \int_{y_1}^{y_2} \int_{x_1}^{x_2} f(x,\ y)\ dx\ dy$$

7. $\displaystyle \int_{1}^{2}\int_{0}^{1} x^2 y\ dy\ dx = \int_{1}^{2} \frac{x^2 y^2}{2}\Big|_0^1\ dx = \int_{1}^{2} \frac{x^2}{2}\ dx = \frac{x^3}{6}\Big|_1^2 = \frac{8}{6} - \frac{1}{6} = \frac{7}{6}$

11. $\displaystyle \int_{0}^{1}\int_{1}^{3} \frac{2xy}{x^2 + 1}\ dy\ dx = \int_{0}^{1} \frac{xy^2}{x^2 + 1}\Big|_1^3\ dx = \int_{0}^{1} \frac{8x}{x^2 + 1}\ dx = 4\int_{0}^{1} \frac{2x\ dx}{x^2 + 1}$

$$= 4\ln(x^2 + 1)\Big|_0^1 = 4\ln 2 = \ln 16$$

15. $\displaystyle \int_{0}^{1}\int_{0}^{2} (2x + 3y)\ dy dx = \int_{0}^{1}\left(2xy + \frac{3y^2}{2}\right)\Big|_0^2\ dx = \int_{0}^{1} (4x + 6)\ dx$

$$= \left(2x^2 + 6x\right)\Big|_0^1 = 8$$

19. $\displaystyle\int_0^1\int_0^4 \sqrt{xy}\,dy\,dx \;=\; \int_0^1 \frac{2x^{1/2}y^{3/2}}{3}\Big|_0^4 dx \;=\; \int_0^1 \frac{16}{3}x^{1/2}dx \;=\; \frac{32}{9}x^{3/2}\Big|_0^1 \;=\; \frac{32}{9}$

23. $\displaystyle\int_0^1\int_0^1 (x+y)^5 dy\,dx \;=\; \int_0^1 \frac{(x+y)^6}{6}\Big|_0^1 dx \;=\; \frac{1}{6}\int_0^1 [(x+1)^6 - x^6]dx$

$$=\; \frac{1}{6}\left[\frac{(x+1)^7}{7} - \frac{x^7}{7}\right]\Big|_0^1 \;=\; \frac{1}{6}\left(\frac{128}{7} - \frac{1}{7} - \frac{1}{7}\right) \;=\; \frac{1}{6}\left(\frac{126}{7}\right) \;=\; 3$$

31. $R \;=\; 1{,}400 \displaystyle\int_R\int f(x,y)\,dy\,dx.$ If $a \le x \le b, \quad c \le y \le d$ then

$$R \;=\; 1{,}400 \int_a^b\int_c^d f(x,y)\,dy\,dx$$

35. $z = f(x,y)$ is a paraboloid opening downward with vertex at $z = 4$ and intercepts of $(\pm 2, 0, 0)$ and $(0, \pm 2, 0)$. The integral represents the volume above the unit square in the first octant. The minimum value for $z = f(1,1) = 2$. Since $z \ge 2$ over the given R, the value of the integral will be greater than the volume of the prism with unit base and height of 2.

39. $\displaystyle\int_R\int\left(1 - \sqrt{x^2 + z^2}\right)dA \;=\; \int_R\int 1\,dA \;-\; \int_R\int\sqrt{x^2 + z^2}\,dA$

If R is the disk $x^2 + z^2 \le 1$, $\displaystyle\int_R\int 1\,dA$ is a cylinder of radius 1 and a

height of 1. $\displaystyle\int_R\int\sqrt{x^2 + z^2}\,dA$ is a cone with radius 1 and a height of 1.

$$V \;=\; \pi r^2 h - \frac{1}{3}\pi r^2 h \;=\; \frac{2}{3}\pi r^2 h \;=\; \frac{2\pi}{3}$$

41. Analysis: $\displaystyle\int_R\int\frac{\partial}{\partial x}\left(\frac{\partial}{\partial y}f(x,y)\right)dA \;=\; \int_{x_1}^{x_2}\int_{y_1}^{y_2}\frac{\partial}{\partial x}\left(\frac{\partial}{\partial y}f(x,y)\right)dy\,dx$

By the Fundamental Theorem of Calculus $\displaystyle\int_a^b \frac{\partial}{\partial x}f(x,y) \;=\; f(x,y)\Big|_a^b$

Integrate and evaluate.

13.2 Double Integration Over Nonrectangular Regions

SURVIVAL HINT

Draw a sketch of the region of integration and decide on vertical or horizontal strips.

For vertical strips you must have numerical values for the x limits and either numerical

or $y = f(x)$ expressions for the y limits:

$$\int_{x_1}^{x_2} \int_{y_1=g(x)}^{y_2=h(x)} f(x,\ y)\ dy\ dx$$

For horizontal strips you must have numerical values for the y limits and either

numerical or $x = f(y)$ expressions for the x limits:

$$\int_{y_1}^{y_2} \int_{x_1=g(y)}^{x_2=h(y)} f(x,\ y)\ dx\ dy$$

3.

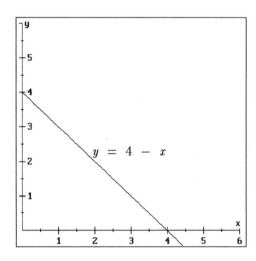

Using vertical strips:

$$\int_0^4 \int_0^{4-x} xy\ dy\ dx\ =\ \int_0^4 \frac{xy^2}{2}\bigg|_0^4\ dx\ =\ \int_0^4 \frac{x}{2}(4\ -\ x)^2\ dx\ =\ \frac{1}{2}\int_0^4 (x^3\ -\ 8x^2\ +\ 16x)\ dx$$

$$=\ \frac{1}{2}\left(\frac{x^4}{4}\ -\ \frac{8x^3}{3}\ +\ 8x^2\right)\bigg|_0^4\ =\ \frac{1}{2}\left[64\ -\ \frac{8}{3}(64)\ +\ 2(64)\right]$$

$$=\ 32\left(1\ -\ \frac{8}{3}\ +\ 2\right)\ =\ \frac{32}{3}$$

7.

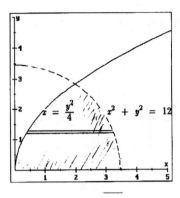

Using horizontal strips: $\displaystyle\int_0^{2\sqrt{3}}\int_{y^2/4}^{\sqrt{12-y^2}} dx\ dy = \int_0^{2\sqrt{3}}\left(\sqrt{12-y^2}-\frac{y^2}{4}\right)dy$

$= \left(\frac{y}{2}\sqrt{12-y^2}+6\sin^{-1}\frac{y}{2\sqrt{3}}-\frac{y^3}{12}\right)\Big|_0^{2\sqrt{3}}$ (formula #231)

$= \sqrt{3}\,(0)+6\sin^{-1}1-\frac{24\sqrt{3}}{12}=3\pi-2\sqrt{3}\ \approx\ 5.9607$

11.

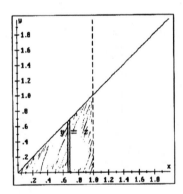

Using vertical strips: $\displaystyle\int_0^1\int_0^x (x^2+2y^2)\ dy\ dx = \int_0^1\left(x^2y+\frac{2y^3}{3}\right)\Big|_0^x dx$

$= \int_0^1\frac{5x^3}{3}\ dx = \frac{5x^4}{12}\Big|_0^1 = \frac{5}{12}$

15. Using vertical strips: $\displaystyle\int_0^1 \int_x^1 (x + y)\, dy\, dx = \int_0^1 \left(xy + \frac{y^2}{2}\right)\Big|_x^1\, dx$

$\displaystyle = \int_0^1 \left(x + \frac{1}{2} - \frac{3x^2}{2}\right) dx = \left(\frac{x^2}{2} + \frac{x}{2} - \frac{x^3}{2}\right)\Big|_0^1 = \frac{1}{2}(1 + 1 - 1) = \frac{1}{2}$

19. Finding the intersection of the curves: $\sqrt{x} = 2 - x$, $x = 4 - 4x + x^2$,

$x^2 - 5x + 4 = 0$, $x = 1, 4$. The intersection is $(1, 1)$.

Using horizontal strips to avoid splitting the region:

$\displaystyle\int_0^1 \int_{y^2}^{2-y} y\, dx\, dy = \int_0^1 xy\Big|_{y^2}^{2-y}\, dy = \int_0^1 (2y - y^2 - y^3)\, dy = \left(y^2 - \frac{y^3}{3} - \frac{y^4}{4}\right)\Big|_0^1$

$\displaystyle = 1 - \frac{1}{3} - \frac{1}{4} = \frac{5}{12}$

23. Finding the intersection of the curves: $x^3 = 1$, $x = 1$, $y = 1$.

Using vertical strips and two integrals since the upper bounds are different:

$\displaystyle\int_0^1 \int_0^x 2x\, dy\, dx + \int_1^2 \int_0^{1/x^2} 2x\, dy\, dx = \int_0^1 2xy\Big|_0^x\, dx + \int_1^2 2xy\Big|_0^{1/x^2}\, dx$

$\displaystyle = \int_0^1 2x^2\, dx + \int_1^2 \frac{2}{x}\, dx = \frac{2x^3}{3}\Big|_0^1 + 2\ln x\Big|_1^2 = \frac{2}{3} + 2\ln 2 = \frac{2}{3} + \ln 4$

≈ 2.0530

27.

a. The $dy\, dx$ indicates vertical strips: $\displaystyle\int_0^4 \int_0^{4-x} xy\, dy\, dx = \int_0^4 \frac{xy^2}{2}\Big|_0^{4-x}\, dx$

$\displaystyle = \frac{1}{2}\int_0^4 (x^3 - 8x^2 + 16x)\, dx = \frac{1}{2}\left(\frac{x^4}{4} - \frac{8x^3}{3} + 8x^2\right)\Big|_0^4$

$\displaystyle = \frac{1}{2}[64 - \frac{8}{3}(64) + 2(64)] = 32(1 - \frac{8}{3} + 2) = \frac{32}{3}$

27. (con't.)

b.   Reversing the order of integration gives $dx\ dy$ and horizontal strips:

$$\int_0^4 \int_0^{4-y} xy\ dx\ dy = \int_0^4 \frac{x^2 y}{2} \Big|_0^{4-y} dy = \frac{1}{2} \int_0^4 (y^3 - 8y^2 + 16y)\ dy$$

$$= \frac{1}{2}\left(\frac{y^4}{4} - \frac{8y^3}{3} + 8y^2\right)\Big|_0^4 = \frac{1}{2}[64 - \frac{8}{3}(64) + 2(64)]$$

$$= 32(1 - \frac{8}{3} + 2) = \frac{32}{3}$$

31.

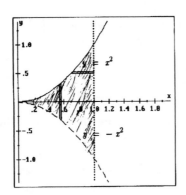

a. $$\int_0^1 \int_{-x^2}^{x^2} dy\ dx = \int_0^1 y \Big|_{-x^2}^{x^2} dx = \int_0^1 2x^2\ dx = \frac{2x^3}{3}\Big|_0^1 = \frac{2}{3}$$

b.   Using the symmetry about the x-axis: $2\int_0^1 \int_{\sqrt{y}}^1 dx\ dy = 2\int_0^1 x\Big|_{\sqrt{y}}^1 dy$

$$= 2\int_0^1 (1 - \sqrt{y})\ dy = 2\left(y - \frac{2y^{3/2}}{3}\right)\Big|_0^1 = 2(1 - \frac{2}{3}) = \frac{2}{3}$$

35.

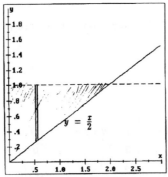

Reversing the order gives $dy\ dx$ and vertical strips: $\displaystyle\int_0^2 \int_{x/2}^1 f(x,\ y)\,dy\ dx$

39.

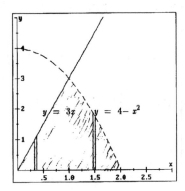

Reversing the order gives $dy\ dx$ and vertical strips. Since the upper boundary
has two different curves, we will need two regions:

$$\int_0^1 \int_0^{3x} f(x,\ y)\ dy\ dx\ +\ \int_1^2 \int_0^{4-x^2} f(x,\ y)\ dy\ dx$$

43.

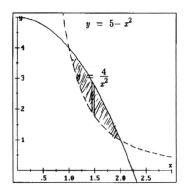

Finding the intersection: $\dfrac{4}{x^2} = 5 - x^2,\ 4 = 5x^2 - x^4,\ x^4 - 5x^2 + 4 = 0,$
$(x^2 - 4)(x^2 - 1) = 0,\ x = \pm 2,\ \pm 1.$ Our intersections are $(1,\ 4),\ (2,\ 1).$

Using horizontal strips: $\displaystyle\int_1^4 \int_{2/\sqrt{y}}^{\sqrt{5-y}} dx\ dy$

Using vertical strips: $\displaystyle\int_1^2 \int_{4/x^2}^{5-x^2} dy\ dx = \int_1^2 \left(5 - x^2 - \frac{4}{x^2}\right) dx$

$$= \left(5x - \frac{x^3}{3} + \frac{4}{x}\right)\Big|_1^2 = 10 - \frac{8}{3} + 2 - 5 + \frac{1}{3} - 4 = 3 - \frac{7}{3} = \frac{2}{3}$$

47.

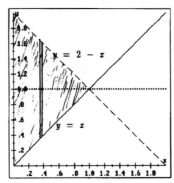

Reversing the order of integration gives a single region: $\displaystyle\int\limits_{0}^{1}\int\limits_{x}^{2-x} (x^2 + y^2)\ dy\ dx$

$$= \int\limits_{0}^{1}\left(x^2 y + \frac{y^3}{3}\right)\Big|_{x}^{2-x} dx = \int\limits_{0}^{1}\left(2x^2 - x^3 + \frac{(2 - x)^3}{3} - x^3 - \frac{x^3}{3}\right) dx$$

$$= \int\limits_{0}^{1}\left(2x^2 - \frac{7x^3}{3} + \frac{(2 - x)^3}{3}\right) dx = \left(\frac{2x^3}{3} - \frac{7x^4}{12} - \frac{(2 - x)^4}{12}\right)\Big|_{0}^{1}$$

$$= \frac{2}{3} - \frac{7}{12} - \frac{1}{12} + \frac{4}{3} = \frac{4}{3}$$

51.

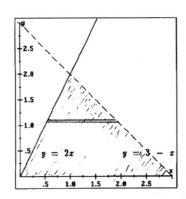

The point of intersection is $(1, 2)$. Horizontal strips will give a single region:

$$\int\limits_{0}^{2}\int\limits_{y/2}^{3-y} (x + 2y + 4)\ dx\ dy = \int\limits_{0}^{2}\left(\frac{x^2}{2} + 2xy + 4x\right)\Big|_{y/2}^{3-y} dy$$

$$= \int\limits_{0}^{2}\left(\frac{(3 - y)^2}{2} + 2y(3 - y) + 4(3 - y) - \frac{y^2}{8} - y^2 - 2y\right) dy$$

$$= \int\limits_{0}^{2}\left(-\frac{21}{8}y^2 - 3y + \frac{33}{2}\right) dy = \left(-\frac{7y^3}{8} - \frac{3y^2}{2} + \frac{33y}{2}\right)\Big|_{0}^{2}$$

$$= -7 - 6 + 33 = 20$$

55.   Analysis: The minimum value for $e^{y\,\sin\,x}$ over the specified region is ≈ 0.7865 somewhere around $(-0.48, 0.52)$, and the maximum value is ≈ 1.54 in the vicinity of $(0.855.\ 0.5725)$. A graphing calculator or computer will be required to find these values. Now since the base region is $\frac{3}{2}$ instead of 1 as specified in problem **54**, $\frac{3}{2}(0.78) \leq \int\int \leq \frac{3}{2}(1.54)$. The value of the integral is between 1.17 and 2.31.

13.3 Double Integrals in Polar Coordinates

3.

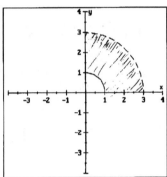

Set up $e^u\ du$ form: $-\dfrac{1}{2}\displaystyle\int_0^{\pi/2}\int_1^3 e^{-r^2}(-2r\,dr)\,d\theta = -\dfrac{1}{2}\displaystyle\int_0^{\pi/2} e^{-r^2}\Big|_1^3\,d\theta$

$= -\dfrac{1}{2}\displaystyle\int_0^{\pi/2}(e^{-9}-e^{-1})\,d\theta = -\dfrac{\pi}{4}(e^{-9}-e^{-1}) = \dfrac{\pi}{4}\Big(\dfrac{1}{e}-\dfrac{1}{e^9}\Big)$

7.

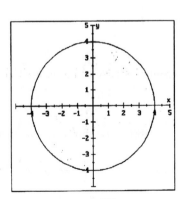

$\displaystyle\int_0^{2\pi}\int_0^4 2\,r^2\cos\theta\,dr\,d\theta = \int_0^{2\pi}(\cos\theta)\Big(\dfrac{2r^3}{3}\Big)\Big|_0^4\,d\theta = \dfrac{128}{3}\int_0^{2\pi}\cos\theta\,d\theta$

$= \dfrac{128}{3}\sin\theta\Big|_0^{2\pi} = 0$

SURVIVAL HINT

Since polar equations are often periodic [e.g. $f(0) = f(2\pi)$], it is sometimes a good idea to make use of any symmetry in the graph. Find the area of the smallest symmetric piece and multiply by the number of such pieces.

11. Using the symmetry about the polar axis: $2\displaystyle\int_0^{\pi} \int_0^{2(1-\cos\theta)} r\, dr\, d\theta$

$$= 2\int_0^{\pi} \frac{r^2}{2}\Big|_0^{2(1-\cos\theta)} d\theta = 4\int_0^{\pi}\left(1 - 2\cos\theta + \frac{1 + \cos 2\theta}{2}\right) d\theta$$

$$= 4\left(\frac{3\theta}{2} - 2\sin\theta + \frac{1}{4}\sin 2\theta\right)\Big|_0^{\pi} = 4\left(\frac{3\pi}{2}\right) = 6\pi$$

15. Finding the area of $\frac{1}{2}$ of one leaf and multiplying by 8: $8\displaystyle\int_0^{\pi/4} \int_0^{\cos 2\theta} r\, dr\, d\theta$

$$= 8\int_0^{\pi/4} \frac{r^2}{2}\Big|_0^{\cos 2\theta} d\theta = 4\int_0^{\pi/4} \frac{1 + \cos 4\theta}{2} d\theta = 4\left(\frac{\theta}{2} + \frac{\sin 4\theta}{8}\right)\Big|_0^{\pi/4}$$

$$= 4\left(\frac{\pi}{8}\right) = \frac{\pi}{2}$$

19. Using the symmetry about the polar axis, and the quarter circle:

$$2(\tfrac{1}{4}\pi r^2) + 2\int_{\pi/2}^{\pi} \int_0^{1+\cos\theta} r\, dr\, d\theta = \frac{\pi}{2} + 2\int_{\pi/2}^{\pi} \frac{r^2}{2}\Big|_0^{1+\cos\theta} d\theta$$

$$= \frac{\pi}{2} + \int_{\pi/2}^{\pi}\left(1 + 2\cos\theta + \frac{1 + \cos 2\theta}{2}\right) d\theta$$

$$= \frac{\pi}{2} + \left(\frac{3\theta}{2} + 2\sin\theta + \frac{\sin 4\theta}{4}\right)\Big|_{\pi/2}^{\pi} = \frac{\pi}{2} + \left(\frac{3\pi}{2} - \frac{3\pi}{4} - 2\right) = \frac{5\pi}{4} - 2$$

SURVIVAL HINT

One of the most common errors when converting from cartesian to polar integration is forgetting the factor of "r". $dy\, dx = dA = r\, dr\, d\theta$.

23.
$$\int_0^{2\pi}\int_0^a r^2 \sin^2\theta \; r \; dr \; d\theta = \int_0^{2\pi} (\sin^2\theta) \left.\frac{r^4}{4}\right|_0^a \; d\theta = \frac{a^4}{4}\int_0^{2\pi}\frac{1-\cos 2\theta}{2}\; d\theta$$

$$= \frac{a^4}{4}\left(\frac{\theta}{2}-\frac{\sin 2\theta}{4}\right)\Big|_0^{2\pi} = \frac{a^4\pi}{4}$$

27. Recall that $x^2 + y^2 = r^2$. $\displaystyle\int_0^{2\pi}\int_0^a \frac{1}{a+r} r \; dr \; d\theta$. Dividing:

$$= \int_0^{2\pi}\int_0^a \left(1-\frac{a}{r+a}\right)dr\;d\theta = \int_0^{2\pi}\left(r-a\ln(r+a)\right)\Big|_0^a \; d\theta$$

$$= \int_0^{2\pi}(a-a\ln 2a + a\ln a)\;d\theta = 2\pi a(1 + \ln a - \ln 2a)$$

$$= 2\pi a(1 + \ln \tfrac{1}{2}) \; \text{ or } \; 2\pi a(1-\ln 2)$$

31. The circle $x^2 + y^2 = 9$ in polar form is $r = 3$.

$$\int_0^{2\pi}\int_0^3 e^{r^2}r\;dr\;d\theta = \frac{1}{2}\int_0^{2\pi}e^{r^2}\Big|_0^3\;d\theta = \frac{e^9-1}{2}\int_0^{2\pi}d\theta = \pi(e^9-1)$$

35. When $z = 4$, $x^2 + y^2 = 4$ which is $r = 2$.

$$\int_0^{2\pi}\int_0^2 r^2 \; r \; dr \; d\theta = \int_0^{2\pi}\left.\frac{r^4}{4}\right|_0^2 \; d\theta = 4\int_0^{2\pi}d\theta = 8\pi$$

39.

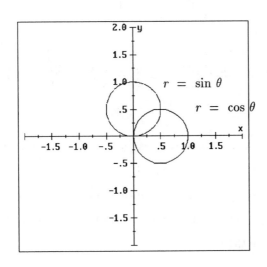

39. (con't.)

The curves intersect at $\theta = \frac{\pi}{4}$. Making use of the symmetry we will integrate the $\sin \theta$ curve from 0 to $\frac{\pi}{4}$ to get half of the desired area.

$$A = 2 \int_0^{\pi/4} \int_0^{a \sin \theta} r \, dr \, d\theta = 2 \int_0^{\pi/4} \frac{a^2 \sin^2 \theta}{2} \, d\theta = a^2 \int_0^{\pi/4} \frac{1 - \cos 2\theta}{2} \, d\theta$$

$$= a^2 \left(\frac{\theta}{2} - \frac{\sin 2\theta}{4} \right) \Big|_0^{\pi/4} = a^2 \left(\frac{\pi}{8} - \frac{1}{4} \right) = \frac{a^2}{8} (\pi - 2)$$

43. $\lim_{n \to \infty} \int_0^{\pi/2} \int_0^n \frac{1}{(1 + r^2)^2} r \, dr \, d\theta = -\frac{1}{2} \lim_{n \to \infty} \int_0^{\pi/2} \frac{1}{1 + r^2} \Big|_0^n \, d\theta$

$$= -\frac{1}{2} \lim_{n \to \infty} \int_0^{\pi/2} \left(\frac{1}{1 + n^2} - 1 \right) d\theta = -\frac{\pi}{4} \lim_{n \to \infty} \left(\frac{1}{1 + n^2} - 1 \right) = \frac{\pi}{4}$$

47. $x^2 + y^2$ and the circles are symmetric about $\theta = \frac{\pi}{4}$ so

$$V = 2 \int_0^{\pi/4} \int_0^{2 \sin \theta} r^2 \, r \, dr \, d\theta = 2 \int_0^{\pi/4} \frac{r^4}{4} \Big|_0^{2 \sin \theta} \, d\theta = 8 \int_0^{\pi/4} \left(\frac{1 - \cos 2\theta}{2} \right)^2 d\theta$$

$$= 2 \int_0^{\pi/4} \left(1 - 2 \cos 2\theta + \frac{1 + \cos 4\theta}{2} \right) d\theta = 2 \left(\frac{3\theta}{2} - \sin 2\theta + \frac{\sin 4\theta}{8} \right) \Big|_0^{\pi/4}$$

$$= 2 \left(\frac{3\pi}{8} - 1 + 0 \right) = \frac{3\pi}{4} - 2$$

51. Analysis: Let $u = \sqrt{2} \, x$, $du = \sqrt{2} \, dx$ then our integral becomes:

$$\int_0^\infty e^{-u^2} \frac{du}{\sqrt{2}} = \frac{1}{\sqrt{2}} \int_0^\infty e^{-u^2} du \quad \text{which by Problem } \mathbf{50c} = \frac{1}{\sqrt{2}} \left(\frac{\sqrt{\pi}}{2} \right) = \frac{\sqrt{2\pi}}{4}$$

13.4 Surface Area

SURVIVAL HINT

Do not try to evaluate any but the simplest of the radical expressions "in your head". You will make fewer errors if you take the time to write down the values of f_x, f_y, square them, add 1, and then take the root.

3.   When $z = 0$, $x^2 + y^2 = 4$ or $r = 2$. $f_x = -2x$, $f_y = -2y$.

$\sqrt{f_x^2 + f_y^2 + 1} = \sqrt{4x^2 + 4y^2 + 1} = \sqrt{4r^2 + 1}$.

$$S = \int_0^{2\pi} \int_0^2 \sqrt{4r^2 + 1}\, r\, dr\, d\theta. \text{ Set up } u^{1/2} du \text{ form:}$$

$$= \frac{1}{8} \int_0^{2\pi} \int_0^2 \sqrt{4r^2 + 1}\, (8r\, dr)\, d\theta = \frac{1}{8} \int_0^{2\pi} \frac{2(4r^2 + 1)^{3/2}}{3} \Big|_0^2 \, d\theta$$

$$= \frac{1}{12} \int_0^{2\pi} (17\sqrt{17} - 1)\, d\theta = \frac{\pi}{6}(17\sqrt{17} - 1)$$

7. $z = x^2 - 9$, $f_x = 2x$, $f_y = 0$, $\sqrt{f_x^2 + f_y^2 + 1} = \sqrt{4x^2 + 1}$.

$$S = \int_0^2 \int_0^2 \sqrt{4x^2 + 1}\, dy\, dx = \int_0^2 \sqrt{4x^2 + 1}\, (2\, dx) \text{ Using formula } \#168 \text{ with}$$

$u = 2x$ and $du = 2\, dx$:

$$= \left(x\sqrt{4x^2 + 1} + \frac{1}{2}\ln|2x + \sqrt{4x^2 + 1}| \right)\Big|_0^2$$

$$= 2\sqrt{17} + \frac{1}{2}\ln(4 + \sqrt{17})$$

SURVIVAL HINT

In any expression where you find a factor of $x^2 + y^2$ you should consider changing to polar coordinates.

11. $\sqrt{f_x^2 + f_y^2 + 1} = \sqrt{4r^2 + 1}$. In polar coordinates $x^2 + y^2 = 1$ is $r = 1$.

$$S = \int_0^{2\pi} \int_0^1 \sqrt{4r^2 + 1}\, r\, dr\, d\theta \text{ Setting up } u^{1/2} du \text{ form: } \frac{1}{8} \int_0^{2\pi} \int_0^1 \sqrt{4r^2 + 1}\, (8r\, dr)\, d\theta$$

$$= \frac{1}{8} \int_0^{2\pi} \frac{2(4r^2 + 1)^{3/2}}{3} \Big|_0^1 \, d\theta = \frac{1}{12} \int_0^{2\pi} (5\sqrt{5} - 1)\, d\theta = \frac{\pi}{6}(5\sqrt{5} - 1)$$

15. $z = \sqrt{4 - x^2 - y^2}$, $f_x = \dfrac{-x}{\sqrt{4 - x^2 - y^2}}$, $f_y = \dfrac{-y}{\sqrt{4 - x^2 - y^2}}$,

$$\sqrt{f_x^{\,2} + f_y^{\,2} + 1} = \sqrt{\frac{x^2 + y^2 + (4 - x^2 - y^2)}{4 - x^2 - y^2}} = \frac{2}{\sqrt{4 - x^2 - y^2}}$$

Changing to polar: $\sqrt{4 - x^2 - y^2}$ becomes $\sqrt{4 - r^2}$ and $x^2 + y^2 = 2y$ becomes $r^2 = 2\,r\sin\theta$ or $r = 2\sin\theta$. Using symmetry, the cylinder cuts through 4 octants of the sphere. We will multiply the first octant surface by 4.

$$S = 4\int_0^{\pi/2}\int_0^{2\sin\theta} \frac{2}{\sqrt{4 - r^2}}\, r\, dr\, d\theta = -4\int_0^{\pi/2}\int_0^{2\sin\theta} \frac{-2r\, dr}{\sqrt{4 - r^2}}\, d\theta$$

$$= -4\int_0^{\pi/2} \frac{(4 - r^2)^{1/2}}{\frac{1}{2}}\Bigg|_0^{2\sin\theta}\, d\theta = -8\int_0^{\pi/2} \left(\sqrt{4 - 4\sin^2\theta} - 2\right) d\theta$$

$$= -8\int_0^{\pi/2} (2\cos\theta - 2)\, d\theta = -16(\sin\theta - \theta)\Big|_0^{\pi/2}$$

$$= -16\left(1 - \frac{\pi}{2}\right) = 8\pi - 16$$

19. $z = 9 - x^2 - y^2$. $\sqrt{f_x^{\,2} + f_y^{\,2} + 1} = \sqrt{4x^2 + 4y^2 + 1}$.
If $z = 0$ then $x^2 + y^2 = 3^2$. Changing to polar: $z = \sqrt{4r^2 + 1}$ and the region in the xy-plane is the circle $r = 3$.

$$S = \int_0^{2\pi}\int_0^3 \sqrt{4r^2 + 1}\, r\, dr\, d\theta = \frac{1}{8}\int_0^{2\pi}\int_0^3 \sqrt{4r^2 + 1}\,(8r\, dr)\, d\theta$$

$$= \frac{1}{8}\int_0^{2\pi} \frac{2(4r^2 + 1)^{3/2}}{3}\Big|_0^3\, d\theta = \frac{1}{12}\int_0^{2\pi} (37\sqrt{37} - 1)\, d\theta = \frac{\pi}{6}(37\sqrt{37} - 1)$$

23. Finding the intercepts: $x = \dfrac{D}{A}$, $y = \dfrac{D}{B}$, $z = \dfrac{D}{C}$. The equation of the line in the first quadrant is $y = -\dfrac{A}{B}x + \dfrac{D}{B}$. $z = \dfrac{D - Ax - By}{C}$

$$\sqrt{f_x^{\,2} + f_y^{\,2} + 1} = \sqrt{\frac{A^2}{C^2} + \frac{B^2}{C^2} + 1} = \frac{\sqrt{A^2 + B^2 + C^2}}{C}$$

23. (con't.)

$$S = \int_0^{D/A} \int_0^{-\frac{A}{B}x+\frac{D}{B}} \frac{\sqrt{A^2 + B^2 + C^2}}{C}\, dy\, dx$$

$$= \frac{\sqrt{A^2 + B^2 + C^2}}{C} \int_0^{D/A} \left(-\frac{A}{B}x + \frac{D}{B}\right) dx$$

$$= \frac{\sqrt{A^2 + B^2 + C^2}}{C} \left(-\frac{Ax^2}{2B} + \frac{Dx}{B}\right)\Big|_0^{D/A}$$

$$= \frac{\sqrt{A^2 + B^2 + C^2}}{C}\left(-\frac{D^2}{2AB} + \frac{D^2}{AB}\right) = \frac{D^2}{2ABC}\sqrt{A^2 + B^2 + C^2}$$

27. $x^2 + y^2 + z^2 = a^2. \quad z = \sqrt{a^2 - x^2 - y^2}$

$$f_x = \frac{-x}{\sqrt{a^2 - x^2 - y^2}}, \quad f_y = \frac{-y}{\sqrt{a^2 - x^2 - y^2}},$$

$$\sqrt{f_x^2 + f_y^2 + 1} = \sqrt{\frac{x^2 + y^2 + (a^2 - x^2 - y^2)}{a^2 - x^2 - y^2}} = \frac{a}{\sqrt{a^2 - x^2 - y^2}}$$

Using polar coordinates and symmetry with just the first octant:

$$S = 8\int_0^{\pi/2}\int_0^a \frac{a}{\sqrt{a^2 - r^2}}\, r\, dr\, d\theta = -4a\int_0^{\pi/2}\int_0^a \frac{(-2\,r)\, dr}{\sqrt{a^2 - r^2}}\, d\theta$$

$$= -4a\int_0^{\pi/2} \frac{(a^2 - r^2)^{1/2}}{\frac{1}{2}}\Big|_0^a\, d\theta = -8a\int_0^{\pi/2}(-a)\, d\theta = 8a^2\left(\frac{\pi}{2}\right) = 4\pi a^2$$

31. Find the intersection of the two surfaces by substituting $x^2 + y^2 = 4z$:

$4z + z^2 = 9z$, $z^2 - 5z = 0$, $z = 0, 5$.

When $z = 0$ the paraboloid and the sphere are tangent. (The sphere

has center at $(0, 0, \frac{9}{2})$ and a radius of $\frac{9}{2}$.) When $z = 5$ the intersection is the

circle $x^2 + y^2 = 20$. The surface we wish to find lies above $x^2 + y^2 = 20$

In polar form $r = 2\sqrt{5}$. $x^2 + y^2 + (z - \frac{9}{2})^2 = \frac{81}{4}$,

31. (con't.)

$$z = \sqrt{\tfrac{81}{4} - x^2 - y^2} + \tfrac{9}{2}. \quad f_x = \frac{-x}{\sqrt{\tfrac{81}{4} - x^2 - y^2}} \quad f_y = \frac{-y}{\sqrt{\tfrac{81}{4} - x^2 - y^2}}$$

$$\sqrt{f_x^{\,2} + f_y^{\,2} + 1} = \sqrt{\frac{x^2 + y^2 + \tfrac{81}{4} - x^2 - y^2}{\tfrac{81}{4} - x^2 - y^2}} = \sqrt{\frac{\tfrac{81}{4}}{\frac{81 - 4x^2 - 4y^2}{4}}}$$

$$= \frac{9}{\sqrt{81 - 4x^2 - 4y^2}}. \quad \text{Using polar coordinates:}$$

$$S = \int_0^{2\pi}\int_0^{2\sqrt{5}} \frac{9}{\sqrt{81 - 4r^2}}\, r\, dr\, d\theta = -\frac{9}{8}\int_0^{2\pi}\int_0^{2\sqrt{5}} \frac{(-8\,r)\, dr}{\sqrt{81 - 4r^2}}\, d\theta$$

$$= -\frac{9}{8}\int_0^{2\pi} 2\sqrt{81 - 4r^2}\,\Big|_0^{2\sqrt{5}}\, d\theta = -\frac{9}{4}\int_0^{2\pi} (1 - 9)\, d\theta = 18(2\pi) = 36\pi$$

35. $z = \cos(x^2 + y^2)$. $f_x = -2x\sin(x^2 + y^2)$, $f_y = -2y\sin(x^2 + y^2)$,

$$\sqrt{f_x^{\,2} + f_y^{\,2} + 1} = \sqrt{(4x^2 + 4y^2)\sin^2(x^2 + y^2) + 1}$$

In polar coordinates the region is a circle with radius of $\sqrt{\tfrac{\pi}{2}}$ and the surface integral is:

$$S = \int_0^{2\pi}\int_0^{\sqrt{\pi/2}} \sqrt{4r^2\sin^2 r^2 + 1}\ r\, dr\, d\theta$$

39. $\boldsymbol{R_u} \times \boldsymbol{R_v} = \begin{vmatrix} \mathbf{i} & \mathbf{j} & \mathbf{k} \\ 2\sin v & 2\cos v & 2u \\ 2u\cos v & -2u\sin v & 0 \end{vmatrix}$

$$= 4u^2\sin v\,\mathbf{i} + 4u^2\cos v\,\mathbf{j} + (-4u\sin^2 v - 4u\cos^2 v)\,\mathbf{k}$$
$$= 4u^2\sin v\,\mathbf{i} + 4u^2\cos v\,\mathbf{j} - 4u\,\mathbf{k}$$

$$\|\boldsymbol{R_u} \times \boldsymbol{R_v}\| = \sqrt{16u^4\sin^2 v + 16u^4\cos^2 v + 16u^2} = \sqrt{16u^4 + 16u^2}$$
$$= 4|u|\sqrt{u^2 + 1}$$

43. $R(u, v) = uv \mathbf{i} + (u - v)\mathbf{j} + (u + v)\mathbf{k}$ over the disk $u^2 + v^2 \leq 1$.

$R_u = v\mathbf{i} + \mathbf{j} + \mathbf{k}, \ R_v = u\mathbf{i} - \mathbf{j} + \mathbf{k}$

$$R_u \times R_v = \begin{vmatrix} \mathbf{i} & \mathbf{j} & \mathbf{k} \\ v & 1 & 1 \\ u & -1 & 1 \end{vmatrix} = 2\mathbf{i} + (u - v)\mathbf{j} - (u + v)\mathbf{k}$$

$\|R_u \times R_v\| = \sqrt{4 + (u - v)^2 + (-u - v)^2} = \sqrt{4 + 2u^2 + 2v^2}$

Using polar coordinates over the unit disk we have:

$$S = \int_0^{2\pi} \int_0^1 \sqrt{4 + 2r^2} \ r \, dr \, d\theta = \frac{1}{4} \int_0^{2\pi} \int_0^1 \sqrt{4 + 2r^2} \ (4 \, r \, dr) \, d\theta$$

$$= \frac{1}{4} \int_0^{2\pi} \frac{2(4 + 2r^2)^{3/2}}{3} \Big|_0^1 \, d\theta = \frac{1}{6} \int_0^{2\pi} (6\sqrt{6} - 8) \, d\theta = \frac{\pi}{3}(6\sqrt{6} - 8)$$

$$= \frac{2\pi}{3}(3\sqrt{6} - 4) \approx 7.01$$

45. Analysis: The intersection of the torus and the xy-plane is the ring with inner radius of $a - b$ and outer radius of $a + b$. Doubling the surface above the plane:

$$S = 2 \int_0^{2\pi} \int_0^{2\pi} \|R_u \times R_v\| \, du \, dv. \text{ This is tedious, but straight-forward. The result}$$

is easily verified with the theorem of Pappus.

47. Analysis: A patch on the surface is approximated by a rectangle on the tangent plane over the corresponding partition of R. If N is the normal to the tangent plane, and $\mathbf{u}$ is a normal to R, then $\dfrac{R_\Delta}{S_\Delta} = \cos\theta$ where θ is the angle between N and $\mathbf{u}$. (Draw a figure and play some geometry with the angles.) Since $N = \nabla f$, $N \bullet \mathbf{u} = \nabla f \bullet \mathbf{u} = \cos\theta$. The ratio is a Riemann Sum of $S = \dfrac{R}{\cos\theta}$.

13.5 Triple Integrals

3. $\displaystyle\int_1^4 \int_{-2}^3 \int_2^5 dx \, dy \, dz = \int_1^4 \int_{-2}^3 x \Big|_1^4 \, dy \, dz = 3 \int_1^4 \int_{-2}^3 dy \, dz = 3 \int_1^4 y \Big|_{-2}^3 \, dz = 3(5) \int_1^4 dz$

$$= 3(5)z \Big|_1^4 = 3(5)(3) = 45$$

7.
$$\int_0^2 \int_0^x \int_0^{x+y} xyz \; dz \; dy \; dx \;=\; \int_0^2 \int_0^x xy\frac{z^2}{2}\Big|_0^{x+y} dy \; dx$$

$$= \frac{1}{2}\int_0^2 \int_0^x (x^3 y + 2x^2 y^2 + xy^3) \; dy \; dx = \frac{1}{2}\int_0^2 \left(\frac{x^3 y^2}{2} + \frac{2x^2 y^3}{3} + \frac{xy^4}{4}\right)\Big|_0^x dx$$

$$= \frac{1}{2}\int_0^2 \left(\frac{x^5}{2} + \frac{2x^5}{3} + \frac{x^5}{4}\right) dx = \frac{1}{2}\Big(\frac{17}{12}\Big)\int_0^2 x^5 dx = \frac{17}{24}\frac{x^6}{6}\Big|_0^2 = \frac{68}{9}$$

11.
$$\int_0^1 \int_0^y \int_0^{\ln y} e^{z+2x} \; dz \; dx \; dy \;=\; \int_0^1 \int_0^y e^{z+2x}\Big|_0^{\ln y} dx \; dy \;=\; \int_0^1 \int_0^y (ye^{2x} - e^{2x}) \; dx \; dy$$

$$\int_0^1 \int_0^y e^{2x}(y - 1) \; dx \; dy \;=\; \int_0^1 (y - 1)\frac{e^{2x}}{2}\Big|_0^y dy \;=\; \frac{1}{2}\int_0^1 (ye^{2y} - e^{2y} - y + 1) \; dy$$

$$= \frac{1}{2}\Big(\frac{ye^{2y}}{2} - \frac{e^{2y}}{4} - \frac{e^{2y}}{2} - \frac{y^2}{2} + y\Big)\Big|_0^1$$

$$= \frac{1}{2}\Big(\frac{e^2}{2} - \frac{e^2}{4} - \frac{e^2}{2} - \frac{1}{2} + 1 + \frac{1}{4} + \frac{1}{2}\Big) = -\frac{e^2}{8} + \frac{5}{8} = \frac{1}{8}(5 - e^2)$$

15.
$$\int_1^3 \int_{-1}^1 \int_2^4 (x^2 y + y^2 z) \; dz \; dy \; dx \;=\; \int_1^3 \int_{-1}^1 \Big(x^2 yz + \frac{y^2 z^2}{2}\Big)\Big|_2^4 dy \; dx$$

$$\int_1^3 \int_{-1}^1 (2x^2 y + 6y^2) \; dy \; dx \;=\; \int_1^3 (x^2 y^2 + 2y^3)\Big|_{-1}^1 dx \;=\; \int_1^3 4 \; dx \;=\; 4x\Big|_1^3 = 8$$

19.   This is the two octants of the unit sphere in which y and z are positive.

$$\int_0^1 \int_0^{\sqrt{1-z^2}} \int_{-\sqrt{1-x^2-y^2}}^{\sqrt{1-x^2-y^2}} xyz \; dx \; dy \; dz \;=\; \int_0^1 \int_0^{\sqrt{1-z^2}} \frac{x^2 yz}{2}\Big|_{-\sqrt{1-x^2-y^2}}^{\sqrt{1-x^2-y^2}} dy \; dz \;=\; 0$$

This is an expected answer since the base regions are symmetric and the function is positive in one and negative in the other.

SURVIVAL HINT

It takes considerable skill and practice to determine which of the six possible permutations of dV will be "easiest" for a given problem. The most common situation is two functions of the form $z = f(x, y)$; in which case you would use $dz\ dA$, where the region A is determined by the intersection of the two functions. For other iterations be certain that the limits for the first integration are either constants or functions of the remaining two variables, the limits of the second integration are constants or functions of the remaining variable, and the limits for the final integration are constants.

23.

$$\int_0^1 \int_0^{-x+1} \int_0^{1-x-y} dz\ dy\ dx = \int_0^1 \int_0^{-x+1} (1 - x - y)\ dy\ dx$$

$$= \int_0^1 \int_0^{-x+1} \left(y - xy - \frac{y^2}{2}\right)\Big|_0^{-x+1} dx$$

$$= \int_0^1 \left(-x + 1 - x(-x + 1) - \frac{(-x + 1)^2}{2}\right) dx$$

$$= \int_0^1 \left(\frac{x^2}{2} - x + \frac{1}{2}\right) dx = \left(\frac{x^3}{6} - \frac{x^2}{2} + \frac{x}{2}\right)\Big|_0^1 = \frac{1}{6}$$

This result is easily verified with the formula for the volume of a pyramid.

27. Find the intersection of the parabolic cylinder and the elliptic paraboloid:

$$4 - y^2 = x^2 + 3y^2, \quad x^2 + 4y^2 = 4, \quad x = 2\sqrt{1 - y^2}$$

Using the first octant volume and symmetry:

$$V = 4 \int_0^1 \int_0^{2\sqrt{1-y^2}} \int_{x^2+3y^2}^{4-y^2} dz\ dx\ dy = 4 \int_0^1 \int_0^{2\sqrt{1-y^2}} (4 - x^2 - 4y^2)\ dx\ dy$$

$$= 4 \int_0^1 \left(4x - \frac{x^3}{3} - 4y^2 x\right)\Big|_0^{2\sqrt{1-y^2}} dy$$

$$= 4 \int_0^1 \left(8\sqrt{1 - y^2} - \frac{8(1 - y^2)^{3/2}}{3} - 8y^2\sqrt{1 - y^2}\right) dy$$

27. (con't.)

Make the substitution $y = \sin \theta$, $dy = \cos \theta \, d\theta$:

$$= 4 \int_0^{\pi/2} \left(8 \cos \theta - \frac{8}{3} \cos^3\theta - 8 \sin^2\theta \cos \theta \right) \cos \theta \, d\theta$$

$$= 32 \int_0^{\pi/2} \left(\cos^2\theta - \frac{1}{3} \cos^4\theta - \sin^2\theta \cos^2\theta \right) d\theta$$

$$= 32 \int_0^{\pi/2} \left((1 - \sin^2\theta)\cos^2\theta - \frac{1}{3} \cos^4\theta \right) d\theta = 32 \int_0^{\pi/2} \frac{2}{3} \cos^4\theta \, d\theta$$

$$= \frac{64}{3} \int_0^{\pi/2} \left(\frac{1 + \cos 2\theta}{2} \right)^2 d\theta = \frac{16}{3} \int_0^{\pi/2} \left(1 + 2 \cos 2\theta + \frac{1 + \cos 4\theta}{2} \right) d\theta$$

$$= \frac{16}{3} \left(\frac{3\theta}{2} + \sin 2\theta + \frac{\sin 4\theta}{8} \right) \Big|_0^{\pi/2} = \frac{16}{3} \left(\frac{3\pi}{4} \right) = 4\pi$$

31. This is the first octant (since all limits are positive) of the sphere
$x^2 + y^2 + z^2 = 4$. For the iteration $dx \, dy \, dz$, z goes from 0 to 2,
in the yz-plane y goes from 0 to the quarter circle $y = \sqrt{4 - z^2}$,
and x goes from the yz-plane to the surface of the sphere $x = \sqrt{4 - y^2 - z^2}$.

$$\int_0^2 \int_0^{\sqrt{4 - z^2}} \int_0^{\sqrt{4 - y^2 - z^2}} f(x, y, z) \, dx \, dy \, dz$$

35. Finding the intersection of the surfaces, and thus the projected R:
$16 - x^2 - 2y^2 = 3x^2 + 2y^2$, $4x^2 + 4y^2 = 16$, $x^2 + y^2 = 4$.
$z = 3x^2 + 2y^2$ is an elliptic paraboloid opening upward and is the lower bound.
$z = 16 - x^2 - 2y^2$ is an elliptic paraboloid opening downward and is the
upper bound. Using the first octant and symmetry:

$$V = 4 \int_0^2 \int_0^{\sqrt{4-x^2}} \int_{3x^2+2y^2}^{16-x^2-2y^2} dz \, dy \, dx = 4 \int_0^2 \int_0^{\sqrt{4-x^2}} (16 - 4x^2 - 4y^2) \, dy \, dx$$

$$= 16 \int_0^2 \int_0^{\sqrt{4-x^2}} (4 - x^2 - y^2) \, dy \, dx. \text{ Switching to polar coordinates:}$$

$$= 16 \int_0^{\pi/2} \int_0^2 (4 - r^2) \, r \, dr \, d\theta = 16 \int_0^{\pi/2} \left(2r^2 - \frac{r^4}{4} \right) \Big|_0^2 d\theta = 16 \int_0^{\pi/2} 4 \, d\theta = 32\pi$$

39. Analysis: For $x^2 + y^2 + z^2 = R^2$ (upper case R to avoid confusion with r in polar coordinates), the region in the xy-palne is the circle $x^2 + y^2 = R^2$.

Using the first octant and polar coordinates:

$$V = 8 \int_0^{\pi/2} \int_0^R \int_0^{\sqrt{R^2-r^2}} dz \, r \, dr \, d\theta, \text{ which should give the standard formula for}$$

the volume of a sphere.

41. Analysis: In the xy-plane $\left(\frac{y}{b}\right)^2 = 1 - \left(\frac{x}{a}\right)^2 = \frac{a^2 - x^2}{a^2}$, $y = \frac{b}{a}\sqrt{a^2 - x^2}$.

Using the first octant volume and symmetry:

$$V = 8 \int_0^a \int_0^{\frac{b}{a}\sqrt{a^2 - x^2}} \int_0^{\frac{c}{ab}\sqrt{a^2b^2 - b^2x^2 - a^2y^2}} dz \, dy \, dx$$

This will require trigonometric substitution and/or the table of formulas.

43. Analysis: The intercepts are: $x = a$, $y = b$, and $z = c$. The boundary line in the xy-plane is: $y = -\frac{b}{a}x + b$ or $x = -\frac{a}{b}y + a$. In the xz-plane: $z = -\frac{c}{a}x + c$ or $x = -\frac{a}{c}z + a$. In the yz-plane: $z = -\frac{c}{b}y + c$ or $y = -\frac{b}{c}z + b$. One possible iteration:

$$\int_0^a \int_0^{-\frac{b}{a}x + b} \int_0^{c\left(1 - \frac{x}{a} - \frac{y}{b}\right)} dz \, dy \, dx.$$

45. Analysis: Let u be the horizontal axis, v be the vertical axis, and consider the triangular region determined by $u = 0$, $v = x$, and $v = u$: $\int_0^x \int_0^v f(u) \, du \, dv$

Switching the order of integration this becomes: $\int_0^x \int_u^x f(u) \, dv \, du$

$$= \int_0^x (x - u) \, f(u) \, du. \text{ But } u \text{ is a dummy variable, so this can be written as:}$$

$\int_0^x (x - t) \, f(t) \, dt.$ In like manner, consider the tetrahedral region determined by the above triangle and the line $u = w$: $\int_0^x \int_0^v \int_0^u f(w) \, dw \, du \, dv$

45. (con't.)

Switching the order of integration this becomes: $\displaystyle\int_0^x \int_w^x \int_w^v f(w) \; du \; dv \; dw$

$$= \int_0^x \int_w^x (v - w) \, f(w) \; dv \; dw = \int_0^x \frac{1}{2}(v - w)^2 \Big|_w^x \; dw = \int_0^x \frac{1}{2} \, f(w) \, (x - w)^2 \; dw$$

But w is a dummy variable, so this can be written as: $\displaystyle\frac{1}{2}\int_0^x (x - t)^2 \, f(t) \; dt$

47.   Analysis: $\displaystyle\int_1^2 \int_{-1}^1 \int_0^2 \int_0^1 xyz^2w^2 \; dx \; dy \; dz \; dw$

Since all of the limits are constants, each variable can be integrated one at a time, in effect treating the other variables as constants.

13.6 Mass, Moments, and Probability Density Functions

SURVIVAL HINT

When finding the center of mass for a lamina you probably thought of the moment about a particular axis as a rotational force. The moment about the y-axis is the same as all of the mass at the centroid: $M_y = m\bar{x}$. The algebraic extension of this concept to volumes is exactly the same. The moment about the xz-plane is the same as all of the mass at the centroid: $M_{yz} = m\bar{x}$. However, the geometric visualization of "rotation" is impossible. A lamina can be rotated about a line by moving in $\mathbb{R}^3$. To "rotate" our volume about a plane we have to move into $\mathbb{R}^4$. Fortunately the algebra is not restricted by our $\mathbb{R}^3$ world.

7. $\bar{x} = \dfrac{M_y}{M}, \; \bar{y} = \dfrac{M_x}{M}, \; \rho = 2$ or homogeneous, the ρ will cancel.

The curves intersect at $(2, 4)$.

$$M = 2 \int_0^2 \int_{x^2}^{2x} dy \; dx = 2 \int_0^2 (2x - x^2) \; dx = 2 \Big(x^2 - \frac{x^3}{3} \Big) \Big|_0^2 = 2 \Big(\frac{4}{3} \Big) = \frac{8}{3}$$

$$M_x = 2 \int_0^2 \int_{x^2}^{2x} y \; dy \; dx = 2 \int_0^2 \frac{y^2}{2} \Big|_{x^2}^{2x} \; dx = \int_0^2 (4x^2 - x^4) \; dx = \Big(\frac{4x^3}{3} - \frac{x^5}{5} \Big) \Big|_0^2$$

$$= 32 \Big(\frac{1}{3} - \frac{1}{5} \Big) = \frac{64}{15}$$

7. (con't.)

$$M_y = 2\int_0^2 \int_{x^2}^{2x} x \, dy \, dx = 2\int_0^2 xy \Big|_{x^2}^{2x} \, dx = 2\int_0^2 (2x^2 - x^3) \, dx = \left(\frac{4x^3}{3} - \frac{x^4}{2}\right)\Big|_0^2$$

$$= \frac{32}{3} - 8 = \frac{8}{3}$$

$$\bar{y} = \frac{\frac{64}{15}}{\frac{8}{3}} = \frac{8}{5} \qquad \bar{x} = \frac{\frac{8}{3}}{\frac{8}{3}} = 1 \quad \text{The centroid is at } \left(1, \frac{8}{5}\right)$$

11. $$\bar{x} = \frac{M_{yz}}{M} = \frac{4\displaystyle\int_0^4 \int_0^{-x+4} \int_0^{4-x-y} x \, dz \, dy \, dx}{4\displaystyle\int_0^4 \int_0^{-x+4} \int_0^{4-x-y} dz \, dy \, dx}$$

$$= \frac{\displaystyle\int_0^4 \int_0^{-x+4} (4x - x^2 - xy) \, dy \, dx}{\text{Vol. of a tetrahedron}}$$

$$= \frac{\displaystyle\int_0^4 \left((4x - x^2)(-x + 4) - \frac{x(-x+4)^2}{2}\right) dx}{\frac{32}{3}}$$

$$= \frac{3}{32}\int_0^4 \left(\frac{x^3}{2} - 4x^2 + 8x\right) dx = \frac{3}{32}\left(\frac{x^4}{8} - \frac{4x^3}{3} + 4x^2\right)\Big|_0^4$$

$$= \frac{3}{32}\left(\frac{256}{8} - \frac{256}{3} + 64\right) = \frac{3}{32}(256)\left(\frac{1}{8} - \frac{1}{3} + \frac{1}{4}\right) = \frac{3}{32}(256)\left(\frac{1}{24}\right) = 1$$

By symmetry $\bar{x} = \bar{y} = \bar{z} = 1$ The centroid is at $(1, 1, 1)$

15.   Using two regions, the equations of the upper boundary lines are:

$$y = \frac{5}{6}x \text{ or } x = \frac{6}{5}y \text{ and } y = -\frac{5}{6}x + 10 \text{ or } x = -\frac{6}{5}y + 12.$$

$$M = \int_0^6 \int_0^{5x/6} 7x \, dy \, dx + \int_6^{12} \int_0^{-5x/6+10} 7x \, dy \, dx$$

$$= \int_0^6 \frac{35}{6}x^2 \, dx + \int_6^{12}\left(-\frac{35x^2}{6} + 70x\right) \cdot dx$$

$$= \frac{35}{6}\left(\frac{x^3}{3}\right)\Big|_0^6 + \left(-\frac{35}{6}\frac{x^3}{3} + \frac{70x^2}{2}\right)\Big|_6^{12}$$

$$= 420 - 3{,}360 + 5{,}040 + 420 - 1{,}260 = 1{,}260$$

15. (con't.)

$$M_x = \int_0^6 \int_0^{5x/6} 7xy \, dy \, dx \; + \; \int_6^{12} \int_0^{-5x/6+10} 7xy \, dy \, dx$$

$$= \int_0^6 \frac{7xy^2}{2}\Big|_0^{5x/6} dx \; + \; \int_6^{12} \frac{7xy^2}{2}\Big|_0^{-5x/6+10} dx$$

$$= \int_0^6 \frac{7x}{2}\Big(\frac{5x}{6}\Big)^2 dx \; + \; \int_6^{12} \frac{7x}{2}\Big(-\frac{5x}{6} + 10\Big)^2 dx$$

$$= \int_0^6 \frac{175}{72}x^3 \; + \; \int_6^{12}\Big(\frac{175}{72}x^3 - \frac{175}{3}x^2 + 350x\Big) dx$$

$$= \frac{175}{72}\Big(\frac{x^4}{4}\Big)\Big|_0^6 \; + \; 175\Big(\frac{1}{72}\frac{x^4}{4} - \frac{1}{3}\frac{x^3}{3} + x^2\Big)\Big|_6^{12}$$

$$= \frac{1{,}575}{2} \; + \; 175(72 - 192 + 144 - \frac{9}{2} + 24 - 36) = 2{,}100$$

For M_y we will use horizontal strips and a single region.

$$M_y = 7\int_0^5 \int_{6y/5}^{-\frac{6}{5}y+12} x^2 \, dx \, dy = 7\int_0^5 \frac{x^3}{3}\Big|_{6y/5}^{-\frac{6}{5}y+12} dy$$

$$= \frac{7}{3}\int_0^5 \Big\{\Big(-\frac{6y}{5} + 12\Big)^3 - \Big(\frac{6y}{5}\Big)^3\Big\} dy = \frac{7(6^3)}{3}\int_0^5 \Big\{\Big(-\frac{y}{5} + 2\Big)^3 - \Big(\frac{y}{5}\Big)^3\Big\} dy$$

$$= \frac{7(6^3)}{3}\int_0^5 \Big(-\frac{2y^3}{5^3} + \frac{6y^2}{5^2} - \frac{12y}{5} + 8\Big) dy$$

$$= \frac{7(6^3)}{3}\Big(-\frac{y^4}{2(5^3)} + \frac{2y^3}{5^2} - \frac{6y^2}{5} + 8y\Big)\Big|_0^5$$

$$= \frac{7(6^3)}{3}\Big(-\frac{5}{2} + 10 - 30 + 40\Big) = \frac{7(6^3)}{3}\Big(\frac{35}{2}\Big) = 8{,}820$$

$$\bar{y} = \frac{2{,}100}{1{,}260} = \frac{5}{3} \quad \bar{x} = \frac{8{,}820}{1{,}260} = 7 \quad \text{The centroid is at } (7, \frac{5}{3}).$$

19.   **a.**   By symmetry $\bar{x} = 0$. $\rho = kr$.

$$M = k \int\limits_0^\pi \int\limits_0^a (r)\, r\, dr\, d\theta = \frac{ka^3}{3} \int\limits_0^\pi d\theta = \frac{k\pi a^3}{3}$$

$$M_x = k \int\limits_0^\pi \int\limits_0^a (r)(r\sin\theta)\, r\, dr\, d\theta = \frac{ka^4}{4} \int\limits_0^\pi \sin\theta\, d\theta = \frac{ka^4}{4}(-\cos\theta)\Big|_0^\pi$$

$$= \frac{ka^4}{4}(1 + 1) = \frac{ka^4}{2}$$

$$\bar{y} = \frac{\dfrac{ka^4}{2}}{\dfrac{k\pi a^3}{3}} = \frac{3a}{2\pi} \quad \text{The centroid is at } (0, \tfrac{3a}{2\pi})$$

b. $\rho = k\theta$ (use $\rho = \theta$ as the k will cancel in the quotient).

$$M = \int\limits_0^\pi \int\limits_0^a (\theta)\, r\, dr\, d\theta = \frac{a^2}{2} \int\limits_0^\pi \theta\, d\theta = \frac{\pi^2 a^2}{4}$$

$$M_x = \int\limits_0^\pi \int\limits_0^a (\theta)(r\sin\theta)\, r\, dr\, d\theta = \frac{a^3}{3} \int\limits_0^\pi \theta\sin\theta\, d\theta$$

For this use integration by parts; $u = \theta,\ dv = \sin\theta\, d\theta$

$$= \frac{a^3}{3}(\ -\theta\cos\theta + \sin\theta)\Big|_0^\pi = \frac{a^3}{3}(\pi - 0) = \frac{\pi a^3}{3}$$

$$M_y = \int\limits_0^\pi \int\limits_0^a (\theta)(r\cos\theta)\, r\, dr\, d\theta = \frac{a^3}{3} \int\limits_0^\pi \theta\cos\theta\, d\theta$$

Again use integration by parts; $u = \theta,\ dv = \cos\theta\, d\theta$

$$= \frac{a^3}{3}(\theta\sin\theta + \cos\theta)\Big|_0^\pi = \frac{a^3}{3}(-1 - 1) = -\frac{2a^3}{3}$$

$$\bar{x} = \frac{-\dfrac{2a^3}{3}}{\dfrac{\pi^2 a^2}{4}} = -\frac{8a}{3\pi^2} \qquad \bar{y} = \frac{\dfrac{\pi a^3}{3}}{\dfrac{\pi^2 a^2}{4}} = \frac{4a}{3\pi}$$

Thus, the centroid is at $\left(-\dfrac{8a}{3\pi^2},\ \dfrac{4a}{3\pi}\right)$

23. $\quad \rho = x^2.\ I_x = 2\displaystyle\int_0^1 \int_0^{1-x^2} x^2 y^2\ dy\ dx = \frac{2}{3}\int_0^1 x^2(1 - x^2)^3\ dx$

$\quad\quad = \dfrac{2}{3}\displaystyle\int_0^1 (-x^8 + 3x^6 - 3x^4 + x^2)\ dx = \dfrac{2}{3}\left(-\dfrac{x^9}{9} + \dfrac{3x^7}{7} - \dfrac{3x^5}{5} + \dfrac{3x^3}{3}\right)\Big|_0^1$

$\quad\quad = \dfrac{2}{3}\left(\dfrac{-35 + 135 - 189 + 105}{9(7)(5)}\right) = \dfrac{2}{3}\left(\dfrac{16}{9(7)(5)}\right) = \dfrac{32}{945}$

27.   By symmetry $\bar{y} = 0$. $\bar{x}$ for the upper half and the lower half are equal, so consider the upper half. $1 + 2\cos\theta = 0$ at $\cos\theta = -\frac{1}{2}$, $\theta = \frac{2\pi}{3}$.

$M = \displaystyle\int_0^{2\pi/3}\int_0^{1+2\cos\theta} r\ dr\ d\theta = \frac{1}{2}\int_0^{2\pi/3}(1 + 2\cos\theta)^2\ d\theta$

$\quad = \dfrac{1}{2}\displaystyle\int_0^{2\pi/3}(1 + 4\cos\theta + 4\cos^2\theta)\ d\theta$

$\quad = \dfrac{1}{2}\displaystyle\int_0^{2\pi/3}(1 + 4\cos\theta + 2 + 2\cos 2\theta)\ d\theta$

$\quad = \dfrac{1}{2}(3\theta + 4\sin\theta + \sin 2\theta)\Big|_0^{2\pi/3} = \dfrac{1}{2}\left(2\pi + 2\sqrt{3} - \dfrac{\sqrt{3}}{2}\right) = \pi + \dfrac{3\sqrt{3}}{4}$

$M_y = \displaystyle\int_0^{2\pi/3}\int_0^{1+2\cos\theta}(r\cos\theta)\ r\ dr\ d\theta = \frac{1}{3}\int_0^{2\pi/3}(\cos\theta)(1 + 2\cos\theta)^3\ d\theta$

$\quad = \dfrac{1}{3}\displaystyle\int_0^{2\pi/3}(\cos\theta)(1 + 6\cos\theta + 12\cos^2\theta + 8\cos^3\theta)\ d\theta$

$\quad = \dfrac{1}{3}\displaystyle\int_0^{2\pi/3}\left\{\cos\theta + \dfrac{6(1 + \cos 2\theta)}{2} + 12(1 - \sin^2\theta)\cos\theta + 8\left(\dfrac{1 + \cos 2\theta}{2}\right)^2\right\}\ d\theta$

$\quad = \dfrac{1}{3}\Big\{\sin\theta + 3\theta + \dfrac{3}{2}\sin 2\theta + 12\sin\theta - \dfrac{12\sin^3\theta}{3}$

$\quad\quad\quad\quad + 2\left(\dfrac{3}{2}\theta + \sin 2\theta + \dfrac{\sin 4\theta}{8}\right)\Big\}\Big|_0^{2\pi/3}$

$\quad = \dfrac{1}{3}\left(6\theta + 13\sin\theta - \dfrac{7}{2}\sin 2\theta - 4\sin^3\theta + \dfrac{1}{4}\sin 4\theta\right)\Big|_0^{2\pi/3}$

$\quad = \dfrac{1}{3}\left(4\pi + \dfrac{13\sqrt{3}}{2} - \dfrac{7\sqrt{3}}{4} - \dfrac{3\sqrt{3}}{2} + \dfrac{\sqrt{3}}{8}\right)$

$\quad = \dfrac{1}{3}\left(4\pi + \dfrac{27\sqrt{3}}{8}\right) = \dfrac{4\pi}{3} + \dfrac{9\sqrt{3}}{8}$

$\bar{x} = \dfrac{M_y}{M} = \dfrac{\dfrac{4\pi}{3} + \dfrac{9\sqrt{3}}{8}}{\pi + \dfrac{3\sqrt{3}}{4}} = \dfrac{32\pi + 27\sqrt{3}}{6(4\pi + 3\sqrt{3})} \approx 1.3821$

31.   Completing the square: $x^2 - ax + y^2 = 0$, $\left(x - \frac{a}{2}\right)^2 + y^2 = \frac{a^2}{4}$.

This is a translated circle with center at $(\frac{a}{2}, 0)$ and radius of $\frac{a}{2}$.

Its equation in polar form is $r = a \cos \theta$. The x-axis is a diameter so the

moment about a diameter is I_x.

$$I_x = \int_0^\pi \int_0^{a \cos \theta} (r \sin \theta)^2 \, r \, dr \, d\theta = \frac{a^4}{4} \int_0^\pi \cos^4\theta \, \sin^2\theta \, d\theta$$

$$= \frac{a^4}{4} \int_0^\pi \left(\frac{1 + \cos 2\theta}{2}\right)^2 \left(\frac{1 - \cos 2\theta}{2}\right) d\theta = \frac{a^4}{32} \int_0^\pi (1 - \cos^2 2\theta)(1 + \cos 2\theta) \, d\theta$$

$$= \frac{a^4}{32} \int_0^\pi (1 + \cos 2\theta - \cos^2 2\theta - \cos^3 2\theta) \, d\theta$$

$$= \frac{a^4}{32} \int_0^\pi \left(1 + \cos 2\theta - \frac{1 + \cos 4\theta}{2} - (1 - \sin^2 2\theta)\cos 2\theta\right) d\theta$$

$$= \frac{a^4}{32}\left(\frac{\theta}{2} - \frac{\sin 4\theta}{8} + \frac{\sin^3 2\theta}{6}\right)\Big|_0^\pi = \frac{a^4}{32}\left(\frac{\pi}{2}\right) = \frac{a^4 \pi}{64}$$

35.   If $\rho = 1$ then $M = A = \pi ab$.   Using symmetry:

$$I_x = 4 \int_0^a \int_0^{\frac{b}{a}\sqrt{a^2 - x^2}} y^2 \, dy \, dx = \frac{4}{3} \int_0^a \left(\frac{b}{a}\sqrt{a^2 - x^2}\right)^3 dx$$

$$= \frac{4b^3}{3a^3} \int_0^a (a^2 - x^2)^{3/2} dx \quad \text{Using formula \#245:}$$

$$= \frac{4b^3}{3a^3}\left\{\frac{1}{4}\left(x(a^2 - x^2)^{3/2} + \frac{3a^2 x}{2}(a^2 - x^2)^{1/2} + \frac{3a^4}{2}\sin^{-1}\frac{x}{a}\right)\right\}\Big|_0^a$$

$$= \frac{b^3}{3a^3}\left(\frac{3a^4 \pi}{4}\right) = \frac{ab^3 \pi}{4} \quad \text{Substituting } m = \pi ab: \ I_x = \frac{mb^2}{4}$$

39. $P[(x, y) \text{ is in } R] = \int_0^1 \int_0^{1-x} xe^{-(x+y)} \, dy \, dx = -\int_0^1 xe^{-(x+y)}\Big|_0^{1-x} dx$

$$= -\int_0^1 \left(xe^{-1} - xe^{-x}\right) dx \quad \text{Use integration by parts on the second term:}$$

$$= -\left(\frac{x^2}{2e} + xe^{-x} + e^{-x}\right)\Big|_0^1 = -\frac{1}{2e} - \frac{1}{e} - \frac{1}{e} + 1 = 1 - \frac{5}{2e}$$

43. $$P(<30) = \frac{1}{300} \int_0^{30} \int_0^{30-x} e^{-x/30} e^{-y/10} \, dy \, dx$$

$$= \frac{1}{300} \int_0^{30} e^{-x/30} (-10) \, e^{-y/10} \Big|_0^{30-x} \, dx = -\frac{1}{30} \int_0^{30} e^{-x/30} \Big(e^{(x-30)/10} - 1 \Big) \, dx$$

$$= -\frac{1}{30} \int_0^{30} \Big(e^{x/15-3} - e^{-x/30} \Big) \, dx = -\frac{1}{30} \Big(15 \, e^{x/15-3} + 30 \, e^{-x/30} \Big) \Big|_0^{30}$$

$$= -\frac{1}{2} \Big(e^{-1} + 2 \, e^{-1} - e^{-3} - 2 \Big) = -\frac{1}{2} \Big(\frac{3}{e} - \frac{1}{e^3} - 2 \Big) \approx 0.4731$$

47.   Since the sphere is symmetric in each of the eight octants, and xyz is positive in four and negative in four, the average value is 0.

53.   Analysis: Put the origin at the center of the rectangle and use symmetry.

$$I = 4 \int_0^{\ell/2} \int_0^{h/2} (x^2 + y^2) \, dy \, dx$$

55.   Analysis: $S = 2\pi L h$. The torus has outer radius of $a + b$ and inner radius of $a - b$. The centroid of the circle is at a and its radius is b.

13.7 Cylindrical and Spherical Coordinates

SURVIVAL HINT

The best way to convert from one notation to another is to understand the derivations and "visualize" the graph. Lacking that skill you will find it necessary to memorize the transformation equations.

3.   Draw a figure and use your basic trigonometry as well as referring to the formulas.

    a.   Exact: $(\sqrt{2}, \frac{\pi}{4}, 1)$ Approximate: $(1.41, 0.79, 1.00)$

    b.   Exact: $\Big(\sqrt{3}, \frac{\pi}{4}, \cos^{-1}\frac{1}{\sqrt{3}} \Big)$ Approximate: $(1.73, \ 0.79, \ 0.96)$

7.   **a.**   Exact: $\Big(\sqrt{5}, \tan^{-1}2, 3 \Big)$ Approximate: $(2.24, 1.11, 3.00)$

    b.   Exact: $\Big(\sqrt{14}, \tan^{-1}2, \cos^{-1}\frac{3}{\sqrt{14}} \Big)$ Approximate: $(3.74, 1.11, 0.64)$

11.   **a.**   Exact: $(\sqrt{2}, \sqrt{2}, \pi)$ Approximate: $(1.41, 1.41, 3.14)$

    b.   Exact: $\Big(\sqrt{\pi^2 + 4}, \frac{\pi}{4}, \cos^{-1}\frac{\pi}{\sqrt{\pi^2 + 4}} \Big)$ Approximate: $(3.72, 0.79, 0.57)$

15. On these the formulas are easier than visualizing a sketch.

    a. Exact: $\left(\frac{3}{2}, \frac{\sqrt{3}}{2}, -1\right)$ Approximate: $(1.50,\ 0.87,\ -1.00)$

    b. Exact: $\left(\sqrt{3},\ \tan^{-1}\frac{\sqrt{3}}{3}, -1\right)$ or $\left(\sqrt{3},\ \frac{\pi}{6}, -1\right)$

    Approximate: $(1.73,\ 0.52,\ -1.00)$

19.   **a.** $(1\sin 3\cos 2,\ 1\sin 3\sin 2,\ 1\cos 3) \approx (-0.06,\ 0.13,\ -0.99)$

    b. $(1\sin 3,\ 2,\ 1\cos 3) \approx (0.14,\ 2.00,\ -0.99)$

23. $\dfrac{x^2}{4} - \dfrac{y^2}{9} + z^2 = 0$ becomes $\dfrac{r^2\cos^2\theta}{4} - \dfrac{r^2\sin^2\theta}{9} + z^2 = 0$

or $9r^2\cos^2\theta - 4r^2\sin^2\theta + 36z^2 = 0$ which is an elliptical cone.

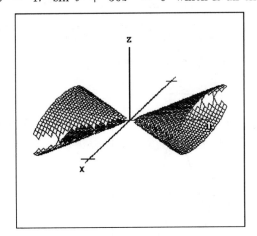

27. $z^2 = x^2 + y^2$ becomes $\rho^2\cos^2\phi = \rho^2\sin^2\phi(\cos^2\theta + \sin^2\theta)$

$\dfrac{\sin^2\phi}{\cos^2\phi} = 1,\ \tan^2\phi = 1,\ \tan\phi = \pm 1,\ \phi = \frac{\pi}{4}, \frac{3\pi}{4}$

But it was specified that $z \geq 0$, so $\phi = \frac{\pi}{4}$ which is a circular cone

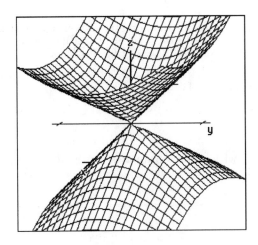

31. $x^2 + y^2 + z^2 = 4$, $\rho^2 = 4$, $\rho = 2$; a sphere.

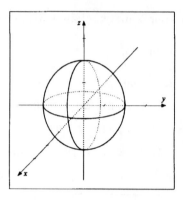

35. $z = r^2 \cos 2\theta$, $z = r^2(\cos^2\theta - \sin^2\theta)$, $z = r^2 \cos^2\theta - r^2 \sin^2\theta$,

$z = x^2 - y^2$; a hyperbolic paraboloid (saddle).

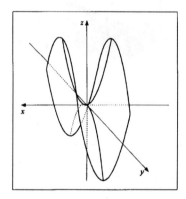

39. $\displaystyle \int_0^\pi \int_0^2 \int_0^{\sqrt{4-r^2}} r \sin\theta \; dz \; dr \; d\theta = -\frac{1}{2} \int_0^\pi \int_0^2 (\sin\theta)\sqrt{4 - r^2} \; (-2r \; dr) \; d\theta$

$\displaystyle = -\frac{1}{2} \int_0^\pi (\sin\theta) \frac{2(4 - r^2)^{3/2}}{3} \Big|_0^2 \; d\theta = -\frac{1}{3} \int_0^\pi (\sin\theta)(-8) \; d\theta$

$\displaystyle = \frac{8}{3} (-\cos\theta) \Big|_0^\pi = \frac{8}{3}(1 + 1) = \frac{16}{3}$

SURVIVAL HINT

Cylindrical coordinates are dz with polar coordinates for the dA. Do not forget the "r".

$dz \; dy \; dx = dV = dz \; r \; dr \; d\theta$.

43. $\displaystyle\int_0^{2\pi}\int_0^4\int_0^1 z\,r\,dz\,dr\,d\theta = \frac{1}{2}\int_0^{2\pi}\int_0^4 r\,dr\,d\theta = \frac{1}{2}\int_0^{2\pi}\frac{r^2}{2}\Big|_0^4\,d\theta = 4\int_0^{2\pi} d\theta = 8\pi$

47. $\displaystyle\iint_R\int (x^2 + y^2)^2\,dx\,dy\,dz$ This is the four octants where $z > 0$.

$\displaystyle 4\int_0^{1/\pi}\int_0^{\pi/2}\int_0^a r^4\,r\,dr\,d\theta\,dz = \frac{4a^6}{6}\int_0^{1/\pi}\int_0^{\pi/2} d\theta\,dz = \frac{4a^6}{6}\left(\frac{\pi}{2}\right)\left(\frac{1}{\pi}\right) = \frac{a^6}{3}$

SURVIVAL HINT

Anytime you find $x^2 + y^2 + z^2$ in a function you should consider changing to spherical coordinates.

51. $\displaystyle\iint_R\int (x^2 + y^2 + z^2)\,dx\,dy\,dz$ Using symmetry and the first octant:

$\displaystyle = 8\int_0^{\pi/2}\int_0^{\pi/2}\int_0^{\sqrt2} \rho^2\,\rho^2\sin\phi\,d\rho\,d\theta\,d\phi = 8\int_0^{\pi/2}\int_0^{\pi/2}\frac{4\sqrt2}{5}\sin\phi\,d\theta\,d\phi$

$\displaystyle = \frac{32\sqrt2}{5}\left(\frac{\pi}{2}\right)\int_0^{\pi/2}\sin\phi\,d\phi = \frac{16\pi\sqrt2}{5}(-\cos\phi)\Big|_0^{\pi/2} = \frac{16\sqrt2\pi}{5}$

55. Since R is a cylinder we will use cylindrical coordinates.

$\displaystyle V = \int_0^{2\pi}\int_0^1\int_0^{4-r^2} dz\,r\,dr\,d\theta = \int_0^{2\pi}\int_0^1 (4 - r^2)\,r\,dr\,d\theta = \int_0^{2\pi}\left(2r^2 - \frac{r^4}{4}\right)\Big|_0^1 d\theta$

$\displaystyle = \int_0^{2\pi}\frac{7}{4}\,d\theta = \frac{7\pi}{2}$

59.

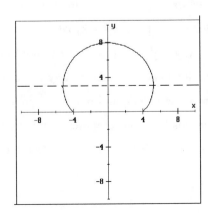

59. (con't.)

The cave is a cardioid of revolution about the z-axis. The spy will begin to drown when the water level reaches 3 ft above the floor (the xy plane); that is when $z = 3$.

$$\rho \cos \phi = 3$$

$$4(1 + \cos \phi) \cos \phi = 3$$

$$4 \cos^2\phi + 4 \cos \phi - 3 = 0$$

$$(2 \cos \phi - 1)(2 \cos \phi + 3) = 0$$

$$\cos \phi = \tfrac{1}{2}, -\tfrac{3}{2} \text{ so in this application } \phi = \tfrac{\pi}{3}$$

The critical water level is $z = 3 = \rho \cos \theta$ or $\rho = 3 \sec \phi$

We need to compute the amount of water in the cave when the depth is 3 ft and use the rate of 25 cu ft/min to determine the time necessary for it to reach that level. In spherical coordinates the volume can be found as the sum of two integrals:

$$V = \int_0^{2\pi} \int_0^{\pi/3} \int_0^{3 \sec\phi} \rho^2\sin \phi \, d\rho \, d\phi \, d\theta + \int_0^{2\pi} \int_{\pi/3}^{\pi/2} \int_0^{4(1+\cos\phi)} \rho^2\sin \phi \, d\rho \, d\phi \, d\theta$$

$$= \int_0^{2\pi} \int_0^{\pi/3} \frac{(3 \sec \phi)^3}{3} \sin \phi \, d\phi \, d\theta + \int_0^{2\pi} \int_{\pi/3}^{\pi/2} \frac{[4(1 + \cos \phi)]^3}{3} \sin \phi \, d\phi \, d\theta$$

$$= \int_0^{2\pi} \frac{9}{2} \tan^2\phi \Big|_0^{\pi/3} d\theta + \int_0^{2\pi} \frac{-64}{3} \frac{(1 + \cos \phi)^4}{4} \Big|_{\pi/3}^{\pi/2} d\theta$$

$$= \frac{9}{2} (3)(2\pi) - \frac{64}{3}\left(\frac{1}{4} - \frac{81}{64}\right)(2\pi)$$

$$= \pi\left(27 + \frac{130}{3}\right) \approx 221 \text{ cu ft of water at the incoming rate of 25 cu ft/min}$$

$$\approx 8.84 \text{ min}$$

So he drowns, you say – nonsense! Any spy worth his salt can hold his breath for a little more than a minute. He frees his hands, stands (to buy more time), hops to the door, pulls up the lever (to stop the water), and opens the door. As the water drains from the room, he unties his feet, and prepares to pursue Purity.

63.   Analysis: Notice that this ellipsoid is a sphere; use spherical coordinates.

$$I_x + I_y + I_z = \iiint_S (y^2 + z^2)\, dV + \iiint_S (x^2 + z^2)\, dV$$

$$+ \iiint_S (x^2 + y^2)\, dV$$

$$= \iiint_S (2x^2 + 2y^2 + 2z^2)\, dV$$

$$= 2 \int_0^\pi \int_0^{2\pi} \int_0^1 \rho^2\, \rho^2 \sin\phi\, d\rho\, d\theta\, d\phi \text{ which is easily integrated.}$$

13.8 Jacobians: Change of Variables

1. $x = u + v,\ y = uv$

$$\frac{\partial(x, y)}{\partial(u, v)} = \begin{vmatrix} \dfrac{\partial x}{\partial u} & \dfrac{\partial x}{\partial v} \\[2mm] \dfrac{\partial y}{\partial u} & \dfrac{\partial y}{\partial v} \end{vmatrix} = \begin{vmatrix} 1 & 1 \\ v & u \end{vmatrix} = u - v$$

5. $x = u^2 v^2,\ y = v^2 - u^2$

$$\frac{\partial(x, y)}{\partial(u, v)} = \begin{vmatrix} \dfrac{\partial x}{\partial u} & \dfrac{\partial x}{\partial v} \\[2mm] \dfrac{\partial y}{\partial u} & \dfrac{\partial y}{\partial v} \end{vmatrix} = \begin{vmatrix} 2uv^2 & 2u^2 v \\ -2u & 2v \end{vmatrix} = 4uv^3 + 4u^3 v = 4uv(u^2 + v^2)$$

9. $x = u + v - w,\ y = 2u - v + 3w,\ z = -u + 2v - w$

$$\frac{\partial(x, y, z)}{\partial(u, v, w)} = \begin{vmatrix} 1 & 1 & -1 \\ 2 & -1 & 3 \\ -1 & 2 & -1 \end{vmatrix}$$

Using the diagonals shortcut:

$$= 1 - 3 - 4 + 1 + 2 - 6 = -9$$

13. In the xy-plane the vertices are: $(0, 0)$, $(0, 5)$, $(6, 0)$, $(6, 5)$.

In the uv-plane these become: $(0, 0)$, $(5, -5)$, $(6, 6)$, and $(11, 1)$.

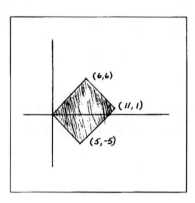

17. $x = u - uv,\; y = uv$

$$\frac{\partial(x, y)}{\partial(u, v)} = \begin{vmatrix} \dfrac{\partial x}{\partial u} & \dfrac{\partial x}{\partial v} \\[2mm] \dfrac{\partial y}{\partial u} & \dfrac{\partial y}{\partial v} \end{vmatrix} = \begin{vmatrix} 1 - v & -u \\ v & u \end{vmatrix} = u - uv + uv = u.$$

So $dx\, dy = u\, du\, dv$.

21. $u = x - y,\; v = x + y$, adding: $x = \dfrac{u + v}{2}$, subtracting: $y = \dfrac{v - u}{2}$

$$\frac{\partial(x, y)}{\partial(u, v)} = \begin{vmatrix} \dfrac{\partial x}{\partial u} & \dfrac{\partial x}{\partial v} \\[2mm] \dfrac{\partial y}{\partial u} & \dfrac{\partial y}{\partial v} \end{vmatrix} = \begin{vmatrix} \dfrac{1}{2} & \dfrac{1}{2} \\[2mm] -\dfrac{1}{2} & \dfrac{1}{2} \end{vmatrix} = \frac{1}{4} + \frac{1}{4} = \frac{1}{2}$$

$(0, 0)$ becomes $(0, 0)$, $(1, 0)$ becomes $(1, 1)$, $(0, 1)$ becomes $(-1, 1)$.

The new boundary lines are: $v = 0$, $v = -u$, and $v = -\frac{1}{2}u + 1$

Using vertical strips: $\displaystyle\int_0^1 \int_0^{1-x} (x - y)^5 (x + y)^3\, dy\, dx$ becomes

(using horizontal strips): $\displaystyle\int_0^1 \int_{-v}^{v} u^5 v^3 \left(\frac{1}{2}\right) du\, dv = \frac{1}{2}\int_0^1 v^3 \left(\frac{u^6}{6}\right)\Big|_{-v}^{v} dv$

$$= \frac{1}{12}\int_0^1 v^3 (v^6 - v^6)\, dv = 0$$

25. $u = \dfrac{2x + y}{5}, \quad v = \dfrac{x - 2y}{5}, \quad$ solving for x and y:

5 times the first minus 10 times the second gives: $y = u - 2v$.

20 times the first plus 10 times the second gives: $x = 2u + v$.

$$\frac{\partial(x,\, y)}{\partial(u,\, v)} = \begin{vmatrix} \dfrac{\partial x}{\partial u} & \dfrac{\partial x}{\partial v} \\[2mm] \dfrac{\partial y}{\partial u} & \dfrac{\partial y}{\partial v} \end{vmatrix} = \begin{vmatrix} 2 & 1 \\[1mm] 1 & -2 \end{vmatrix} = |-5| = 5$$

For the new region: $(0, 0)$ becomes $(0, 0)$, $(1, -2)$ becomes $(0, 1)$,

$(2, 1)$ becomes $(1, 0)$, and $(3, -1)$ becomes $(1, 1)$. R is the unit square.

$$\int\int_{S} \left(\frac{2x + y}{x - 2y + 5}\right)^2 dy\, dx \text{ becomes } \int_0^1\int_0^1 \left(\frac{u}{v + 1}\right)^2 5\, du\, dv$$

$$= \frac{5}{3}\int_0^1 \frac{1}{(v + 1)^2}\, u^3\Big|_0^1\, dv = \frac{5}{3}\int_0^1 (v + 1)^{-2}dv = \frac{5}{3}\left(-\frac{1}{v + 1}\right)\Big|_0^1$$

$$= \frac{5}{3}\left(-\frac{1}{2} + 1\right) = \frac{5}{6}$$

29. $u = 2x + y, \quad v = \tan^{-1}(x - 2y)$

Solving for x and y: $x = \frac{1}{5}(2u + \tan v), \quad y = \frac{1}{5}(u - 2\tan v)$

$$\frac{\partial(x,\, y)}{\partial(u,\, v)} = \begin{vmatrix} \dfrac{\partial x}{\partial u} & \dfrac{\partial x}{\partial v} \\[2mm] \dfrac{\partial y}{\partial u} & \dfrac{\partial y}{\partial v} \end{vmatrix} = \begin{vmatrix} \dfrac{2}{5} & \dfrac{1}{5}\sec^2 v \\[2mm] \dfrac{1}{5} & -\dfrac{2}{5}\sec^2 v \end{vmatrix} = \left|-\frac{1}{5}\sec^2 v\right| = \frac{1}{5}\sec^2 v$$

For the new region: $(0, 0)$ becomes $(0, 0)$, $(1, -2)$ becomes $(0, \tan^{-1}5)$,

$(2, 1)$ becomes $(5, 0)$, and $(3, -1)$ becomes $(5, \tan^{-1}5)$.

$$\int\int_{S} (2x + y)\tan^{-1}(x - 2y)\, dx\, dy \text{ becomes } \frac{1}{5}\int_0^5\int_0^{\tan^{-1}5} u\, v\, \sec^2 v\, dv\, du$$

Integrating by parts: $\dfrac{1}{5}\int_0^5 u[v\tan v + \ln \sec v]\Big|_0^{\tan^{-1}5}\, du$

$$= \frac{1}{5}\int_0^5 u[5\tan^{-1}5 - \ln \sqrt{26}\,]\, du$$

$$= \frac{5}{2}[5\tan^{-1}5 - \frac{1}{2}\ln 26]$$

$$= \frac{25}{2}\tan^{-1}5 - \frac{5}{4}\ln 26 \approx 13.09$$

33. $x = s^2 - t^2$, $y = 2st$. Squaring and adding:

$x^2 + y^2 = (s^2 - t^2)^2 + 4s^2t^2 = s^4 + 2s^2t^2 + t^4 = (s^2 + t^2)^2$

So $s^2 + t^2 = \sqrt{x^2 + y^2}$ and the new region is $\sqrt{x^2 + y^2} \leq 1$, which represents a semicircle. Using polar coordinates:

$$\int_0^\pi \int_0^1 \frac{1}{r} r \, dr \, d\theta = \pi$$

37. $x = r \cos \theta$, $y = r \sin \theta$, $z = z$

$$\frac{\partial(x, y, z)}{\partial(r, \theta, z)} = \begin{vmatrix} \cos \theta & \sin \theta & 0 \\ -r \sin \theta & r \cos \theta & 0 \\ 0 & 0 & 1 \end{vmatrix}$$

$= r \cos^2\theta + 0 + 0 - 0 + r \sin^2\theta - 0 = r(\cos^2\theta + \sin^2\theta) = r$

So $dx \, dy \, dz$ becomes $r \, dr \, d\theta \, dz$.

Chapter 13 Review

PRACTICE PROBLEMS

23.
$$\int_0^{\pi/3} \int_0^{\sin y} e^{-x} \cos y \, dx \, dy = \int_0^{\pi/3} (\cos y)(-e^{-x}) \Big|_0^{\sin y} dy$$

$$= \int_0^{\pi/3} \left((e^{-\sin y})(-\cos y) + \cos y \right) dy = \left(e^{-\sin y} + \sin y \right) \Big|_0^{\pi/3}$$

$$= e^{-\sqrt{3}/2} + \frac{\sqrt{3}}{2} - 1$$

24.
$$\int_{-1}^1 \int_0^z \int_y^{y-z} (x + y - z) \, dx \, dy \, dz = \int_{-1}^1 \int_0^z \left(\frac{x^2}{2} + (y - z)x \right) \Big|_y^{y-z} dy \, dz$$

$$= \int_{-1}^1 \int_0^z \left(\frac{(y - z)^2}{2} + (y - z)y - (y - z)z - \frac{y^2}{2} - y^2 + yz \right) dy \, dz$$

$$= \int_{-1}^1 \int_0^z \left(-2yz + \frac{3z^2}{2} \right) dy \, dz = \int_{-1}^1 \left(-y^2z + \frac{3z^2y}{2} \right) \Big|_0^z dz$$

$$= \int_{-1}^1 \left(-z^3 + \frac{3}{2}z^3 \right) dz = \frac{z^4}{8} \Big|_{-1}^1 = 0$$

25. The x-intercepts are ± 3. Use symmetry about the y-axis.

$$2\int_0^3\int_0^{9-x^2} dy\,dx = 2\int_0^3 (9 - x^2)\,dx = 2\left(9x - \frac{x^3}{3}\right)\Big|_0^3 = 2(27 - 9) = 36$$

26. $y = \sqrt{1 - x^2}$ or $x^2 + y^2 = 1$ and $0 \le x \le 1$ is the first quadrant of the unit circle.

$$A = \int_0^{\pi/2}\int_0^1 \cos r^2\, r\,dr\,d\theta = \frac{1}{2}\int_0^{\pi/2}\int_0^1 (\cos r^2)(2r\,dr)\,d\theta = \frac{1}{2}\int_0^{\pi/2} \sin r^2\Big|_0^1 d\theta$$

$$= \frac{1}{2}\int_0^{\pi/2} \sin 1\, d\theta = \frac{\pi}{4}\sin 1$$

27. This is the four z positive octants of the sphere with radius of 2.

Using polar coordinates: $S = \int_0^{2\pi}\int_0^2 \sqrt{f_x^2 + f_y^2 + 1}\, r\,dr\,d\theta$

$$z = \sqrt{4 - x^2 - y^2}, \quad f_x = \frac{-x}{\sqrt{4 - x^2 - y^2}}, \quad f_y = \frac{-y}{\sqrt{4 - x^2 - y^2}}$$

$$\sqrt{f_x^2 + f_y^2 + 1} = \sqrt{\frac{x^2}{4 - x^2 - y^2} + \frac{y^2}{4 - x^2 - y^2} + \frac{4 - x^2 - y^2}{4 - x^2 - y^2}}$$

$$= \frac{2}{\sqrt{4 - x^2 - y^2}}$$

$$S = \int_0^{2\pi}\int_0^2 \frac{2}{\sqrt{4 - r^2}}\, r\,dr\,d\theta = -2\int_0^{2\pi} \sqrt{4 - r^2}\Big|_0^2 d\theta = -2(-2)\int_0^{2\pi} d\theta = 8\pi$$

28. Finding the intersection of the plane and the paraboloid:

$$x^2 + 2y^2 = 4x, \quad (x - 2)^2 + 2y^2 = 4, \quad \frac{(x - 2)^2}{4} + \frac{y^2}{2} = 1$$

R is a translated ellipse with x intercepts of 0 and 4.

$$V = \int_0^4\int_0^{\sqrt{4x-x^2}/\sqrt{2}}\int_{x^2+2y^2}^{4x} dz\,dy\,dx = \int_0^4\int_0^{\sqrt{4x-x^2}/\sqrt{2}} (4x - x^2 - 2y^2)\,dy\,dx$$

$$= \int_0^4 \left((4x - x^2)y - \frac{2y^3}{3}\right)\Big|_0^{\sqrt{4x-x^2}/\sqrt{2}} dx$$

$$= \int_0^4 \left\{\frac{(4x - x^2)^{3/2}}{\sqrt{2}} - \frac{2}{3}\frac{(4x - x^2)^{3/2}}{2\sqrt{2}}\right\} dx = \frac{2}{3\sqrt{2}}\int_0^4 (4x - x^2)^{3/2}dx$$

28. (con't.)

Completing the square to have $a^2 - u^2$ form:

$$= \frac{\sqrt{2}}{3} \int_0^4 \left(2^2 - (x - 2)^2\right)^{3/2} dx \quad \text{Now using formula \#245:}$$

$$= \frac{\sqrt{2}}{3}\left(\frac{(x - 2)(4x - x^2)^{3/2}}{4} + \frac{12(x - 2)(4x - x^2)^{1/2}}{8} + \frac{3}{8}(16)\sin^{-1}\frac{x - 2}{2}\right)\Big|_0^4$$

$$= \frac{\sqrt{2}}{3}\left(0 + 0 + 6(\tfrac{\pi}{2}) - 0 - 0 - 6(-\tfrac{\pi}{2})\right) = \frac{\sqrt{2}}{3}(6\pi) = 2\pi\sqrt{2}$$

29. The intersection is $x^2 + y^2 = 4$, so R is a circle with radius of 2.

Using cylindrical coordinates and $\rho = r$:

$$M = \int_0^{2\pi}\int_0^2\int_{r^2}^4 (r)\, r\, dz\, dr\, d\theta = \int_0^{2\pi}\int_0^2 r^2(4 - r^2)\, dr\, d\theta = \int_0^{2\pi}\left(\frac{4r^3}{3} - \frac{r^5}{5}\right)\Big|_0^2 d\theta$$

$$= \int_0^{2\pi}\left(\frac{32}{3} - \frac{32}{5}\right) d\theta = \frac{64}{15}\int_0^{2\pi} d\theta = \frac{128\pi}{15}$$

30. Let $u = x + y$ and $v = x - 2y$. Then $u - v = 3y$, $y = \frac{v - u}{3}$ and $2u + v = 3x$, $x = \frac{2u + v}{3}$. For the region transformation $(0, 0)$ becomes $(0, 0)$, $(2, 0)$ becomes $(2, 2)$ and $(1, 1)$ becomes $(2, -1)$. The equations of the boundary lines for the uv region triangle are: $v = u$ and $v = -\frac{1}{2}u$ or $u = -2v$.

The Jacobian:

$$\frac{\partial(x,y)}{\partial(u,v)} = \begin{vmatrix} \frac{2}{3} & \frac{1}{3} \\ \frac{1}{3} & -\frac{1}{3} \end{vmatrix} = |-\tfrac{1}{3}| = \frac{1}{3} \quad \text{For a single region use vertical slices.}$$

$$\int_0^2\int_{-u/2}^u u\, e^v\, (\tfrac{1}{3})\, dv\, du = \frac{1}{3}\int_0^2 u\, e^v\Big|_{-u/2}^u du = \frac{1}{3}\int_0^2\left(u\, e^u - u\, e^{-u/2}\right) du$$

Using integration by parts: $= \frac{1}{3}\left(u\, e^u - e^u + 2u\, e^{-u/2} + 4\, e^{-u/2}\right)\Big|_0^2$

$$= \frac{1}{3}\left(2e^2 - e^2 + 4e^{-1} + 4e^{-1} + 1 - 4\right) = \frac{1}{3}\left(e^2 + \frac{8}{e} - 3\right)$$

Chapter 14
Vector Analysis

14.1 Properties of a Vector Field: Divergence and Curl

3. $\mathbf{F} = x\mathbf{i} + y\mathbf{j}$

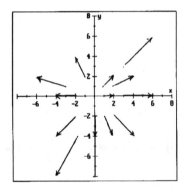

7. $\mathbf{F} = y\mathbf{i} + x\mathbf{j}$

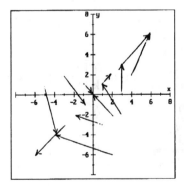

SURVIVAL HINT

Remember that div $\mathbf{F}$ is always a scalar function and curl $\mathbf{F}$ is always a vector function.

11. $\mathbf{F} = 2y\mathbf{j}$

div $\mathbf{F} = 2$

curl $\mathbf{F} = \begin{vmatrix} \mathbf{i} & \mathbf{j} & \mathbf{k} \\ \frac{\partial}{\partial x} & \frac{\partial}{\partial y} & \frac{\partial}{\partial z} \\ 0 & 2y & 0 \end{vmatrix} = -\frac{\partial}{\partial z}(2y)\mathbf{i} + \frac{\partial}{\partial x}(2y)\mathbf{k} = 0$

15. $\mathbf{F} = xyz\mathbf{i} + y\mathbf{j} + x\mathbf{k}$ at $(1, 2, 3)$

div $\mathbf{F} = \frac{\partial}{\partial x}(xyz) + \frac{\partial}{\partial y}(y) + \frac{\partial}{\partial z}(x) = yz + 1$

div $\mathbf{F}(1, 2, 3) = 7$

curl $\mathbf{F} = \begin{vmatrix} \mathbf{i} & \mathbf{j} & \mathbf{k} \\ \frac{\partial}{\partial x} & \frac{\partial}{\partial y} & \frac{\partial}{\partial z} \\ xyz & y & x \end{vmatrix}$

$\quad = \left(\frac{\partial}{\partial y}x - \frac{\partial}{\partial z}y\right)\mathbf{i} + \left(\frac{\partial}{\partial z}xyz - \frac{\partial}{\partial x}x\right)\mathbf{j} + \left(\frac{\partial}{\partial x}y - \frac{\partial}{\partial y}xyz\right)\mathbf{k}$

$\quad = (xy - 1)\mathbf{j} - xz\mathbf{k}$

curl $\mathbf{F}(1, 2, 3) = \mathbf{j} - 3\mathbf{k}$

19. div $\mathbf{F} = \cos x - \sin y$

curl $\mathbf{F} = \begin{vmatrix} \mathbf{i} & \mathbf{j} & \mathbf{k} \\ \frac{\partial}{\partial x} & \frac{\partial}{\partial y} & \frac{\partial}{\partial z} \\ \sin x & \cos y & 0 \end{vmatrix}$

$\quad = -\frac{\partial}{\partial z}(\cos y)\mathbf{i} + \frac{\partial}{\partial z}(\sin x)\mathbf{j} + \left(\frac{\partial}{\partial x}\cos y - \frac{\partial}{\partial y}\sin x\right)\mathbf{k}$

$\quad = \mathbf{0}$

23. div $\mathbf{F} = \dfrac{\sqrt{x^2 + y^2} - \dfrac{x^2}{\sqrt{x^2 + y^2}}}{x^2 + y^2} + \dfrac{\sqrt{x^2 + y^2} - \dfrac{y^2}{\sqrt{x^2 + y^2}}}{x^2 + y^2}$

$\quad = \dfrac{y^2 + x^2}{(x^2 + y^2)^{3/2}} = \dfrac{1}{\sqrt{x^2 + y^2}}$

curl $\mathbf{F} = \begin{vmatrix} \mathbf{i} & \mathbf{j} & \mathbf{k} \\ \frac{\partial}{\partial x} & \frac{\partial}{\partial y} & \frac{\partial}{\partial z} \\ \dfrac{x}{\sqrt{x^2 + y^2}} & \dfrac{y}{\sqrt{x^2 + y^2}} & 0 \end{vmatrix}$

$\quad = \left(\dfrac{\partial}{\partial x}\dfrac{y}{\sqrt{x^2 + y^2}} - \dfrac{\partial}{\partial y}\dfrac{x}{\sqrt{x^2 + y^2}}\right)\mathbf{k}$

$\quad = \left(\dfrac{-xy}{(x^2 + y^2)^{3/2}} - \dfrac{-xy}{(x^2 + y^2)^{3/2}}\right)\mathbf{k} = \mathbf{0}$

27.   div $\mathbf{F} = 2x + 2y + 2z = 2(x + y + z)$

curl $\mathbf{F} = \begin{vmatrix} \mathbf{i} & \mathbf{j} & \mathbf{k} \\ \dfrac{\partial}{\partial x} & \dfrac{\partial}{\partial y} & \dfrac{\partial}{\partial z} \\ x^2 & y^2 & z^2 \end{vmatrix} = \mathbf{0}$

31.   div $\mathbf{F} = yz + 2x^2yz^2 + 3y^2z^2$

curl $\mathbf{F} = \begin{vmatrix} \mathbf{i} & \mathbf{j} & \mathbf{k} \\ \dfrac{\partial}{\partial x} & \dfrac{\partial}{\partial y} & \dfrac{\partial}{\partial z} \\ xyz & x^2y^2z^2 & y^2z^3 \end{vmatrix}$

$\qquad = (2yz^3 - 2x^2y^2z)\mathbf{i} + xy\mathbf{j} + (2xy^2z^2 - xz)\mathbf{k}$

35. $u = e^{-x}(\cos y - \sin y)$

$u_x = -e^{-x}(\cos y - \sin y), \; u_{xx} = e^{-x}(\cos y - \sin y)$

$u_y = e^{-x}(-\sin y - \cos y), \; u_{yy} = e^{-x}(-\cos y + \sin y)$

$u_{xx} + u_{yy} = 0 \quad u$ is harmonic

39. $\mathbf{F} \times \mathbf{G} = \begin{vmatrix} \mathbf{i} & \mathbf{j} & \mathbf{k} \\ 2 & 2x & 3y \\ x & -y & z \end{vmatrix} = (2xz + 3y^2)\mathbf{i} + (3xy - 2z)\mathbf{j} + (-2y - 2x^2)\mathbf{k}$

curl $\mathbf{F} \times \mathbf{G} = \begin{vmatrix} \mathbf{i} & \mathbf{j} & \mathbf{k} \\ \dfrac{\partial}{\partial x} & \dfrac{\partial}{\partial y} & \dfrac{\partial}{\partial z} \\ 2xz + 3y^2 & 3xy - 2z & -2y - 2x^2 \end{vmatrix}$

$\qquad = (-2 + 2)\mathbf{i} + (2x + 4x)\mathbf{j} + (3y - 6y)\mathbf{k}$

$\qquad = 6x\mathbf{j} - 3y\mathbf{k}$

43. $f(x, y, z) = xy^3z^2$

$\nabla f = y^3z^2\mathbf{i} + 3xy^2z^2\mathbf{j} + 2xy^3z\mathbf{k}$

div $\mathbf{F} = 6xyz^2 + 2xy^3$

47.   Analysis: Let $\mathbf{A} = a\mathbf{i} + b\mathbf{j} + c\mathbf{k}$ Find $\mathbf{A} \times \mathbf{R}$, and then div $\mathbf{A} \times \mathbf{R}$.

49.   Analysis:

curl $\mathbf{F} = \begin{vmatrix} \mathbf{i} & \mathbf{j} & \mathbf{k} \\ \dfrac{\partial}{\partial x} & \dfrac{\partial}{\partial y} & \dfrac{\partial}{\partial z} \\ u(x, y) & v(x, y) & 0 \end{vmatrix}$

$\qquad = -\dfrac{\partial}{\partial z}v(x, y)\mathbf{i} + \dfrac{\partial}{\partial z}u(x, y)\mathbf{j} + \left(\dfrac{\partial}{\partial x}v(x, y) - \dfrac{\partial}{\partial y}u(x, y)\right)\mathbf{k}$

$\qquad = \mathbf{0}$ if and only if the coefficient of $\mathbf{k} = 0$

51. Analysis: If the angular velocity is ω, let the speed be ω_0, then

$$\omega = \omega_0\left(\frac{dx}{dt}\mathbf{i} + \frac{dy}{dt}\mathbf{j} + 0\mathbf{k}\right)$$

a. $\mathbf{V} = \omega \times \mathbf{R} = \begin{vmatrix} \mathbf{i} & \mathbf{j} & \mathbf{k} \\ 0 & 0 & \omega_0 \\ x & y & z \end{vmatrix} = -\omega_0 y\mathbf{i} + \omega_0 x\mathbf{j}$

b. $\operatorname{div}\mathbf{V} = z - z + 0 = 0$

computation of curl $\mathbf{V}$ is straightforward:

$$\operatorname{curl}\mathbf{V} = \begin{vmatrix} \mathbf{i} & \mathbf{j} & \mathbf{k} \\ \dfrac{\partial}{\partial x} & \dfrac{\partial}{\partial y} & \dfrac{\partial}{\partial z} \\ -\omega_0 y & \omega_0 x & 0 \end{vmatrix}$$

$$= 2\omega_0\mathbf{k}$$

$$= 2\omega$$

53. Analysis: Let $\mathbf{F} = F_1\mathbf{i} + F_2\mathbf{j} + F_3\mathbf{k}$ and $\mathbf{G} = G_1\mathbf{i} + G_2\mathbf{j} + G_3\mathbf{k}$

$$\mathbf{F} \times \mathbf{G} = \begin{vmatrix} \mathbf{i} & \mathbf{j} & \mathbf{k} \\ F_1 & F_2 & F_3 \\ G_1 & G_2 & G_3 \end{vmatrix}$$

$$= (F_2 G_3 - F_3 G_2)\mathbf{i} + (F_3 G_1 - F_1 G_3)\mathbf{j} + (F_1 G_2 - F_2 G_1)\mathbf{k}$$

Now work out each of I through IV to see which are equivalent.

55. Analysis:

$$\operatorname{div}\mathbf{F} + \operatorname{div}\mathbf{G} = \frac{\partial}{\partial x}F_1 + \frac{\partial}{\partial y}F_2 + \frac{\partial}{\partial z}F_3 + \frac{\partial}{\partial x}G_1 + \frac{\partial}{\partial y}G_2 + \frac{\partial}{\partial z}G_3$$

Now commute and associate.

57. Analysis: $c\mathbf{F} = cF_1\mathbf{i} + cF_2\mathbf{j} + cF_3\mathbf{k}$

Find curl $c\mathbf{F}$ and factor the c.

59. Analysis: $f\mathbf{F} = fF_1\mathbf{i} + fF_2\mathbf{j} + fF_3\mathbf{k}$

Find curl $f\mathbf{F}$ and show that it is the same as the sum of the two vectors

$f(\operatorname{curl}\mathbf{F})$ and $(\nabla f \times \mathbf{F})$.

61. Analysis: Let $\nabla f = f_x\mathbf{i} + f_y\mathbf{j} + f_z\mathbf{k}$ and $\mathbf{F} = F_1\mathbf{i} + F_2\mathbf{j} + F_3\mathbf{k}$

Do the required computations and equate.

63. Analysis:

Compute $\nabla \times \nabla f = \begin{vmatrix} \mathbf{i} & \mathbf{j} & \mathbf{k} \\ \dfrac{\partial}{\partial x} & \dfrac{\partial}{\partial y} & \dfrac{\partial}{\partial z} \\ f_x & f_y & f_z \end{vmatrix}$

$$= (f_y f_z - f_z f_y)\mathbf{i} + (f_z f_x - f_x f_z)\mathbf{j} + (f_x f_y - f_y f_x)\mathbf{k}$$

14.2 Line Integrals

3.   Let $x = t$, $y = 4t^2$ on $0 \leq t \leq 1$

$(-y\,dx + x\,dy) = -4t^2\,dt + t(8t)\,dt = 4t^2\,dt$

$$\int_C (-y\,dx + x\,dy) = 4\int_0^1 t^2\,dt = \frac{4}{3}$$

SURVIVAL HINT

If C is not a smooth curve (continuous and differentiable) then the line integral needs to be computed for each smooth segment.

7. C needs to be considered as two regions: let $x = t$ then $y = -2t$ on $-1 \leq t \leq 0$

and $y = 2t$ on $0 \leq t \leq 1$

On $[-1, 0]$: $[(x + y)^2\,dx - (x - y)^2\,dy] = (-t)^2\,dt - (3t)^2(-2dt) = 19t^2\,dt$

On $[0, 1]$: $[(x + y)^2\,dx - (x - y)^2\,dy] = (3t)^2\,dt - (-t)^2(2dt) = 7t^2\,dt$

$$\int_C [(x + y)^2\,dx - (x - y)^2\,dy] = \int_{-1}^0 19t^2\,dt + \int_0^1 7t^2\,dt = \frac{19}{3} + \frac{7}{3} = \frac{26}{3}$$

11.   Since this is not a smooth curve use two regions.

On $[0, 2]$: $x = t$, $y = 0$ and on $[2, 4]$ $x = 2$. $y = t$

For the horizontal segment: $[x^2 y\,dx + (x^2 - y^2)\,dy] = 0$

For the vertical segment: $[x^2 y\,dx + (x^2 - y^2)\,dy] = (4 - t^2)\,dt$

$$\int_C [x^2 y\,dx + (x^2 - y^2)\,dy] = \int_0^4 (4 - t^2)\,dt = \left(4t - \frac{t^3}{3}\right)\Big|_0^4 = -\frac{16}{3}$$

15.   For the first piece let $0 \leq t \leq 1$, $x = t$, $y = t^2$

or the second piece let $0 \leq t \leq 1$, $x = 1 - t$, $y = 1 - t$

1st piece: $(x^2 y\,dx - xy\,dy) = t^4\,dt - 2t^4\,dt = -t^4\,dt$

2nd piece: $(x^2 y\,dx - xy\,dy) = -(1 - t)^3\,dt + (1 - t)^2\,dt = (t^3 - 2t^2 + t)\,dt$

$$\int_C (x^2 y\,dx - xy\,dy) = \int_0^1 (-t^4)\,dt + \int_0^1 (t^3 - 2t^2 + t)\,dt$$

$$= -\frac{1}{5} + \frac{1}{4} - \frac{2}{3} + \frac{1}{2} = -\frac{7}{60}$$

19.   Since C is not smooth use two regions.

On $(0, 0)$ to $(0, 1)$: $\mathbf{R} = t\mathbf{j}$, for $0 \le t \le 1$ $d\mathbf{R} = \mathbf{j}$, $\mathbf{F} \bullet d\mathbf{R} = x = 0$

On $(0, 1)$ to $(2, 1)$: $\mathbf{R} = t\mathbf{i} + \mathbf{j}$, for $0 \le t \le 2$ $d\mathbf{R} = \mathbf{i}$,

$\mathbf{F} \bullet d\mathbf{R} = 5x + y = 5t + 1$

$$\int_C \mathbf{F} \bullet d\mathbf{R} = \int_0^2 (5t + 1)dt = \frac{5t^2}{2} + t \Big|_0^2 = 12$$

23.   **a.**   For the arc: $x = t^2$, $y = t$, $z = 0$, $0 \le t \le 1$

$(5xy\,dx + 10yz\,dy + z\,dz) = 10t^4\,dt$

For the line: $x = 1$, $y = 1$, $z = t$, $0 \le t \le 1$

$(5xy\,dx + 10yz\,dy + z\,dz) = t\,dt$

$$\int_C (5xy\,dx + 10yz\,dy + z\,dz) = \int_0^1 (10t^4 + t)\,dt$$

$$= \left(2t^5 + \frac{t^2}{2}\right)\Big|_0^1 = \frac{5}{2}$$

b. $x = y = z = t$ for $0 \le t \le 1$

$(5xy\,dx + 10yz\,dy + z\,dz) = 5t^2 dt + 10t^2 dt + t dt = (15t^2 + t)dt$

$$\int_C (5xy\,dx + 10yz\,dy + z\,dz) = \int_0^1 (15t^2 + t)\,dt$$

$$= \left(5t^3 + \frac{t^2}{2}\right)\Big|_0^1 = \frac{11}{2}$$

27. $\mathbf{R} = t\mathbf{i} + t^2\mathbf{j} - \mathbf{k}$, $d\mathbf{R} = \mathbf{i} + 2t\mathbf{j}$

Using the parameters: $x = t, y = t^2, z = -1$,

$\mathbf{F} = (y - 2z)\mathbf{i} + x\mathbf{j} - 2xy\mathbf{k}$ becomes $\mathbf{F} = (t^2 + 2)\mathbf{i} + t\mathbf{j} - 2t^3\mathbf{k}$

$\mathbf{F} \bullet d\mathbf{R} = t^2 + 2 + 2t^2 = 3t^2 + 2$

$$\int_C \mathbf{F} \bullet d\mathbf{R} = \int_1^2 (3t^2 + 2)dt = \left(t^3 + 2t\right)\Big|_1^2 = 12 - 3 = 9$$

31. $\mathbf{F} = -x\mathbf{i} + 2\mathbf{k}$, for $0 \le t \le 1$:

   $(0, 0)$ to $(0, 1)$, $x = 0$, $y = t$, $\mathbf{R} = t\mathbf{j}$, $d\mathbf{R} = \mathbf{j}$, $\mathbf{F} = 2\mathbf{k}$

   $(0, 1)$ to $(2, 1)$, $x = 2t$, $y = 1$, $\mathbf{R} = 2t\mathbf{i}$, $d\mathbf{R} = 2\mathbf{i}$, $\mathbf{F} = 2t\mathbf{i} + 2\mathbf{k}$

   $(2, 1)$ to $(1, 0)$, $x = 2 - t$, $y = 1 - t$, $\mathbf{R} = (2 - t)\mathbf{i} + (1 - t)\mathbf{j}$,

   $\qquad d\mathbf{R} = -\mathbf{i} - \mathbf{j}$, $\mathbf{F} = (t - 2)\mathbf{i} + 2\mathbf{k}$

   $(1, 0)$ to $(0, 0)$, $x = 1 - t$, $y = 0$, $\mathbf{R} = (1 - t)\mathbf{i}$,

   $\qquad d\mathbf{R} = -\mathbf{i}$, $\mathbf{F} = (t - 1)\mathbf{i} + 2\mathbf{k}$

   $$\int_C \mathbf{F} \cdot \mathbf{T} \, ds = \int_C \mathbf{F} \cdot d\mathbf{R} = W = \int_0^1 [0 - 4t + (2 - t) + (1 - t)] \, dt$$

   $$= \int_0^1 (-6t + 3) \, dt = (-3t^2 + 3t)\Big|_0^1 = 0$$

35. For $0 \le t \le 2\pi$, $x = \cos t$, $y = \sin t$

   $$\frac{x \, dy - y \, dx}{x^2 + y^2} = \frac{\cos^2 t + \sin^2 t}{\cos^2 t + \sin^2 t} \, dt = 1 \, dt$$

   $$\int_C \frac{x \, dy - y \, dx}{x^2 + y^2} = \int_0^{2\pi} 1 \, dt = 2\pi$$

39. For $0 \le t \le 1$, $x = 1 - t$, $y = t$, $\mathbf{R} = (1 - t)\mathbf{i} + t\mathbf{j}$, $d\mathbf{R} = -\mathbf{i} + \mathbf{j}$

   $\mathbf{F} = y\mathbf{i} + 2x\mathbf{j} = t\mathbf{i} + 2(1 - t)\mathbf{j}$, $\mathbf{F} \cdot d\mathbf{R} = -t + 2(1 - t) = -3t + 2$

   $$W = \int_C \mathbf{F} \cdot d\mathbf{R} = \int_0^1 (-3t + 2) \, dt = \left(-\frac{3t^2}{2} + 2t\right)\Big|_0^1 = \frac{1}{2}$$

43. $W = \int_C \mathbf{F} \cdot d\mathbf{R}$ $\mathbf{R} = t^2\mathbf{i} + 2t\mathbf{j} + 4t^3\mathbf{k}$, $d\mathbf{R} = 2t\mathbf{i} + 2\mathbf{j} + 12t^2\mathbf{k}$

   For $0 \le t \le 1$, $x = t^2$, $y = 2t$, $z = 4t^3$

   $\mathbf{F} = x\mathbf{i} + y\mathbf{j} + (xz - y)\mathbf{k} = t^2\mathbf{i} + 2t\mathbf{j} + (4t^5 - 2t)\mathbf{k}$

   $\mathbf{F} \cdot d\mathbf{R} = 2t^3 + 4t + 48t^7 - 24t^3$

   $$\int_C \mathbf{F} \cdot d\mathbf{R} = \int_0^1 (48t^7 - 22t^3 + 4t) \, dt = 6 - \frac{11}{2} + 2 = \frac{5}{2}$$

47. Analysis: Work backward from the line integral:

$$\int_C \mathbf{E} \cdot d\mathbf{R} \;=\; \int_C \mathbf{E} \cdot \frac{d\mathbf{R}}{dt}\, dt, \;\text{ but } \mathbf{E} \cdot \frac{d\mathbf{R}}{dt} \;=\; \frac{m}{Q}\frac{d\mathbf{V}}{dt} \cdot \mathbf{V}$$

Some rather tricky substitutions should give the desired result:

$$\mathbf{F} \;=\; m\mathbf{A} \;=\; m\frac{d\mathbf{V}}{dt},$$

$$\mathbf{F} \cdot \mathbf{V} \;=\; Q(\mathbf{E} + \mathbf{V}\times\mathbf{B}) \cdot \mathbf{V} \;=\; Q\mathbf{E} \cdot \mathbf{V} + (\mathbf{V}\times\mathbf{B}) \cdot \mathbf{V} \;=\; Q\mathbf{E} \cdot \mathbf{V} \;=\; Q\mathbf{E} \cdot \frac{d\mathbf{R}}{dt}$$

So $\mathbf{F} \cdot \mathbf{V} \;=\; m\frac{d\mathbf{V}}{dt} \;=\; Q\mathbf{E} \cdot \frac{d\mathbf{R}}{dt}$

14.3 Independence of Path

5. For $\mathbf{F}$ to be conservative we must have $\dfrac{\partial}{\partial y}\, 2xy^3 \;=\; \dfrac{\partial}{\partial x}\, 3y^2x^2$.

Both equal $6xy^2$, so $\mathbf{F}$ is conservative.

To find f: $\dfrac{\partial f}{\partial x} = 2xy^3$ and $\dfrac{\partial f}{\partial y} = 3y^2x^2$. Doing the "partial integration" of $\dfrac{\partial f}{\partial x}$:

$f = x^2y^3 + C(y)$. But $\dfrac{\partial f}{\partial y}$ must $= 3y^2x^2$, so $3x^2y^2 + \dfrac{\partial C(y)}{\partial y} = 3x^2y^2$.

So $C(y) = 0$ and $f = x^2y^3$

9. $\dfrac{\partial}{\partial y}\, e^{2x}\sin y \;=\; e^{2x}\cos y, \quad \dfrac{\partial}{\partial x}\, e^{2x}\cos y \;=\; 2e^{2x}\cos y$

They are not equal so $\mathbf{F}$ is not conservative.

13. $\displaystyle\int_C (2xy\, dx + x^2\, dy)$

    a. $\dfrac{\partial}{\partial y}\, 2xy = 2x, \quad \dfrac{\partial}{\partial x}\, x^2 = 2x, \;\; \mathbf{F}$ is conservative on a closed curve C so

$$\int_C (2xy\, dx + x^2\, dy) = 0$$

    b. For $0 \le t \le 2$, $x = t$, $y = 2t^2$, $\mathbf{R} = t\mathbf{i} + 2t^2\mathbf{j}$, $d\mathbf{R} = \mathbf{i} + 4t\mathbf{j}$

$\mathbf{F} = 2xy\mathbf{i} + x^2\mathbf{j} = 4t^3\mathbf{i} + t^2\mathbf{j}$, $\mathbf{F} \cdot d\mathbf{R} = 4t^3 + 4t^3 = 8t^3$

$$\int_C (2xy\, dx + x^2\, dy) = \int_0^2 8t^3\, dt = 2t^4 \Big|_0^2 = 32$$

    c. Since $\mathbf{F}$ is conservative the value of the line integral is independent of the path. This path also gives 32.

SURVIVAL HINT

The fundamental theorem of calculus provides a simple method for the evaluation of a definite integral. Likewise the fundamental theorem on line integrals provides a simple method for the evaluation of a line integral. Just as the hypothesis of the fundamental theorem of calculus requires a continuous f', the hypothesis of the fundamental theorem on line integrals requires a conservative vector field. If the hypothesis is met we have independence of path, and the value of the line integral is the difference in the scalar potentials at the ending and beginning points. So the essence of a problem becomes verifying that $\mathbf{F}$ is conservative, and finding f.

17. $\frac{\partial}{\partial y}(2x - y) = -1$, $\frac{\partial}{\partial x}(y^2 - x) = -1$, $\mathbf{F}$ is conservative

The partial integral with respect to x of $(2x - y)$ equals $x^2 - yx + C(y)$,

and $\frac{\partial}{\partial y}[x^2 - yx + C(y)] = y^2 - x$. So $-x + \frac{\partial}{\partial y}C(y) = y^2 - x$,

$\frac{\partial}{\partial y}C(y) = y^2$, $C(y) = \frac{1}{3}y^3 + C$ (let $C = 0$). $f = x^2 - xy + \frac{1}{3}y^3$

By the fundamental theorem:

$$\int_C \mathbf{F} \cdot d\mathbf{R} = f(1, 1) - f(0, 0) = \frac{1}{3}$$

21. $\frac{\partial}{\partial y}(3x^2 + 2x + y^2) = 2y$, $\frac{\partial}{\partial x}(2xy + y^3) = 2y$, f is conservative and the value of the line integral is independent of the path.

The partial integral with respect to x of $(3x^2 + 2x + y^2)$ is $x^3 + x^2 + y^2x + C(y)$

$\frac{\partial}{\partial y}[x^3 + x^2 + y^2x + C(y)] = 2xy + y^3$, so $2xy + \frac{\partial}{\partial y}C(y) = 2xy + y^3$,

$\frac{\partial}{\partial y}C(y) = y^3$, $C(y) = \frac{y^4}{4} + C$ (let $C = 0$). $f = x^3 + x^2 + y^2x + \frac{y^4}{4}$

By the fundamental theorem $\int_C \mathbf{F} \cdot d\mathbf{R} = f(1, 1) - f(0, 0) = \frac{13}{4}$

25. $\frac{\partial}{\partial y}\sin y = \cos y$, $\frac{\partial}{\partial x}(3 + x \cos y) = \cos y$, f is conservative and the value of the line integral is independent of the path.

The partial integral with respect to x of $\sin y$ equals $x \sin y + C(y)$ and

$\frac{\partial}{\partial y}[x \sin y + C(y)] = 3 + x \cos y$, $x \cos y + \frac{\partial}{\partial y}C(y) = 3 + x \cos y$,

$\frac{\partial}{\partial y}C(y) = 3$, $C(y) = 3y + C$ (let $C = 0$), $f = x \sin y + 3y$

25. (con't.)

For $0 \le t \le 1$ we need equivalent initial and terminal values.

For $t = 0$, $(2 \sin \frac{\pi}{2} t \cos \pi t, \sin^{-1} t) = (0, 0)$

For $t = 1$, $(2 \sin \frac{\pi}{2} t \cos \pi t, \sin^{-1} t) = (-2, \frac{\pi}{2})$

The fundamental theorem now gives:

$$\int_C \mathbf{F} \cdot d\mathbf{R} = f(-2, \frac{\pi}{2}) - f(0, 0) = \frac{3\pi}{2} - 2$$

29. $\frac{\partial}{\partial y} 2y = -2$, $\frac{\partial}{\partial x} 2x = 2$, $\mathbf{F}$ is conservative

The partial integral with respect to x of $2y$ equals $2xy + C(y)$ and

$\frac{\partial}{\partial y}\left(2xy + C(y)\right) = 2x$, $2x + \frac{\partial}{\partial y} C(y) = 2x$, $\frac{\partial}{\partial y} C(y) = 0$,

$C(y) = c$ (any constant, let $c = 0$). $f = 2xy$

By the fundamental theorem $\int_C (2y \, dx + 2x \, dy) = f(4, 4) - f(0, 0) = 32$

33. $\mathbf{F} = M\mathbf{i} + N\mathbf{j} + P\mathbf{k} = e^{xy} yz\mathbf{i} + e^{xy} xz\mathbf{j} + e^{xy}\mathbf{k}$

$\frac{\partial P}{\partial y} = e^{xy} x = \frac{\partial N}{\partial z}$, $\frac{\partial M}{\partial z} = e^{xy} y = \frac{\partial P}{\partial x}$, and $\frac{\partial N}{\partial x} = e^{xy} z = \frac{\partial M}{\partial y}$,

so $\mathbf{F}$ is conservative

37. $\mathbf{F} = M\mathbf{i} + N\mathbf{j} + P\mathbf{k} = (xy^2 + yz)\mathbf{i} + (x^2 y + xz + 3y^2 z)\mathbf{j} + (xy + y^3)\mathbf{k}$

$\frac{\partial P}{\partial y} = x + 3y^2 = \frac{\partial N}{\partial z}$, $\frac{\partial M}{\partial z} = y = \frac{\partial P}{\partial x}$, and $\frac{\partial N}{\partial x} = 2xy + z = \frac{\partial M}{\partial y}$,

so $\mathbf{F}$ is conservative

41. $\mathbf{F}$ is constant, and therefore, conservative. $\mathbf{R}$ is a simple closed curve,

so $W = 0$. If you want to work it out; $\mathbf{R} = (3 \cos t)\mathbf{i} + (3 \sin t)\mathbf{j}$,

$F = \frac{mv^2}{r} = \frac{\frac{30}{32}[3(2\pi)]^2}{3} = \frac{45\pi^2}{4}$, $\mathbf{F}$ is always in a direction that is normal

to $\mathbf{R}$, so $\mathbf{F} = \frac{45\pi^2}{4}(-\mathbf{R})$. So $\mathbf{F} \cdot \mathbf{R}$, and the line integral, $= 0$.

45. Analysis: An alternative way to verify the three necessary equalities is to show:

$$\begin{vmatrix} \mathbf{i} & \mathbf{j} & \mathbf{k} \\ \frac{\partial}{\partial x} & \frac{\partial}{\partial y} & \frac{\partial}{\partial z} \\ M & N & P \end{vmatrix} = 0$$

a. The partial integral of f_x with respect to $x = \frac{\partial}{\partial x}(y^2 - 2xz)$

$f = xy^2 - x^2 z + c(y, z)$

45. (con't.)

b.   Now $f_y = \frac{\partial}{\partial y}[xy^2 - x^2z + c(y, z)] = 2xy + \frac{\partial}{\partial y} c(y, z)$,

so $\frac{\partial}{\partial y} c(y, z) = z$, and $c(y, z) = yz + c(z)$

c.   Find $\frac{\partial}{\partial z}f$ to determine $c(z)$:

$\frac{\partial}{\partial z}f = -x^2 + y + c'(z)$, so $c'(z) = 0$ and $c(0) = k$

$f = xy^2 - x^2z + yz + k$

47.   Analysis: As suggested in the hint, pick any two points P and Q on C.

Let P to Q be C_1 and Q to P be C_2. If $\int_C = 0$ then $\int_{C_1} - \int_{C_2} = 0$

so two different paths have the same value and $\int_C$ is independent of path and,

therefore, conservative.

14.4 Green's Theorem

3.   By Green's theorem: $\int_C (M\,dx + N\,dy) = \int\int_D \left(\frac{\partial N}{\partial x} - \frac{\partial M}{\partial y}\right) dA$

Using horizontal strips and $(-)$ due to clockwise orientation:

$$\int_C [(2x^2 + 3y)\,dx + (-3y^2)\,dy] = -\int_0^1 \int_y^{-y+2} (0 - 3)\,dx\,dy$$

$$= 3\int_0^1 (-2y + 2)\,dy = 3(-y^2 + 2y)\Big|_0^1 = 3(-1 + 2) = 3$$

By parametrization: using $0 \le t \le 1$,

C_1: $x = t$, $y = t$, $F_1 = 2t^2 + 3t - 3t^2 = 3t - t^2$

C_2: $x = t + 1$, $y = 1 - t$,

$\quad F_2 = [2(t + 1)^2 + 3(1 - t)] - 3(1 - t)^2(-1) = 5t^2 - 5t + 8$

C_3: $x = 2 - 2t$, $y = 0$, $F_3 = 2(2 - 2t)^2(-2) = -16(t^2 - 2t + 1)$

$$\int_0^1 (F_1 + F_2 + F_3)\,dt = \int_0^1 (-12t^2 + 30t - 8)\,dt$$

$$= (-4t^3 + 15t^2 - 8t)\Big|_0^1 = 3$$

7. $\displaystyle\int_C (2y\ dx\ -\ x\ dy)\ =\ \int_D\int (-1\ -\ 2)\,dA$

$$=\ -\ 3(\text{the area of a semicircle with radius 2})$$

$$=\ -\ 3(2\pi)\ =\ -\ 6\pi$$

11. $\displaystyle\int_C [(x\ -\ y^2)\,dx\ +\ 2xy\ dy]\ =\ \int_D\int (2y\ +\ 2y)\,dA\ =\ \int_0^2\int_0^2 4y\ dy\ dx$

$$=\ \int_0^2 2y^2\,\Big|_0^2\ dx\ =\ \int_0^2 8\ dx\ =\ 16$$

SURVIVAL HINT

When using Green's theorem to find area, take the time to write out the values of dx and dy before substituting (except in the simplest of problems), and do not forget the $\frac{1}{2}$.

15. Parametrizing a circle with radius of 2: $x\ =\ 2\cos t,\ \ y\ =\ 2\sin t,$

By Theorem 14.6, $\displaystyle A\ =\ \frac{1}{2}\int_C (-\ y\ dx\ +\ x\ dy)$

$$=\ \frac{1}{2}\int_C [(-\ 2\sin t)(-\ 2\sin t)\ +\ (2\cos\theta)(2\cos t)]\,dt$$

$$=\ \frac{1}{2}\int_C 4\ dt\ =\ \frac{1}{2}\int_0^{2\pi} 4\ dt\ =\ 4\pi$$

Area of a circle with radius of 2 is $\pi r^2\ =\ 4\pi$

19. $\displaystyle\int_C (x^2 y\ dx\ -\ y^2 x\ dy)\ =\ \int_D\int (-\ y^2\ -\ x^2)\,dA\ =\ -\int_D\int r^2\ dA$

$$=\ -\int_0^{\pi}\int_0^a r^3\ dr\ d\theta\ =\ -\frac{\pi a^4}{4}$$

23. By Green's theorem $\displaystyle\int_C x^2\ dy\ =\ \int_D\int (2x\ -\ 0)\ dA\ =\ 2\int_D\int x\ dA$

$$=\ 2\,\bar{x}\,A$$

$$A\bar{x}\ =\ \frac{1}{2}\int_C x^2\ dy$$

27. $M = \dfrac{-y}{(x - 1)^2 + y^2}$, $N = \dfrac{x - 1}{(x - 1)^2 + y^2}$,

$\dfrac{\partial N}{\partial x} = \dfrac{(x - 1)^2 + y^2 - (x - 1)2(x - 1)}{[(x - 1)^2 + y^2]^2} = \dfrac{y^2 - (x - 1)^2}{[(x - 1)^2 + y^2]^2}$

$\dfrac{\partial M}{\partial y} = \dfrac{[(x - 1)^2 + y^2](- 1) - (- y)(2y)}{[(x - 1)^2 + y^2]^2} = \dfrac{y^2 - (x - 1)^2}{[(x - 1)^2 + y^2]^2}$

So $\displaystyle\int_C \dfrac{- y\, dx + (x - 1)\, dy}{(x - 1)^2 + y^2}$ is conservative, and a Jordan curve is closed,

therefore, the value $= 0$

31. The normal derivative, $\dfrac{\partial f}{\partial n} = \nabla f \bullet \mathbf{N}$, so $\dfrac{\partial x}{\partial n} = \nabla x \bullet \mathbf{N} = \mathbf{i} \bullet \left(\dfrac{dy}{ds}\mathbf{i} - \dfrac{dx}{ds}\mathbf{j}\right)$

$\displaystyle\int_C x \dfrac{\partial x}{\partial n}\, ds = \int_C x \dfrac{dy}{ds}\, ds = \int_C x\, dy = \int\int_D 1\, dA$ (by Green's theorem)

but $A = 1$, so $\displaystyle\int_C x \dfrac{\partial x}{\partial n}\, ds = 1$

35. Analysis: $\dfrac{\partial f}{\partial n} = \nabla f \bullet \mathbf{N}$ where $\nabla f = f_x \mathbf{i} + f_y \mathbf{j}$ and $\mathbf{N} = \dfrac{dy}{ds}\mathbf{i} - \dfrac{dx}{ds}\mathbf{j}$

so $\dfrac{\partial f}{\partial n} = f_x \dfrac{dy}{ds} - f_y \dfrac{dx}{ds}$

$\displaystyle\int_C f \dfrac{\partial x}{\partial n}\, ds = \int_C f\left(f_x \dfrac{dy}{ds} - f_y \dfrac{dx}{ds}\right) ds = \int_C f\left(f_x\, dy - f_y\, dx\right)$

$= \displaystyle\int_C (f f_x\, dy - f f_y\, dx)$ Now apply Green's theorem,

using the product rule on the partials of M and N.

37. Analysis: $\dfrac{\partial g}{\partial n} = \nabla g \bullet \mathbf{N}$ where $\nabla g = g_x \mathbf{i} + g_y \mathbf{j}$ and $\mathbf{N} = \dfrac{dy}{ds}\mathbf{i} - \dfrac{dx}{ds}\mathbf{j}$

so $\dfrac{\partial g}{\partial n} = g_x \dfrac{dy}{ds} - g_y \dfrac{dx}{ds}$.

$\displaystyle\int_C f \dfrac{\partial g}{\partial n}\, ds = \int_C f\left(g_x \dfrac{dy}{ds} - g_y \dfrac{dx}{ds}\right) ds = \int_C f\left(g_x\, dy - g_y\, dx\right)$

$= \displaystyle\int_C (- f\, g_y\, dx + f g_x\, dy)$

Now use Green's theorem, using the product rule on the partials of M and N.

14.5 Surface Integrals

SURVIVAL HINT

If you think of a surface integral as a double integral over a curved surface, rather than a flat region of a plane, then the radical is the "slope" factor to transform R into S.
Since there are several steps in finding the value of the radical, taking the time to carefully write them out will save time in the long run as you will have fewer errors.

3.   The hemisphere $x^2 + y^2 + z^2 = 4$ has R: $x^2 + y^2 = 4$ or $r = 2$

Using polar coordinates: $z = \sqrt{4 - r^2}$, $x = r \cos \theta$, $y = r \sin \theta$

$$dS = \sqrt{\left(\frac{x}{\sqrt{4 - r^2}}\right)^2 + \left(\frac{y}{\sqrt{4 - r^2}}\right)^2 + 1} = \sqrt{\frac{4}{4 - r^2}} = \frac{2}{\sqrt{4 - r^2}}$$

$$\int\int_S (x - 2y)\, dS = \int_0^{2\pi}\int_0^2 (2 \cos \theta - 4 \sin \theta)\frac{2}{\sqrt{4 - r^2}}\, r\, dr\, d\theta$$

$$= -2\int_0^{2\pi}\int_0^2 (\cos \theta - 2 \sin \theta)\frac{-2r\, dr}{\sqrt{4 - r^2}}\, d\theta$$

$$= -4\int_0^{2\pi} (\cos \theta - 2 \sin \theta) \sqrt{4 - r^2}\,\Big|_0^2\, d\theta$$

$$= 8\int_0^{2\pi} (\cos \theta - 2 \sin \theta)\, d\theta$$

$$= 8(\sin \theta + 2 \cos \theta)\Big|_0^{2\pi} = 8(2 - 2) = 0$$

7. $\displaystyle\int\int_S xy\, dS$ where $z = 2 - y$ and R is a 2 by 2 square

$$dS = \sqrt{0 + (-1)^2 + 1} = \sqrt{2}$$

$$\int\int_S xy\, dS = \int_0^2\int_0^2 xy \sqrt{2}\, dy\, dx = \sqrt{2}\int_0^2 x\frac{y^2}{2}\Big|_0^2\, dx = 2\sqrt{2}\int_0^2 x\, dx$$

$$= 2\sqrt{2}\,\frac{x^2}{2}\Big|_0^2 = 4\sqrt{2}$$

11. $\displaystyle\iint_S (x^2 + y^2)\,dS$ where $z = 4 - x - 2y$, R: $0 \le x \le 4$, $0 \le y \le 2$

$$dS = \sqrt{(-1)^2 + (-2)^2 + 1} = \sqrt{6}$$

$$\iint_S (x^2 + y^2)\,dS = \sqrt{6} \int_0^4 \int_0^2 (x^2 + y^2)\,dy\,dx = \sqrt{6}\int_0^4 \left(x^2 y + \frac{y^3}{3}\right)\Big|_0^2\,dx$$

$$= \sqrt{6}\int_0^4 \left(2x^2 + \frac{8}{3}\right)dx = \sqrt{6}\left(\frac{2x^3}{3} + \frac{8x}{3}\right)\Big|_0^4$$

$$= \sqrt{6}\left(\frac{128}{3} + \frac{32}{3}\right) = \frac{160\sqrt{6}}{3}$$

15. $z = x^2 + y^2$, $z \le 4$ gives R: $x^2 + y^2 \le 4$

$$dS = \sqrt{(2x)^2 + (2y)^2 + 1} = \sqrt{4(x^2 + y^2) + 1}$$

Using polar coordinates:

$$\iint_S z\,dS = \int_0^{2\pi}\int_0^2 r^2\sqrt{4r^2 + 1}\, r\,dr\,d\theta = 4\pi\int_0^2 r^3\sqrt{r^2 + \left(\tfrac{1}{2}\right)^2}\,dr$$

Using trigonometric substitution or formula #171:

$$4\pi\left[\frac{\left(r^2 + \frac{1}{4}\right)^{5/2}}{5} - \frac{\frac{1}{4}\left(r^2 + \frac{1}{4}\right)^{3/2}}{3}\right]\Big|_0^2 = 4\pi\left[\frac{r^2}{5} - \frac{2\left(\frac{1}{4}\right)}{15}\right]\left(r^2 + \frac{1}{4}\right)^{3/2}\Big|_0^2$$

$$= 4\pi\left[\left(\frac{4}{5} - \frac{1}{30}\right)\frac{17}{4}\sqrt{\frac{17}{4}} + \frac{1}{30}\left(\frac{1}{8}\right)\right] = 4\pi\left(\frac{391\sqrt{17}}{240} + \frac{1}{240}\right) = \frac{\pi}{60}\left(391\sqrt{17} + 1\right)$$

19. When $z = 0$, R: $x^2 + y^2 = 1$

$$dS = \sqrt{\left(\frac{-x}{\sqrt{1 - r^2}}\right)^2 + \left(\frac{-y}{\sqrt{1 - r^2}}\right)^2 + 1} = \sqrt{\frac{1}{1 - r^2}} = \frac{1}{\sqrt{1 - r^2}}$$

Using polar coordinates:

$$\iint_S (x^2 + y^2)\,dS = \int_0^{2\pi}\int_0^1 \frac{r^2}{\sqrt{1 - r^2}}\, r\,dr\,d\theta = 2\pi\int_0^1 \frac{r^3\,dr}{\sqrt{1 - r^2}}$$

Using trigonometric substitution or formula #227:

$$= 2\pi\left[\frac{(1 - r^2)^{3/2}}{3} - (1 - r^2)^{1/2}\right]\Big|_0^1$$

$$= 2\pi\left(-\frac{1}{3} + 1\right) = \frac{4\pi}{3}$$

23. $\displaystyle\iint\limits_{S} \mathbf{F} \bullet \mathbf{N}\, dS$ with $\mathbf{F} = x\mathbf{i} + 2y\mathbf{j} - 3z\mathbf{k}$ and $z = 2 - 5x + 4y$ and

S is the portion of the plane above the unit square $0 \leq x \leq 1,\ 0 \leq y \leq 1$

$dS = \sqrt{(-5)^2 + 4^2 + 1} = \sqrt{42}$

$g(x,\ y,\ z) = z + 5x - 4y - 2,\quad \mathbf{N} = \dfrac{5\mathbf{i} - 4\mathbf{j} + \mathbf{k}}{\sqrt{42}},$

$\mathbf{F} \bullet \mathbf{N} = \dfrac{5x - 8y - 3z}{\sqrt{42}}$

$\qquad = \dfrac{5x - 8y - 3(2 - 5x + 4y)}{\sqrt{42}} = \dfrac{2(10x - 10y - 3)}{\sqrt{42}}$

Now not all of the plane is above the unit square. When $z = 0$, $5x - 4y = 2$.

The line $y = \dfrac{5}{4}x - \dfrac{1}{2}$ cuts off a corner of the square, requiring two regions.

$\displaystyle\iint\limits_{S} \mathbf{F} \bullet \mathbf{N}\, dS$

$\displaystyle = \int_{0}^{2/5}\int_{0}^{1} 2(10x - 10y - 3)\, dy\, dx + \int_{2/5}^{1}\int_{5x/4-1/2}^{1} 2(10x - 10y - 3)\, dy\, dx$

$\displaystyle = 2\int_{0}^{2/5} (10xy - 5y^2 - 3y)\Big|_{0}^{1} + 2\int_{2/5}^{1} (10xy - 5y^2 - 3y)\Big|_{5x/4-1/2}^{1}$

$\displaystyle = \int_{0}^{2/5} (10x - 8)\, dx + 2\int_{2/5}^{1}\left[10x - 8 - \left(\frac{5x}{4} - \frac{1}{2}\right)\left\{10x - 5\left(\frac{5x}{4} - \frac{1}{2}\right) - 3\right\}\right] dx$

$\displaystyle = 2(5x^2 - 8x)\Big|_{0}^{2/5} + 2\int_{2/5}^{1}\left[10x - 8 - \left(\frac{5x}{4} - \frac{1}{2}\right)\left(\frac{15x}{4} - \frac{1}{2}\right)\right] dx$

$\displaystyle = 2\left(-\frac{12}{5}\right) + 2\int_{2/5}^{1}\left(10x - 8 - \frac{75x^2}{16} + \frac{5x}{2} - \frac{1}{4}\right) dx$

$\displaystyle = -\frac{24}{5} + 2\int_{2/5}^{1}\left(\frac{-75x^2}{16} + \frac{25x}{2} - \frac{33}{4}\right) dx$

$\displaystyle = -\frac{24}{5} + 2\left(-\frac{25x^3}{16} + \frac{25x^2}{4} - \frac{33x}{4}\right)\Big|_{2/5}^{1}$

$\displaystyle = -\frac{24}{5} + 2\left[-\frac{25}{16} + \frac{25}{4} - \frac{33}{4} - \left(-\frac{1}{10} + 1 - \frac{33}{10}\right)\right]$

$\displaystyle = -\frac{24}{5} + 2\left(-\frac{25}{16} - 2 + \frac{12}{5}\right) = -\frac{57}{8}$

27.　　$\mathbf{F} = x^2\mathbf{i} + y^2\mathbf{j} + z^2\mathbf{k}, \quad z = y + 1, \quad dS = \sqrt{2}$

$g(x, y, z) = z - y - 1, \quad \mathbf{N} = \dfrac{-\mathbf{j} + \mathbf{k}}{\sqrt{2}}$

$\mathbf{F} \bullet \mathbf{N} = \dfrac{-y^2 + z^2}{\sqrt{2}} = \dfrac{2y + 1}{\sqrt{2}}$

Now $R: x^2 + y^2 = 1$, since the plane intersects the cylinder as a tangent in the

xy-plane. Using polar coordinates:

$$\int_S \int \mathbf{F} \bullet \mathbf{N} \, dS = \int_0^{2\pi} \int_0^1 (2r \sin\theta + 1) \, r \, dr \, d\theta = \int_0^{2\pi} \left(\frac{2r^3}{3} \sin\theta + \frac{r^2}{2} \right)\Big|_0^1 d\theta$$

$$= \int_0^{2\pi} \left(\frac{2}{3} \sin\theta + \frac{1}{2} \right) d\theta = \left(-\frac{2}{3} \cos\theta + \frac{\theta}{2} \right)\Big|_0^{2\pi}$$

$$= -\frac{2}{3} + \pi + \frac{2}{3} = \pi$$

31.　　$\displaystyle\int_S \int (x^2 + y - z) \, dS$ where $\mathbf{R} = u\mathbf{i} - u^2\mathbf{j} + v\mathbf{k}, \quad 0 \le u \le 2, \, 0 \le v \le 1$

$$\int_S \int (x^2 + y - z) \, dS = \int_D \int f(\mathbf{R}) \, \|\mathbf{R}_u \times \mathbf{R}_v\| \, du \, dv$$

$$\mathbf{R}_u \times \mathbf{R}_v = \begin{vmatrix} \mathbf{i} & \mathbf{j} & \mathbf{k} \\ 1 & -2u & 0 \\ 0 & 0 & 1 \end{vmatrix} = -2u\mathbf{i} - \mathbf{j}, \quad \|\mathbf{R}_u \times \mathbf{R}_v\| = \sqrt{4u^2 + 1},$$

$x = u, \quad y = -u^2, \quad z = v, \quad$ so $f(\mathbf{R}) = u^2 - u^2 - v = -v$

$$\int_D \int f(\mathbf{R}) \, \|\mathbf{R}_u \times \mathbf{R}_v\| \, dv \, du = \int_0^2 \int_0^1 -v\sqrt{4u^2 + 1} \, dv \, du = -\frac{1}{2} \int_0^2 2\sqrt{u^2 + \frac{1}{4}} \, du$$

$$= -\int_0^2 \sqrt{u^2 + \frac{1}{4}} \, du \quad \text{Using trigonometric substitution or formula \#168:}$$

$$= -\left[\frac{u\sqrt{u^2 + \frac{1}{4}}}{2} + \frac{1}{8}\ln\left(u + \sqrt{u^2 + \frac{1}{4}}\right) \right]\Big|_0^2$$

$$= -\left[\frac{\sqrt{17}}{2} + \frac{1}{8}\ln\left(2 + \frac{\sqrt{17}}{2}\right) - \frac{1}{8}\ln\frac{1}{2} \right]$$

$$= -\frac{\sqrt{17}}{2} - \frac{1}{8}\ln(4 + \sqrt{17})$$

35. $f(x,\, y,\, z) = 1 - x^2 - y^2 - z,\quad dS = \sqrt{4x^2 + 4y^2 + 1}$

$$m = \int\int_S \rho(x,\, y,\, z)\, dS = \int_0^{2\pi}\int_0^1 1\,\sqrt{4r^2 + 1}\; r\, dr\, d\theta$$

$$= \frac{1}{8}\int_0^{2\pi}\int_0^1 \sqrt{4r^2 + 1}\;(8r\,dr)\, d\theta = \frac{1}{8}\int_0^{2\pi}\left.\frac{2(4r^2 + 1)^{3/2}}{3}\right|_0^1 d\theta$$

$$= \frac{1}{8}\int_0^{2\pi}\left(\frac{10\sqrt{5}}{3} - \frac{2}{3}\right) d\theta = \frac{\pi}{4}\left(\frac{10\sqrt{5} - 2}{3}\right) = \frac{\pi(5\sqrt{5} - 1)}{6}$$

39.   Analysis: $I_z = \int\int_S (x^2 + y^2)\, dS$ for a conical shell with radius $= a$, and vertex at the origin, has a projection in the xy-plane of $\mathbf{R}$: $x^2 + y^2 = a^2$. Since we are not given a vertex angle, the equation of the cone can be: $z = C\sqrt{x^2 + y^2}$. Use this equation to find dS. Now the desired result is in terms of the mass, m. To find m use $\int\int_S \rho(x,\, y,\, z)\, dS$. Integrate I_z with polar coordinates.

14.6 Stokes's Theorem

SURVIVAL HINT

Green's theorem relates a double integral over a flat region of a plane to a line integral over the boundary of the region. Stokes's theorem relates a double integral over a curved surface in $\mathbb{R}^3$ to the line integral over its boundary.

SURVIVAL HINT

As with any other theorem, Stokes's theorem has a hypothesis that must be verified before the conclusion can be applied. In each case verify that the orientation of C is compatable with the orientation on S.

3.   Evaluating the line integral $\displaystyle\int_C \mathbf{F} \bullet d\mathbf{R}$

$\mathbf{F} = (x + 2z)\mathbf{i} + (y - x)\mathbf{j} + (z - y)\mathbf{k}$ and the portion of the plane $x + 2y + z = 3$ cut off by the first octant. The three edges of the boundary triangle are E_1: $x + 2y = 3$, E_2: $2y + z = 3$, E_3: $x + z = 3$.

3. (con't.)

Parametrizing all three with $0 \leq t \leq \frac{3}{2}$: (to avoid fractions)

E_1: $x = 3 - 2t$, $y = t$, $z = 0$, $\mathbf{R} = (3 - 2t)\mathbf{i} + t\mathbf{j}$,

$\quad\quad d\mathbf{R} = -2\mathbf{i} + \mathbf{j}$, $\mathbf{F} \bullet d\mathbf{R} = -2(x + 2z) + (y - x)$

$\quad\quad\quad\quad\quad\quad\quad\quad\quad\quad = -3x + y - 4z$

$\quad\quad\quad\quad\quad\quad\quad\quad\quad\quad = 7t - 9$

E_2: $x = 0$, $y = \frac{3}{2} - t$, $z = 2t$, $\mathbf{R} = (\frac{3}{2} - t)\mathbf{j} + 2t\mathbf{k}$,

$\quad\quad d\mathbf{R} = -\mathbf{j} + 2\mathbf{k}$, $\mathbf{F} \bullet d\mathbf{R} = -(y - x) + 2(z - y)$

$\quad\quad\quad\quad\quad\quad\quad\quad\quad\quad = x - 3y + 2z$

$\quad\quad\quad\quad\quad\quad\quad\quad\quad\quad = 7t - \frac{9}{2}$

E_3: $x = 2t$, $y = 0$, $z = 3 - 2t$, $\mathbf{R} = 2t\mathbf{i} + (3 - 2t)\mathbf{k}$,

$\quad\quad d\mathbf{R} = 2\mathbf{i} - 2\mathbf{k}$, $\mathbf{F} \bullet d\mathbf{R} = 2(x + 2z) - 2(z - y)$

$\quad\quad\quad\quad\quad\quad\quad\quad\quad\quad = 2x + 2y + 2z = 6$

$$\int_{E_1} \mathbf{F} \bullet d\mathbf{R} = \int_0^{3/2} (7t - 9)\,dt = \left(\frac{7t^2}{2} - 9t\right)\Big|_0^{3/2} = \frac{63}{8} - \frac{27}{2} = -\frac{45}{8}$$

$$\int_{E_2} \mathbf{F} \bullet d\mathbf{R} = \int_0^{3/2} \left(7t - \frac{9}{2}\right)dt = \left(\frac{7t^2}{2} - \frac{9t}{2}\right)\Big|_0^{3/2} = \frac{9}{8}$$

$$\int_{E_3} \mathbf{F} \bullet d\mathbf{R} = \int_0^{3/2} 6\,dt = 9$$

$$\int_C \mathbf{F} \bullet d\mathbf{R} = -\frac{45}{8} + \frac{9}{8} + 9 = \frac{9}{2}$$

Using Stokes's theorem: find $\iint_S \text{curl } \mathbf{F} \bullet \mathbf{N}\, dS$

$$\text{curl } \mathbf{F} = \begin{vmatrix} \mathbf{i} & \mathbf{j} & \mathbf{k} \\ \frac{\partial}{\partial x} & \frac{\partial}{\partial y} & \frac{\partial}{\partial z} \\ x + 2z & y - x & z - y \end{vmatrix} = -\mathbf{i} + 2\mathbf{j} - \mathbf{k}$$

The unit normal to the plane: $\mathbf{N} = \dfrac{\mathbf{i} + 2\mathbf{j} + \mathbf{k}}{\sqrt{6}}$

Finally, we need $dS = \sqrt{6}\, dy\, dx$

$$\iint_S \text{curl } \mathbf{F} \bullet \mathbf{N}\, dS = \iint_R \frac{-1 + 4 - 1}{\sqrt{6}} \sqrt{6}\, dy\, dx$$

$$= 2(\text{the area of the triangular region}) = 2\left(\frac{9}{4}\right) = \frac{9}{2}$$

7.

$$\text{curl } \mathbf{F} = \begin{vmatrix} \mathbf{i} & \mathbf{j} & \mathbf{k} \\ \dfrac{\partial}{\partial x} & \dfrac{\partial}{\partial y} & \dfrac{\partial}{\partial z} \\ z & x & y \end{vmatrix} = \mathbf{i} + \mathbf{j} + \mathbf{k}$$

The unit normal to the plane $2x + y + 3z = 6$, $\mathbf{N} = \dfrac{2\mathbf{i} + \mathbf{j} + 3\mathbf{k}}{\sqrt{14}}$

$\text{curl } \mathbf{F} \bullet \mathbf{N} = \dfrac{2 + 1 + 3}{\sqrt{14}} = \dfrac{6}{\sqrt{14}}$, $\quad dS = \dfrac{\sqrt{14}}{3} \, dy \, dx$

$$\int_S \int \text{curl } \mathbf{F} \bullet \mathbf{N} \, dS = \int_R \int \dfrac{6}{\sqrt{14}} \dfrac{\sqrt{14}}{3} \, dy \, dx = -2(\text{the area of the triangular region})$$

$$= -2(9) = -18 \quad (\text{negative due to left-handed orientation})$$

11.

$$\text{curl } \mathbf{F} = \begin{vmatrix} \mathbf{i} & \mathbf{j} & \mathbf{k} \\ \dfrac{\partial}{\partial x} & \dfrac{\partial}{\partial y} & \dfrac{\partial}{\partial z} \\ y & z & x \end{vmatrix} = -\mathbf{i} - \mathbf{j} - \mathbf{k}$$

The unit normal to the plane $x + y = 2$; $\mathbf{N} = \dfrac{\mathbf{i} + \mathbf{j}}{\sqrt{2}}$

$\text{curl } \mathbf{F} \bullet \mathbf{N} = \dfrac{-1 - 1}{\sqrt{2}} = -\sqrt{2}$

To find S and dS we need the intersection of the plane and the sphere:

completing the square for the sphere; $(x - 1)^2 + (y - 1)^2 + z^2 = 2$.

The projection of this sphere in the yz-plane will be a circle with center at $(1, 0)$

and radius of 1. $\quad y = 2 - x$, $\quad dS = \sqrt{f_x^2 + f_z^2 + 1} = \sqrt{2}$

$$\int_S \int \text{curl } \mathbf{F} \bullet \mathbf{N} \, dS = \int_R \int (-\sqrt{2}) \sqrt{2} \, dy \, dx$$

$$= -2 \, (\text{the area of a circle with } r = 1) = -2\pi$$

15. $\quad \mathbf{F} = xy\mathbf{i} - z\mathbf{j},$

$$\text{curl } \mathbf{F} = \begin{vmatrix} \mathbf{i} & \mathbf{j} & \mathbf{k} \\ \dfrac{\partial}{\partial x} & \dfrac{\partial}{\partial y} & \dfrac{\partial}{\partial z} \\ xy & -z & 0 \end{vmatrix} = \mathbf{i} - x\mathbf{k}$$

The surface is not smooth, so there are 5 different representations for the

outward unit normal: $\mathbf{k}, \mathbf{i}, -\mathbf{i}, \mathbf{j}, -\mathbf{j}$. $\text{curl } \mathbf{F} \bullet \mathbf{N}$ for each of these is:

$-x, 1, -1, 0, 0$. The 2nd and the 3rd are inverses, and $dS = 1 \, dy \, dx$, so

$$\int_S \int \text{curl } \mathbf{F} \bullet \mathbf{N} \, dS = \int_0^1 \int_0^1 (-x) \, dy \, dx = -\frac{1}{2}$$

19. $\mathbf{F} = 4y\mathbf{i} + z\mathbf{j} + 2y\mathbf{k}$

$$\text{curl } \mathbf{F} = \begin{vmatrix} \mathbf{i} & \mathbf{j} & \mathbf{k} \\ \dfrac{\partial}{\partial x} & \dfrac{\partial}{\partial y} & \dfrac{\partial}{\partial z} \\ 4y & z & 2y \end{vmatrix} = \mathbf{i} - 4\mathbf{k}$$

$\mathbf{N} = \mathbf{k}$, curl $\mathbf{F} \bullet \mathbf{N} = -4$, R is the circle in the xy-plane with radius of 2.

Using polar coordinates:

$$\int\int_S \text{curl } \mathbf{F} \bullet \mathbf{N} \ dS = \int_0^{2\pi}\int_0^2 (-4) \ r \ dr \ d\theta = -8\pi \int_0^2 r \ dr \ d\theta = -8\pi \left.\frac{r^2}{2}\right|_0^2 = -16\pi$$

23. By Stokes's theorem $\displaystyle\int\int_S \text{curl } \mathbf{F} \bullet \mathbf{N} \ dS = \int_C \mathbf{F} \bullet d\mathbf{R}$

$\mathbf{F} = y^2\mathbf{i} + xy\mathbf{j} + xz\mathbf{k}$, $\mathbf{R} = x^2\mathbf{i} + y^2\mathbf{j}$, $d\mathbf{R} = 2x\mathbf{i} + 2y\mathbf{j}$

$\mathbf{F} \bullet d\mathbf{R} = 2xy^2 + 2xy^2 = 4xy^2$, using polar coordinates on the unit circle:

$$\int_C \mathbf{F} \bullet d\mathbf{R} = \int_0^{2\pi} 4(\cos\theta)(\sin^2\theta) \ d\theta = 4\left.\frac{\sin^3\theta}{3}\right|_0^{2\pi} = 4(0) = 0$$

25. Analysis: Consider the z-positive half of the ellipsoid, S_1, and the z-negative half, S_2, separately. By the symmetry of the surface $\mathbf{N}(S_1) = -\mathbf{N}(S_2)$.

$$\int\int_S \text{curl } \mathbf{F} \bullet \mathbf{N} \ dS = \int\int_{S_1} \text{curl } \mathbf{F} \bullet \mathbf{N} \ dS_1 + \int\int_{S_2} \text{curl } \mathbf{F} \bullet (-\mathbf{N}) \ dS_2 = 0$$

If $\mathbf{F}$ is a vector field whose component functions have continuous partial derivatives and S can be broken into a number of simple curves which are symmetric about the xy-plane, then $\displaystyle\int\int_S \text{curl } \mathbf{F} \bullet \mathbf{N} \ dS = 0$

27. Analysis: By Stokes's theorem

$$\int_C \mathbf{B} \bullet d\mathbf{R} = \int\int_S \text{curl } \mathbf{B} \bullet \mathbf{N} \ dS, \text{ given curl } \mathbf{B} = k\mathbf{J}$$

$$= \int\int_S k\mathbf{J} \bullet \mathbf{N} \ dS = k\int\int_S \mathbf{J} \bullet \mathbf{N} \ dS$$

also $\displaystyle I = \int_C \mathbf{H} \bullet d\mathbf{R} = \int\int_S \text{curl } \mathbf{H} \bullet \mathbf{N} \ dS = \int\int_S \mathbf{J} \bullet \mathbf{N} \ dS$ from Example 5

Substitute for the desired result.

14.7 Divergence Theorem

SURVIVAL HINT

Green's theorem, which found a relationship between a double integral over a region in a plane and the line integral of its boundary, is extended to the divergence theorem.

This gives the relationship between the triple integral over a portion of space, D, and the surface integral over the boundary, S, of that region of $\mathbb{R}^3$.

3. First the surface integral:

$$\mathbf{F} = 2y^2\mathbf{j}, \quad \mathbf{N} = \frac{\mathbf{i} + 4\mathbf{j} + \mathbf{k}}{3\sqrt{2}}, \quad \mathbf{F} \bullet \mathbf{N} = \frac{8y^2}{3\sqrt{2}}$$

$dS = 3\sqrt{2}\, dA,$ and the region in the xy-plane has boundary line $x + 4y = 8$

$$\int\int_S \mathbf{F} \bullet \mathbf{N}\, dS = \int_0^2 \int_0^{8-4y} \frac{8y^2}{3\sqrt{2}} 3\sqrt{2}\, dx\, dy = 8\int_0^2 y^2(8 - 4y)\, dy$$

$$= 8\left(\frac{8y^3}{3} - y^4\right)\Big|_0^2 = 8\left(\frac{64}{3} - 16\right) = \frac{128}{3}$$

Now the volume integral:

$$\text{div } \mathbf{F} = 4y$$

$$\int\int\int_D \text{div } \mathbf{F}\, dV = 4\int_0^2 \int_0^{8-4y} \int_0^{8-x-4y} y\, dz\, dx\, dy$$

$$= 4\int_0^2 \int_0^{8-4y} (8y - xy - 4y^2)\, dx\, dy$$

$$= 4\int_0^2 \left(8xy - \frac{x^2 y}{2} - 4xy^2\right)\Big|_0^{8-4y}$$

$$= 4\int_0^2 (32y - 32y^2 + 8y^3)\, dy = 32\int_0^2 (4y - 4y^2 + y^3)\, dy$$

$$= 32\left(2y^2 - \frac{4y^3}{3} + \frac{y^4}{4}\right)\Big|_0^2 = 32\left(8 - \frac{32}{3} + 4\right) = \frac{128}{3}$$

7. $\mathbf{F} = (\cos yz)\mathbf{i} + e^{xz}\mathbf{j} + 3z^2\mathbf{k}$, div $\mathbf{F} = 6z$, using cylindrical coordinates:

$$\iiint_D \text{div } \mathbf{F} \, dV = 6\int_0^{2\pi}\int_0^2\int_0^{\sqrt{4-r^2}} z \, dz \, r \, dr \, d\theta = 3\int_0^{2\pi}\int_0^2 (4 - r^2) \, r \, dr \, d\theta$$

$$= 3\int_0^{2\pi}\left(2r^2 - \frac{r^4}{4}\right)\Big|_0^2 \, d\theta = 3\int_0^{2\pi}(8 - 4) \, d\theta = 24\pi$$

11. Since this is not a closed surface we will use the divergence theorem for the closed surface of the paraboloid and its disk, then subtract the disk.

$\mathbf{F} = x\mathbf{i} + y\mathbf{j} + z\mathbf{k}$, div $\mathbf{F} = 1 + 1 + 1 = 3$,

Using cylindrical coordinates, $z = x^2 + y^2 = r^2$:

$$\iiint_D \text{div } \mathbf{F} \, dV = 3\int_0^{2\pi}\int_0^3\int_0^{r^2} dz \, r \, dr \, d\theta = 3\int_0^{2\pi}\int_0^3 r^2 \, r \, dr \, d\theta$$

$$= 6\pi\int_0^3 r^3 \, dr = 6\pi \frac{r^4}{4}\Big|_0^3 = \frac{243\pi}{2}$$

For the disk: $x^2 + y^2 = 9$ $g(x, y, z) = 9 - x^2 - y^2$, $\mathbf{F} = -x^2\mathbf{i} - y^2\mathbf{j}$

and $\mathbf{N} = \mathbf{k}$, $\mathbf{F} \cdot \mathbf{N} = 0$, so $\iint_S \mathbf{F} \cdot \mathbf{N} \, dS = 0$, and our value remains $\frac{243\pi}{2}$.

15. $\mathbf{F} = x\mathbf{i} + y\mathbf{j} + (z^2 - 1)\mathbf{k}$, div $\mathbf{F} = 1 + 1 + 2z = 2(z + 1)$

Using cylindrical coordinates:

$$\iiint_D \text{div } \mathbf{F} \, dV = 2\int_0^{2\pi}\int_0^2\int_0^1 (z + 1) \, dz \, r \, dr \, d\theta = 2\int_0^{2\pi}\int_0^2 \left(\frac{z^2}{2} + z\right)\Big|_0^1$$

$$= 2\int_0^{2\pi}\int_0^2 \frac{3}{2} r \, dr \, d\theta = 3\int_0^{2\pi}\frac{r^2}{2}\Big|_0^2 \, d\theta = 6(2\pi) = 12\pi$$

19. $\mathbf{F} = x^3\mathbf{i} + y^3\mathbf{j} + 3a^2z\mathbf{k}$, div $\mathbf{F} = 3x^2 + 3y^2 + 3a^2$

Using cylindrical coordinates: div $\mathbf{F} = 3(r^2 + a^2)$

$$\iiint_D \text{div } \mathbf{F} \, dV = \int_0^{2\pi}\int_0^a\int_0^1 3(r^2 + a^2) \, dz \, r \, dr \, d\theta$$

$$= 3\int_0^{2\pi}\int_0^a (r^3 + a^2r) \, dr \, d\theta = 3\int_0^{2\pi}\left(\frac{r^4}{4} + \frac{a^2r^2}{2}\right)\Big|_0^a \, d\theta$$

$$= 3\int_0^{2\pi}\frac{3a^4}{4} \, d\theta = \frac{9\pi a^4}{2}$$

23. Analysis:

　a. $\dfrac{\partial u}{\partial n}$ can be found by $\nabla u \bullet \mathbf{N}$, (p. 955), substitute and apply Stokes's theorem.

　b.
$$\iint_S (u\nabla v) \bullet \mathbf{N}\; dS = \iiint_V \text{div}\,(u\nabla v)\; dV = \iiint_V \text{div}\,(x^2 + y^2 + z^2)\; dV$$

$$= \iiint_V (\nabla u \bullet \nabla v + u^2 v)\, dV$$

$$= \int_0^1 \int_0^1 \int_0^1 [x + y + z + 3(x + y + z)]\, dz\, dy\, dx = 6$$

25. Analysis: Recall that the Laplacian of f equal to 0 is defined as harmonic. (p. 904) Use this definition with the result of **23 a.**

27. Analysis: Follow the derivation on p. 954, keeping in mind that k now represents a variable, $k(x, y\; z)$, rather than a constant. Also recall that the del operation uses a product-like rule: $\nabla(uv) = v\nabla u + u\nabla v$. So that $\nabla(k\nabla T) = \nabla T \nabla k + k\nabla^2 T$.

29. Analysis: By the divergence theorem $q = \int_S \int \mathbf{D} \bullet \mathbf{N}\; dS = \iiint_D \text{div}\,\mathbf{D}\; dV$

$$= \epsilon \iiint_D \text{div}\,\mathbf{E}\; dV, \quad \text{but the total electric charge, } q, \text{ is also given by}$$

$$q = \iiint_D Q\; dV, \quad \text{where } Q \text{ represents the charge density function.}$$

Chapter 14 Review

PRACTICE PROBLEMS

23. $\mathbf{F} = yz\mathbf{i} + xz\mathbf{j} + xy\mathbf{k} = M\mathbf{i} + N\mathbf{j} + P\mathbf{k}$

$\dfrac{\partial M}{\partial y} = z = \dfrac{\partial N}{\partial x}, \dfrac{\partial M}{\partial z} = y = \dfrac{\partial P}{\partial x}, \dfrac{\partial N}{\partial z} = z = \dfrac{\partial P}{\partial y},$ so $\mathbf{F}$ is conservative

$\dfrac{\partial f}{\partial x} = yz$, so $f = xyz + g(y, z)$ and $\dfrac{\partial f}{\partial y} = xz + \dfrac{\partial g}{\partial y}$, so from $\mathbf{F}$, $\dfrac{\partial g}{\partial y} = 0$,

$g = h(z).$ $f = xyz + h(z)$, but $\dfrac{\partial f}{\partial z} = xy + \dfrac{\partial h}{\partial z}$, so from $\mathbf{F}$, $\dfrac{\partial h}{\partial z} = 0$,

$h = C$ (a constant) and $f = xyz + C$

24. $\mathbf{F} = x^2 y \mathbf{i} - e^{yz}\mathbf{j} + \frac{x}{2}\mathbf{k}$

div $\mathbf{F} = 2xy - ze^{yz}$

$$\text{curl } \mathbf{F} = \begin{bmatrix} \mathbf{i} & \mathbf{j} & \mathbf{k} \\ \frac{\partial}{\partial x} & \frac{\partial}{\partial y} & \frac{\partial}{\partial z} \\ x^2 y & -e^{yz} & \frac{x}{2} \end{bmatrix} = ye^{yz}\mathbf{i} - \frac{1}{2}\mathbf{j} - x^2\mathbf{k}$$

25.   By Green's theorem $\displaystyle\int_C \mathbf{F} \bullet d\mathbf{R} = \int_C (M\, dx + N\, dy) = \int\!\!\int_D \left(\frac{\partial N}{\partial x} - \frac{\partial M}{\partial y}\right) dA$

$$= \int\!\!\int_D (-1)\, dA$$

$$= -\text{(the area of the triangle)} = -2$$

26.   By Stokes's theorem $\displaystyle\int_C \mathbf{F} \bullet d\mathbf{R} = \int\!\!\int_S (\text{curl } \mathbf{F}) \bullet \mathbf{N}\, dS$

$$\text{curl } \mathbf{F} = \begin{bmatrix} \mathbf{i} & \mathbf{j} & \mathbf{k} \\ \frac{\partial}{\partial x} & \frac{\partial}{\partial y} & \frac{\partial}{\partial z} \\ 2y & z & y \end{bmatrix} = (1 - 1)\mathbf{i} - 2\mathbf{k} = -2\mathbf{k}$$

The intersection of the sphere and the plane: $x^2 + y^2 + (x + 2)^2 = 4(x + 2)$,

$2x^2 + y^2 = 4$, is the ellipse: $\dfrac{x^2}{2} + \dfrac{y^2}{4} = 1$

$g(x, y, z) = x - z + 2$, $\mathbf{N} = \dfrac{\mathbf{i} - \mathbf{k}}{\sqrt{2}}$, curl $\mathbf{F} \bullet \mathbf{N} = \dfrac{2}{\sqrt{2}}$, $dS = \sqrt{2}\, dy\, dx$

$$\int\!\!\int_S (\text{curl } \mathbf{F}) \bullet \mathbf{N}\, dS = \int\!\!\int_S \sqrt{2}\, \sqrt{2}\, dy\, dx = 2\text{ (the area of the ellipse, } \pi ab)$$

$$= 2\,(\pi 2\sqrt{2}) = 4\pi\sqrt{2}$$

27.   By the divergence theorem $\displaystyle\int\!\!\int_S \mathbf{F} \bullet \mathbf{N}\, dS = \int\!\!\int\!\!\int_D \text{div } \mathbf{F}\, dV$

$\mathbf{F} = x^2\mathbf{i} + (y + z)\mathbf{j} - 2z\mathbf{k}$, div $\mathbf{F} = 2x + 1 - 2 = 2x - 1$

$$\int\!\!\int\!\!\int_D \text{div } \mathbf{F}\, dV = \int\!\!\int\!\!\int_D (2x - 1)\, dV = \int_0^1\!\!\int_0^1\!\!\int_0^1 (2x - 1)\, dz\, dy\, dx$$

$$= \int_0^1 (2x - 1)\, dx = (x^2 - x)\Big|_0^1 = 0$$

28. $\quad \mathbf{F} = \dfrac{x\,dx}{(x^2 + y^2)^2} + \dfrac{y\,dy}{(x^2 + y^2)^2} = M\,dx + N\,dy$

$\dfrac{\partial N}{\partial x} = \dfrac{-2y(2x)}{(x^2 + y^2)^3} = \dfrac{-4xy}{(x^2 + y^2)^3}, \quad \dfrac{\partial M}{\partial y} = \dfrac{-2x(2y)}{(x^2 + y^2)^3} = \dfrac{-4xy}{(x^2 + y^2)^3}$

$\dfrac{\partial N}{\partial x} = \dfrac{\partial M}{\partial y}$ so $\mathbf{F}$ is conservative and independent of path. Since C is a closed path the value of the line integral is 0.

29. Since $m\omega^2$ is a scalar $\mathbf{F}$ is conservative if $\mathbf{R}$ is conservative. Testing $\mathbf{R}$:

$\dfrac{\partial M}{\partial y} = 0 = \dfrac{\partial N}{\partial x}, \; \dfrac{\partial M}{\partial z} = 0 = \dfrac{\partial P}{\partial x}, \; \dfrac{\partial N}{\partial z} = 0 = \dfrac{\partial P}{\partial y}$, so $\mathbf{F}$ is conservative.

Ignore for the moment the factor $m\omega^2$.

$\dfrac{\partial f}{\partial x} = x$, so $f = \dfrac{x^2}{2} + g(y, z)$ and $\dfrac{\partial f}{\partial y} = \dfrac{\partial g}{\partial y}$, so from $\mathbf{F}$, $\dfrac{\partial g}{\partial y} = y$,

$g = \dfrac{y^2}{2} + h(z)$. $f = \dfrac{x^2}{2} + \dfrac{y^2}{2} + h(z)$, but $\dfrac{\partial f}{\partial z} = \dfrac{\partial h}{\partial z}$, so from $\mathbf{F}$, $\dfrac{\partial h}{\partial z} = z$,

$h = \dfrac{z^2}{2} + C$ (a constant) and $f = \dfrac{x^2}{2} + \dfrac{y^2}{2} + \dfrac{z^2}{2} + C$

Inserting the scalar factor: $f = \dfrac{m\omega^2}{2}(x^2 + y^2 + z^2) + C$

30. $\quad W = \displaystyle\int_C \mathbf{F} \bullet d\mathbf{R}$ where $\mathbf{F} = m\omega^2(x\mathbf{i} + y\mathbf{j} + z\mathbf{k})$ and $\mathbf{R}$ is the circular

path from $(3, 0, 2)$ to $(-3, 0, 2)$. Parametrizing: $x = 3\cos t, \, y = 3\sin t,$

$\mathbf{R} = (3\cos t)\mathbf{i} + (3\sin t)\mathbf{j}$ for $0 \le t \le \pi$. $d\mathbf{R} = (-3\sin t)\mathbf{i} + (3\cos t)\mathbf{j}$,

$\mathbf{F} \bullet d\mathbf{R} = m\omega^2(-9\sin t \cos t + 9\sin t \cos t) = 0$. Therefore, $W = 0$.

This result should have been anticipated as z is constant and $\mathbf{R}$ is symmetric about the y-axis.

SURVIVAL HINT

The Cumulative Review for Chapters 12-14 can be very valuable to refresh some
of the skills and concepts that you may not have been using often. It can also serve
as a valuable tool in preparing for a final exam. If you do not have the time to actually
do all of the problems, try looking at each one to see if you can recall the concept
involved and how to proceed with the solution. If you are confident about your
ability to solve the problem, do not spend the time. If you feel a little uncertain
about the problem, refer back to the appropriate section, review the concepts,
look in your old homework for a similar problem, and then see if you can work it.
Be more concerned about understanding the concept than about getting exactly the
right answer. Do not spend a lot of your time looking for algebra and arithmetic errors.

CONGRATULATIONS!!   YOU ARE A SURVIVOR.